Numerical Methods in Engineering with MATLAB®

Numerical Methods in Engineering with MATLAB® is a text for engineering students and a reference for practicing engineers, especially those who wish to explore the power and efficiency of MATLAB. The choice of numerical methods was based on their relevance to engineering problems. Every method is discussed thoroughly and illustrated with problems involving both hand computation and programming. MATLAB M-files accompany each method and are available on the book web site. This code is made simple and easy to understand by avoiding complex book-keeping schemes, while maintaining the essential features of the method. MATLAB was chosen as the example language because of its ubiquitous use in engineering studies and practice. Moreover, it is widely available to students on school networks and through inexpensive educational versions. MATLAB a popular tool for teaching scientific computation.

Jaan Kiusalaas is a Professor Emeritus in the Department of Engineering Science and Mechanics at the Pennsylvania State University. He has taught numerical methods, including finite element and boundary element methods for over 30 years. He is also the co-author of four other Books—*Engineering Mechanics: Statics, Engineering Mechanics: Dynamics, Mechanics of Materials,* and an alternate version of this work with Python code.

NUMERICAL METHODS IN ENGINEERING WITH
MATLAB®

Jaan Kiusalaas

The Pennsylvania State University

CAMBRIDGE
UNIVERSITY PRESS

CAMBRIDGE UNIVERSITY PRESS
Cambridge, New York, Melbourne, Madrid, Cape Town, Singapore, São Paulo

Cambridge University Press
32 Avenue of the Americas, New York, NY 10013-2473, USA

www.cambridge.org
Information on this title: www.cambridge.org/9780521852883

First published 2005
Reprinted 2006

Printed in the United States of America

A catalog record for this publication is available from the British Library.

Library of Congress Cataloging in Publication Data

Kiusalaas, Jaan.
Numerical methods in engineering with MATLAB / Jaan Kiusalaas.
   p.   cm.
Includes bibliographical references and index.
ISBN-13: 978-0-521-85288-3 (hardback)
ISBN-10: 0-521-85288-9 (hardback)
1. MATLAB.   2. Engineering mathematics – Data processing.   3. Numerical
analysis – Data processing.   I. Title.
TA345.K58   2005
620′.001′518 – dc22   2005011264

ISBN-13   978-0-521-85288-3 hardback
ISBN-10   0-521-85288-9 hardback

Contents

Preface

This book is targeted primarily toward engineers and engineering students of advanced standing (sophomores, seniors and graduate students). Familiarity with a computer language is required; knowledge of basic engineering subjects is useful, but not essential.

The text attempts to place emphasis on numerical methods, not programming. Most engineers are not programmers, but problem solvers. They want to know what methods can be applied to a given problem, what are their strengths and pitfalls and how to implement them. Engineers are not expected to write computer code for basic tasks from scratch; they are more likely to utilize functions and subroutines that have been already written and tested. Thus programming by engineers is largely confined to assembling existing pieces of code into a coherent package that solves the problem at hand.

The "piece" of code is usually a function that implements a specific task. For the user the details of the code are unimportant. What matters is the interface (what goes in and what comes out) and an understanding of the method on which the algorithm is based. Since no numerical algorithm is infallible, the importance of understanding the underlying method cannot be overemphasized; it is, in fact, the rationale behind learning numerical methods.

This book attempts to conform to the views outlined above. Each numerical method is explained in detail and its shortcomings are pointed out. The examples that follow individual topics fall into two categories: hand computations that illustrate the inner workings of the method, and small programs that show how the computer code is utilized in solving a problem. Problems that require programming are marked with ■.

The material consists of the usual topics covered in an engineering course on numerical methods: solution of equations, interpolation and data fitting, numerical differentiation and integration, solution of ordinary differential equations and eigenvalue problems. The choice of methods within each topic is tilted toward relevance to engineering problems. For example, there is an extensive discussion of symmetric, sparsely populated coefficient matrices in the solution of simultaneous equations. In the same vein, the solution of eigenvalue problems concentrates on methods that efficiently extract specific eigenvalues from banded matrices.

An important criterion used in the selection of methods was clarity. Algorithms requiring overly complex bookkeeping were rejected regardless of their efficiency and robustness. This decision, which was taken with great reluctance, is in keeping with the intent to avoid emphasis on programming.

The selection of algorithms was also influenced by current practice. This disqualified several well-known historical methods that have been overtaken by more recent developments. For example, the secant method for finding roots of equations was omitted as having no advantages over Brent's method. For the same reason, the multistep methods used to solve differential equations (e.g., Milne and Adams methods) were left out in favor of the adaptive Runge–Kutta and Bulirsch–Stoer methods.

Notably absent is a chapter on partial differential equations. It was felt that this topic is best treated by finite element or boundary element methods, which are outside the scope of this book. The finite difference model, which is commonly introduced in numerical methods texts, is just too impractical in handling multidimensional boundary value problems.

As usual, the book contains more material than can be covered in a three-credit course. The topics that can be skipped without loss of continuity are tagged with an asterisk (*).

The programs listed in this book were tested with MATLAB® 6.5.0 and under Windows® XP. The source code can be downloaded from the book's website at

www.cambridge.org/0521852889

The author wishes to express his gratitude to the anonymous reviewers and Professor Andrew Pytel for their suggestions for improving the manuscript. Credit is also due to the authors of *Numerical Recipes* (Cambridge University Press) whose presentation of numerical methods was inspirational in writing this book.

1   Introduction to MATLAB

1.1   General Information

Quick Overview

This chapter is not intended to be a comprehensive manual of MATLAB$^{\circledR}$. Our sole aim is to provide sufficient information to give you a good start. If you are familiar with another computer language, and we assume that you are, it is not difficult to pick up the rest as you go.

MATLAB is a high-level computer language for scientific computing and data visualization built around an interactive programming environment. It is becoming the premiere platform for scientific computing at educational institutions and research establishments. The great advantage of an interactive system is that programs can be tested and debugged quickly, allowing the user to concentrate more on the principles behind the program and less on programming itself. Since there is no need to compile, link and execute after each correction, MATLAB programs can be developed in much shorter time than equivalent FORTRAN or C programs. On the negative side, MATLAB does not produce stand-alone applications—the programs can be run only on computers that have MATLAB installed.

MATLAB has other advantages over mainstream languages that contribute to rapid program development:

- MATLAB contains a large number of functions that access proven numerical libraries, such as LINPACK and EISPACK. This means that many common tasks (e.g., solution of simultaneous equations) can be accomplished with a single function call.
- There is extensive graphics support that allows the results of computations to be plotted with a few statements.
- All numerical objects are treated as double-precision arrays. Thus there is no need to declare data types and carry out type conversions.

The syntax of MATLAB resembles that of FORTRAN. To get an idea of the similarities, let us compare the codes written in the two languages for solution of simultaneous equations $\mathbf{Ax} = \mathbf{b}$ by Gauss elimination. Here is the subroutine in FORTRAN 90:

```
subroutine gauss(A,b,n)
    use prec_mod
    implicit none
    real(DP), dimension(:,:), intent(in out) :: A
    real(DP), dimension(:),   intent(in out) :: b
    integer, intent(in)                      :: n
    real(DP) :: lambda
    integer  :: i,k
 ! -------------Elimination phase--------------
    do k = 1,n-1
        do i = k+1,n
            if(A(i,k) /= 0) then
                lambda = A(i,k)/A(k,k)
                A(i,k+1:n) = A(i,k+1:n) - lambda*A(k,k+1:n)
                b(i) = b(i) - lambda*b(k)
            end if
        end do
    end do
 ! -----------Back substitution phase----------
    do k = n,1,-1
        b(k) = (b(k) - sum(A(k,k+1:n)*b(k+1:n)))/A(k,k)
    end do
    return
end subroutine gauss
```

The statement `use prec_mod` tells the compiler to load the module `prec_mod` (not shown here), which defines the word length DP for floating-point numbers. Also note the use of array sections, such as `a(k,k+1:n)`, a feature that was not available in previous versions of FORTRAN.

The equivalent MATLAB function is (MATLAB does not have subroutines):

```
function b = gauss(A,b)
n = length(b);
%----------------Elimination phase-------------
for k = 1:n-1
    for i = k+1:n
```

```
        if A(i,k) ~= 0
            lambda = A(i,k)/A(k,k);
            A(i,k+1:n) = A(i,k+1:n) - lambda*A(k,k+1:n);
            b(i)= b(i) - lambda*b(k);
        end
    end
end
%-------------Back substitution phase-----------
for k = n:-1:1
    b(k) = (b(k) - A(k,k+1:n)*b(k+1:n))/A(k,k);
end
```

Simultaneous equations can also be solved in MATLAB with the simple command A\b (see below).

MATLAB can be operated in the interactive mode through its command window, where each command is executed immediately upon its entry. In this mode MATLAB acts like an electronic calculator. Here is an example of an interactive session for the solution of simultaneous equations:

```
>> A = [2 1 0; -1 2 2; 0 1 4]; % Input 3 x 3 matrix
>> b = [1; 2; 3];              % Input column vector
>> soln = A\b                  % Solve A*x = b by left division
soln =
    0.2500
    0.5000
    0.6250
```

The symbol >> is MATLAB's prompt for input. The percent sign (%) marks the beginning of a comment. A semicolon (;) has two functions: it suppresses printout of intermediate results and separates the rows of a matrix. Without a terminating semicolon, the result of a command would be displayed. For example, omission of the last semicolon in the line defining the matrix A would result in

```
>> A = [2 1 0; -1 2 2; 0 1 4]
A =
     2     1     0
    -1     2     2
     0     1     4
```

Functions and programs can be created with the MATLAB editor/debugger and saved with the .m extension (MATLAB calls them M-files). The file name of a saved function should be identical to the name of the function. For example, if the function for Gauss elimination listed above is saved as gauss.m, it can be called just like any MATLAB function:

```
>> A = [2 1 0; -1 2 2; 0 1 4];
>> b = [1; 2; 3];
>> soln = gauss(A,b)
soln =
    0.2500
    0.5000
    0.6250
```

1.2 Data Types and Variables

Data Types

The most commonly used MATLAB data types, or *classes*, are double, char and logical, all of which are considered by MATLAB as arrays. Numerical objects belong to the class double, which represents double-precision arrays; a scalar is treated as a 1×1 array. The elements of a char type array are strings (sequences of characters), whereas a logical type array element may contain only 1 (true) or 0 (false).

Another important class is function_handle, which is unique to MATLAB. It contains information required to find and execute a function. The name of a function handle consists of the character @, followed by the name of the function; e.g., @sin. Function handles are used as input arguments in function calls. For example, suppose that we have a MATLAB function plot(func,x1,x2) that plots any user-specified function func from x1 to x2. The function call to plot $\sin x$ from 0 to π would be plot(@sin,0,pi).

There are other data types, but we seldom come across them in this text. Additional classes can be defined by the user. The class of an object can be displayed with the class command. For example,

```
>> x = 1 + 3i  % Complex number
>> class(x)
ans =
double
```

Variables

Variable names, which must start with a letter, are *case sensitive*. Hence `xstart` and `xStart` represent two different variables. The length of the name is unlimited, but only the first N characters are significant. To find N for your installation of MATLAB, use the command `namelengthmax`:

```
>> namelengthmax
ans =
    63
```

Variables that are defined within a MATLAB function are local in their scope. They are not available to other parts of the program and do not remain in memory after exiting the function (this applies to most programming languages). However, variables can be shared between a function and the calling program if they are declared `global`. For example, by placing the statement `global X Y` in a function as well as the calling program, the variables X and Y are shared between the two program units. The recommended practice is to use capital letters for global variables.

MATLAB contains several built-in constants and special variables, most important of which are

ans	Default name for results
eps	Smallest number for which `1 + eps > 1`
inf	Infinity
NaN	Not a number
i or j	$\sqrt{-1}$
pi	π
realmin	Smallest usable positive number
realmax	Largest usable positive number

Here are a few of examples:

```
>> warning off   % Suppresses print of warning messages
>> 5/0
ans =
   Inf

>> 0/0
```

```
ans =
   NaN

>> 5*NaN          % Most operations with NaN result in NaN
ans =
   NaN

>> NaN == NaN     % Different NaN's are not equal!
ans =
     0

>> eps
ans =
   2.2204e-016
```

Arrays

Arrays can be created in several ways. One of them is to type the elements of the array between brackets. The elements in each row must be separated by blanks or commas. Here is an example of generating a 3×3 matrix:

```
>> A = [ 2 -1   0
        -1   2 -1
         0 -1   1]
A =
     2    -1     0
    -1     2    -1
     0    -1     1
```

The elements can also be typed on a single line, separating the rows with semicolons:

```
>> A = [2 -1 0; -1 2 -1; 0 -1 1]
A =
     2    -1     0
    -1     2    -1
     0    -1     1
```

Unlike most computer languages, MATLAB differentiates between row and column vectors (this peculiarity is a frequent source of programming and input errors). For example,

```
>> b = [1 2 3]          % Row vector
b =
        1       2       3

>> b = [1; 2; 3]        % Column vector
b =
        1
        2
        3

>> b = [1 2 3]'         % Transpose of row vector
b =
        1
        2
        3
```

The single quote (') is the *transpose operator* in MATLAB; thus b ' is the transpose of b.

The elements of a matrix, such as

$$\mathbf{A} = \begin{bmatrix} A_{11} & A_{12} & A_{13} \\ A_{21} & A_{22} & A_{23} \\ A_{31} & A_{32} & A_{33} \end{bmatrix}$$

can be accessed with the statement A(i,j), where i and j are the row and column numbers, respectively. A section of an array can be extracted by the use of colon notation. Here is an illustration:

```
>> A = [8 1 6; 3 5 7; 4 9 2]
A =
        8       1       6
        3       5       7
        4       9       2

>> A(2,3)       % Element in row 2, column 3
ans =
        7

>> A(:,2)       % Second column
```

```
ans =
      1
      5
      9
```

```
>> A(2:3,2:3)    % The 2 x 2 submatrix in lower right corner
ans =
      5      7
      9      2
```

Array elements can also be accessed with a single index. Thus A(i) extracts the ith element of A, counting the elements down the columns. For example, A(7) and A(1,3) would extract the same element from a 3×3 matrix.

Cells

A cell array is a sequence of arbitrary objects. Cell arrays can be created by enclosing their contents between braces {}. For example, a cell array c consisting of three cells can be created by

```
>> c = {[1 2 3], 'one two three', 6 + 7i}
c =
    [1x3 double]    'one two three'    [6.0000+ 7.0000i]
```

As seen above, the contents of some cells are not printed in order to save space. If all contents are to be displayed, use the celldisp command:

```
>> celldisp(c)
c{1} =
      1      2      3
c{2} =
one two three
c{3} =
    6.0000 + 7.0000i
```

Braces are also used to extract the contents of the cells:

```
>> c{1}                 % First cell
ans =
      1      2      3
```

```
>> c{1}(2)              % Second element of first cell
ans =
     2
>> c{2}                 % Second cell
ans =
one two three
```

Strings

A string is a sequence of characters; it is treated by MATLAB as a character array. Strings are created by enclosing the characters between single quotes. They are concatenated with the function `strcat`, whereas a colon operator (`:`) is used to extract a portion of the string. For example,

```
>> s1 = 'Press return to exit';   % Create a string
>> s2 = ' the program';           % Create another string
>> s3 = strcat(s1,s2)             % Concatenate s1 and s2
s3 =
Press return to exit the program
>> s4 = s1(1:12)                  % Extract chars. 1-12 of s1
s4 =
Press return
```

1.3   Operators

Arithmetic Operators

MATLAB supports the usual arithmetic operators:

+	Addition
−	Subtraction
*	Multiplication
^	Exponentiation

When applied to matrices, they perform the familiar matrix operations, as illustrated below.

```
>> A = [1 2 3; 4 5 6]; B = [7 8 9; 0 1 2];

>> A + B                % Matrix addition
```

```
ans =
        8       10      12
        4        6       8


>> A*B'                    % Matrix multiplication
ans =
       50        8
      122       17


>> A*B                     % Matrix multiplication fails
??? Error using ==> *      % due to incompatible dimensions
Inner matrix dimensions must agree.
```

There are two division operators in MATLAB:

/	Right division
\	Left division

If a and b are scalars, the right division a/b results in a divided by b, whereas the left division is equivalent to b/a. In the case where A and B are matrices, A/B returns the solution of X*A = B and A\B yields the solution of A*X = B.

Often we need to apply the *, / and ^ operations to matrices in an element-by-element fashion. This can be done by preceding the operator with a period (.) as follows:

.*	Element-wise multiplication
./	Element-wise division
.^	Element-wise exponentiation

For example, the computation $C_{ij} = A_{ij} B_{ij}$ can be accomplished with

```
>> A = [1 2 3; 4 5 6]; B = [7 8 9; 0 1 2];
>> C = A.*B
C =
        7       16      27
        0        5      12
```

Comparison Operators

The comparison (relational) operators return 1 for true and 0 for false. These operators are

<	Less than
>	Greater than
<=	Less than or equal to
>=	Greater than or equal to
==	Equal to
~=	Not equal to

The comparison operators always act element-wise on matrices; hence they result in a matrix of `logical` type. For example,

```
>> A = [1 2 3; 4 5 6]; B = [7 8 9; 0 1 2];
>> A > B
ans =
     0     0     0
     1     1     1
```

Logical Operators

The logical operators in MATLAB are

&	AND
|	OR
~	NOT

They are used to build compound relational expressions, an example of which is shown below.

```
>> A = [1 2 3; 4 5 6]; B = [7 8 9; 0 1 2];
>> (A > B) | (B > 5)
ans =
     1     1     1
     1     1     1
```

1.4   Flow Control

Conditionals

if, else, elseif
The `if` construct

```
if condition
    block
end
```

executes the *block* of statements if the *condition* is true. If the condition is false, the block skipped. The `if` conditional can be followed by any number of `elseif` constructs:

```
if condition
    block
elseif condition
    block
    ⋮
end
```

which work in the same manner. The `else` clause

```
    ⋮
else
    block
end
```

can be used to define the block of statements which are to be executed if none of the `if-elseif` clauses are true. The function `signum` below illustrates the use of the conditionals.

```
function sgn = signum(a)
if a > 0
    sgn = 1;
elseif a < 0
    sgn = -1;
else
```

```
        sgn = 0;
end


>> signum (-1.5)
ans =
    -1
```

switch

The switch construct is

```
switch expression
    case value1
        block
    case value2
        block
        ⋮
    otherwise
        block
end
```

Here the *expression* is evaluated and the control is passed to the case that matches the value. For instance, if the value of *expression* is equal to *value2*, the *block* of statements following case *value2* is executed. If the value of *expression* does not match any of the case values, the control passes to the optional otherwise block. Here is an example:

```
function y = trig(func,x)
switch func
    case 'sin'
        y = sin(x);
    case 'cos'
        y = cos(x);
    case 'tan'
        y = tan(x);
    otherwise
        error('No such function defined')
end


>> trig('tan',pi/3)
ans =
    1.7321
```

Loops

while

The `while` construct

```
while condition:
    block
end
```

executes a *block* of statements if the *condition* is true. After execution of the block, *condition* is evaluated again. If it is still true, the block is executed again. This process is continued until the *condition* becomes false.

The following example computes the number of years it takes for a $1000 principal to grow to $10,000 at 6% annual interest.

```
>> p = 1000; years = 0;
>> while p < 10000
       years = years + 1;
       p = p*(1 + 0.06);
   end
>> years
years =
    40
```

for

The `for` loop requires a *target* and a *sequence* over which the target loops. The form of the construct is

```
for target = sequence
    block
end
```

For example, to compute $\cos x$ from $x = 0$ to $\pi/2$ at increments of $\pi/10$ we could use

```
>> for n = 0:5  % n loops over the sequence 0 1 2 3 4 5
       y(n+1) = cos(n*pi/10);
   end
>> y
y =
    1.0000    0.9511    0.8090    0.5878    0.3090    0.0000
```

Loops should be avoided whenever possible in favor of *vectorized* expressions, which execute much faster. A vectorized solution to the last computation would be

```
>> n = 0:5;
>> y = cos(n*pi/10)
y =
    1.0000    0.9511    0.8090    0.5878    0.3090    0.0000
```

break

Any loop can be terminated by the `break` statement. Upon encountering a `break` statement, the control is passed to the first statement outside the loop. In the following example the function `buildvec` constructs a row vector of arbitrary length by prompting for its elements. The process is terminated when an empty element is encountered.

```
function x = buildvec
for i = 1:1000
    elem = input('==> '); % Prompts for input of element
    if isempty(elem)       % Check for empty element
        break
    end
    x(i) = elem;
end
```

```
>> x = buildvec
==> 3
==> 5
==> 7
==> 2
==>
x =
      3     5     7     2
```

continue

When the `continue` statement is encountered in a loop, the control is passed to the next iteration without executing the statements in the current iteration. As an illustration, consider the following function that strips all the blanks from the string s1:

```
function s2 = strip(s1)
s2 = '';                        % Create an empty string
for i = 1:length(s1)
```

```
    if s1(i) == ' '
        continue
    else
        s2 = strcat(s2,s1(i)); % Concatenation
    end
end
```

```
>> s2 = strip('This is too bad')
s2 =
Thisistoobad
```

return

A function normally returns to the calling program when it runs out of statements. However, the function can be forced to exit with the `return` command. In the example below, the function `solve` uses the Newton–Raphson method to find the zero of $f(x) = \sin x - 0.5x$. The input x (guess of the solution) is refined in successive iterations using the formula $x \leftarrow x + \Delta x$, where $\Delta x = -f(x)/f'(x)$, until the change Δx becomes sufficiently small. The procedure is then terminated with the `return` statement. The `for` loop assures that the number of iterations does not exceed 30, which should be more than enough for convergence.

```
function x = solve(x)
for numIter = 1:30
    dx = -(sin(x) - 0.5*x)/(cos(x) - 0.5); % -f(x)/f'(x)
    x = x + dx;
    if abs(dx) < 1.0e-6          % Check for convergence
        return
    end
end
error('Too many iterations')
```

```
>> x = solve(2)
x =
    1.8955
```

error

Execution of a program can be terminated and a message displayed with the `error` function

$$\text{error('}message\text{')}$$

For example, the following program lines determine the dimensions of a matrix and aborts the program if the dimensions are not equal.

```
[m,n] = size(A);   % m = no. of rows; n = no. of cols.
if m ~= n
    error('Matrix must be square')
end
```

1.5   Functions

Function Definition

The body of a function must be preceded by the function definition line

$$\texttt{function } [output\_args] = function\_name(input\_arguments)$$

The input and output arguments must be separated by commas. The number of arguments may be zero. If there is only one output argument, the enclosing brackets may be omitted.

To make the function accessible to other programs units, it must be saved under the file name *function_name* . m. This file may contain other functions, called *subfunctions*. The subfunctions can be called only by the primary function *function_name* or other subfunctions in the file; they are not accessible to other program units.

Calling Functions

A function may be called with fewer arguments than appear in the function definition. The number of input and output arguments used in the function call can be determined by the functions `nargin` and `nargout`, respectively. The following example shows a modified version of the function `solve` that involves two input and two output arguments. The error tolerance `epsilon` is an optional input that may be used to override the default value `1.0e-6`. The output argument `numIter`, which contains the number of iterations, may also be omitted from the function call.

```
function [x,numIter] = solve(x,epsilon)
if nargin == 1              % Specify default value if
    epsilon = 1.0e-6;       % second input argument is
end                         % omitted in function call
for numIter = 1:100
    dx = -(sin(x) - 0.5*x)/(cos(x) - 0.5);
    x = x + dx;
    if abs(dx) < epsilon    % Converged; return to
        return              % calling program
    end
```

```
end
error('Too many iterations')


>> x = solve(2)                    % numIter not printed
x =
    1.8955


>> [x,numIter] = solve(2)   % numIter is printed
x =
    1.8955
numIter =
    4


>> format long
>> x = solve(2,1.0e-12)      % Solving with extra precision
x =
    1.89549426703398
>>
```

Evaluating Functions

Let us consider a slightly different version of the function `solve` shown below. The expression for dx, namely $\Delta x = -f(x)/f'(x)$, is now coded in the function `myfunc`, so that `solve` contains a call to `myfunc`. This will work fine, provided that `myfunc` is stored under the file name `myfunc.m` so that MATLAB can find it.

```
function [x,numIter] = solve(x,epsilon)
if nargin == 1; epsilon = 1.0e-6; end
for numIter = 1:30
    dx = myfunc(x);
    x = x + dx;
    if abs(dx) < epsilon; return; end
end
error('Too many iterations')

function y = myfunc(x)
y = -(sin(x) - 0.5*x)/(cos(x) - 0.5);


>> x = solve(2)
x =
    1.8955
```

In the above version of `solve` the function returning `dx` is stuck with the name `myfunc`. If `myfunc` is replaced with another function name, `solve` will not work unless the corresponding change is made in its code. In general, it is not a good idea to alter computer code that has been tested and debugged; all data should be communicated to a function through its arguments. MATLAB makes this possible by passing the function handle of `myfunc` to `solve` as an argument, as illustrated below.

```
function [x,numIter] = solve(func,x,epsilon)
if nargin == 2; epsilon = 1.0e-6; end
for numIter = 1:30
    dx = feval(func,x);    % feval is a MATLAB function for
    x = x + dx;            % evaluating a passed function
    if abs(dx) < epsilon; return; end
end
error('Too many iterations')

>> x = solve(@myfunc,2)    % @myfunc is the function handle
x =
    1.8955
```

The call `solve(@myfunc,2)` creates a function handle to `myfunc` and passes it to `solve` as an argument. Hence the variable `func` in `solve` contains the handle to `myfunc`. A function passed to another function by its handle is evaluated by the MATLAB function

$$feval\,(function\_handle, arguments)$$

It is now possible to use `solve` to find a zero of any $f(x)$ by coding the function $\Delta x = -f(x)/f'(x)$ and passing its handle to `solve`.

In-Line Functions

If the function is not overly complicated, it can also be represented as an *inline* object:

$$function\_name = inline('expression','var1','var2',\ldots)$$

where *expression* specifies the function and *var1, var2, . . .* are the names of the independent variables. Here is an example:

```
>> myfunc = inline ('x^2 + y^2','x','y');
>> myfunc (3,5)
ans =
    34
```

The advantage of an in-line function is that it can be embedded in the body of the code; it does not have to reside in an M-file.

1.6   Input/Output

Reading Input

The MATLAB function for receiving user input is

$$value = \text{input}('prompt')$$

It displays a prompt and then waits for input. If the input is an expression, it is evaluated and returned in *value*. The following two samples illustrate the use of `input`:

```
>> a = input('Enter expression: ')
Enter expression: tan(0.15)
a =
    0.1511

>> s = input('Enter string: ')
Enter string: 'Black sheep'
s =
Black sheep
```

Printing Output

As mentioned before, the result of a statement is printed if the statement does not end with a semicolon. This is the easiest way of displaying results in MATLAB. Normally MATLAB displays numerical results with about five digits, but this can be changed with the format command:

`format long`	switches to 16-digit display
`format short`	switches to 5-digit display

To print formatted output, use the `fprintf` function:

$$\text{fprintf}('format', \, list)$$

where *format* contains formatting specifications and *list* is the list of items to be printed, separated by commas. Typically used formatting specifications are

%w.df	Floating point notation
%w.de	Exponential notation
\n	Newline character

where w is the width of the field and d is the number of digits after the decimal point. Line break is forced by the newline character. The following example prints a formatted table of sin x vs. x at intervals of 0.2:

```
>> x = 0:0.2:1;
>> for i = 1:length(x)
       fprintf('%4.1f %11.6f\n',x(i),sin(x(i)))
   end
 0.0     0.000000
 0.2     0.198669
 0.4     0.389418
 0.6     0.564642
 0.8     0.717356
 1.0     0.841471
```

1.7   Array Manipulation

Creating Arrays

We learned before that an array can be created by typing its elements between brackets:

```
>> x = [0 0.25 0.5 0.75 1]
x =
         0    0.2500    0.5000    0.7500    1.0000
```

Colon Operator
Arrays with equally spaced elements can also be constructed with the colon operator.

$$x = first\_elem : increment : last\_elem$$

For example,

```
>> x = 0:0.25:1
x =
         0    0.2500    0.5000    0.7500    1.0000
```

linspace

Another means of creating an array with equally spaced elements is the `linspace` function. The statement

$$x = \texttt{linspace}(\mathit{xfirst}, \mathit{xlast}, n)$$

creates an array of n elements starting with *xfirst* and ending with *xlast*. Here is an illustration:

```
>> x = linspace(0,1,5)
x =
        0    0.2500    0.5000    0.7500    1.0000
```

logspace

The function `logspace` is the logarithmic counterpart of `linspace`. The call

$$x = \texttt{logspace}(\mathit{zfirst}, \mathit{zlast}, n)$$

creates n logarithmically spaced elements starting with $x = 10^{z\,first}$ and ending with $x = 10^{z\,last}$. Here is an example:

```
>> x = logspace(0,1,5)
x =
    1.0000    1.7783    3.1623    5.6234    10.0000
```

zeros

The function call

$$X = \texttt{zeros}(m, n)$$

returns a matrix of m rows and n columns that is filled with zeroes. When the function is called with a single argument, e.g., `zeros(n)`, a $n \times n$ matrix is created.

ones

$$X = \texttt{ones}(m, n)$$

The function `ones` works in the manner as `zeros`, but fills the matrix with ones.

rand

$$X = \texttt{rand}(m, n)$$

This function returns a matrix filled with random numbers between 0 and 1.

eye

The function eye

$$X = \text{eye}(n)$$

creates an $n \times n$ identity matrix.

Array Functions

There are numerous array functions in MATLAB that perform matrix operations and other useful tasks. Here are a few basic functions:

length

The length n (number of elements) of a vector x can be determined with the function length:

$$n = \text{length}(x)$$

size

If the function size is called with a single input argument:

$$[m,n] = \text{size}(X)$$

it determines the number of rows m and number of columns n in the matrix X. If called with two input arguments:

$$m = \text{size}(X,dim)$$

it returns the length of X in the specified dimension ($dim = 1$ yields the number of rows, and $dim = 2$ gives the number of columns).

reshape

The reshape function is used to rearrange the elements of a matrix. The call

$$Y = \text{reshape}(X,m,n)$$

returns a $m \times n$ matrix the elements of which are taken from matrix X in the column-wise order. The total number of elements in X must be equal to $m \times n$. Here is an example:

```
>> a = 1:2:11
a =
     1     3     5     7     9    11
>> A = reshape(a,2,3)

A =
     1     5     9
     3     7    11
```

dot

$$a = \texttt{dot}(x, y)$$

This function returns the dot product of two vectors x and y which must be of the same length.

prod

$$a = \texttt{prod}(x)$$

For a vector x, $\texttt{prod}(x)$ returns the product of its elements. If x is a matrix, then a is a row vector containing the products over each column. For example,

```
>> a = [1 2 3 4 5 6];
>> A = reshape(a,2,3)
A =
     1     3     5
     2     4     6

>> prod(a)
ans =
   720

>> prod(A)
ans =
     2    12    30
```

sum

$$a = \texttt{sum}(x)$$

This function is similar to $\texttt{prod}$, except that it returns the sum of the elements.

cross

$$c = \text{cross}(a, b)$$

The function cross computes the cross product: $c = a \times b$, where vectors a and b must be of length 3.

1.8   Writing and Running Programs

MATLAB has two windows available for typing program lines: the *command window* and the *editor/debugger*. The command window is always in the interactive mode, so that any statement entered into the window is immediately processed. The interactive mode is a good way to experiment with the language and try out programming ideas.

MATLAB opens the editor window when a new M-file is created, or an existing file is opened. The editor window is used to type and save programs (called *script files* in MATLAB) and functions. One could also use a text editor to enter program lines, but the MATLAB editor has MATLAB-specific features, such as color coding and automatic indentation, that make work easier. Before a program or function can be executed, it must be saved as a MATLAB M-file (recall that these files have the .m extension). A program can be run by invoking the *run* command from the editor's *debug* menu.

When a function is called for the first time during a program run, it is compiled into P-code (pseudo-code) to speed up execution in subsequent calls to the function. One can also create the P-code of a function and save it on disk by issuing the command

$$\text{pcode } function\_name$$

MATLAB will then load the P-code (which has the .p extension) into the memory rather than the text file.

The variables created during a MATLAB session are saved in the MATLAB *workspace* until they are cleared. Listing of the saved variables can be displayed by the command who. If greater detail about the variables is required, type whos. Variables can be cleared from the workspace with the command

$$\text{clear } a\, b \ldots$$

which clears the variables a, b, ... If the list of variables is omitted, all variables are cleared.

Assistance on any MATLAB function is available by typing

help *function_name*

in the command window.

1.9 Plotting

MATLAB has extensive plotting capabilities. Here we illustrate some basic commands for two-dimensional plots. The example below plots sin x and cos x on the same plot.

```
>> x = 0:0.2:pi;    % Create x-array
>> y = sin(x);      % Create y-array
>> plot(x,y,'k:o')  % Plot x-y points with specified color
                    % and symbol ('k' = black, 'o' = circles)
>> hold on          % Allow overwriting of current plot
>> z = cos(x);      % Create z-array
>> plot(x,z,'k:x')  % Plot x-z points ('x' = crosses)
>> grid on          % Display coordinate grid
>> xlabel('x')      % Display label for x-axis
>> ylabel('y')      % Display label for y-axis
>> gtext('sin x')   % Create mouse-movable text
>> gtext('cos x')
```

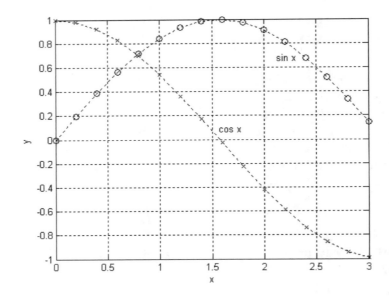

A function stored in a M-file can be plotted with a single command, as shown below.

```
function y = testfunc(x)      % Stored function
y = (x.^3).*sin(x) - 1./x;


>> fplot(@testfunc,[1 20])    % Plot from x = 1 to 20
>> grid on
```

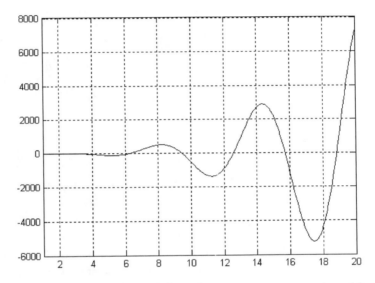

The plots appearing in this book from here on were not produced by MATLAB. We used the copy/paste operation to transfer the numerical data to a spreadsheet and then let the spreadsheet create the plot. This resulted in plots more suited for publication.

2 Systems of Linear Algebraic Equations

> Solve the simultaneous equations $\mathbf{Ax} = \mathbf{b}$

2.1 Introduction

In this chapter we look at the solution of n linear, algebraic equations in n unknowns. It is by far the longest and arguably the most important topic in the book. There is a good reason for this—it is almost impossible to carry out numerical analysis of any sort without encountering simultaneous equations. Moreover, equation sets arising from physical problems are often very large, consuming a lot of computational resources. It usually possible to reduce the storage requirements and the run time by exploiting special properties of the coefficient matrix, such as sparseness (most elements of a sparse matrix are zero). Hence there are many algorithms dedicated to the solution of large sets of equations, each one being tailored to a particular form of the coefficient matrix (symmetric, banded, sparse, etc.). A well-known collection of these routines is LAPACK – Linear Algebra PACKage, originally written in Fortran77[1].

We cannot possibly discuss all the special algorithms in the limited space available. The best we can do is to present the basic methods of solution, supplemented by a few useful algorithms for banded and sparse coefficient matrices.

Notation

A system of algebraic equations has the form

---

[1] LAPACK is the successor of LINPACK, a 1970s and 80s collection of Fortran subroutines.

$$A_{11}x_1 + A_{12}x_2 + \cdots + A_{1n}x_n = b_1$$

$$A_{21}x_1 + A_{22}x_2 + \cdots + A_{2n}x_n = b_2$$

$$A_{31}x_1 + A_{32}x_2 + \cdots + A_{3n}x_n = b_3 \tag{2.1}$$

$$\vdots$$

$$A_{n1}x_1 + A_{n2}x_2 + \cdots + A_{nn}x_n = b_n$$

where the coefficients A_{ij} and the constants b_j are known, and x_i represent the unknowns. In matrix notation the equations are written as

$$\begin{bmatrix} A_{11} & A_{12} & \cdots & A_{1n} \\ A_{21} & A_{22} & \cdots & A_{2n} \\ \vdots & \vdots & \ddots & \vdots \\ A_{n1} & A_{n2} & \cdots & A_{nn} \end{bmatrix} \begin{bmatrix} x_1 \\ x_2 \\ \vdots \\ x_n \end{bmatrix} = \begin{bmatrix} b_1 \\ b_2 \\ \vdots \\ b_n \end{bmatrix} \tag{2.2}$$

or, simply

$$\mathbf{Ax} = \mathbf{b} \tag{2.3}$$

A particularly useful representation of the equations for computational purposes is the *augmented coefficient matrix*, obtained by adjoining the constant vector $\mathbf{b}$ to the coefficient matrix $\mathbf{A}$ in the following fashion:

$$\begin{bmatrix} \mathbf{A} \mid \mathbf{b} \end{bmatrix} = \begin{bmatrix} A_{11} & A_{12} & \cdots & A_{1n} & b_1 \\ A_{21} & A_{22} & \cdots & A_{2n} & b_2 \\ \vdots & \vdots & \ddots & \vdots & \vdots \\ A_{n1} & A_{n2} & \cdots & A_{nn} & b_n \end{bmatrix} \tag{2.4}$$

Uniqueness of Solution

A system of n linear equations in n unknowns has a unique solution, provided that the determinant of the coefficient matrix is *nonsingular*, i.e., if $|\mathbf{A}| \neq 0$. The rows and columns of a nonsingular matrix are *linearly independent* in the sense that no row (or column) is a linear combination of other rows (or columns).

If the coefficient matrix is *singular*, the equations may have an infinite number of solutions, or no solutions at all, depending on the constant vector. As an illustration, take the equations

$$2x + y = 3 \qquad 4x + 2y = 6$$

Since the second equation can be obtained by multiplying the first equation by two, any combination of x and y that satisfies the first equation is also a solution of the second equation. The number of such combinations is infinite. On the other hand, the equations

$$2x + y = 3 \qquad 4x + 2y = 0$$

have no solution because the second equation, being equivalent to $2x + y = 0$, contradicts the first one. Therefore, any solution that satisfies one equation cannot satisfy the other one.

Ill-Conditioning

An obvious question is: what happens when the coefficient matrix is almost singular; i.e., if $|A|$ is very small? In order to determine whether the determinant of the coefficient matrix is "small," we need a reference against which the determinant can be measured. This reference is called the *norm* of the matrix, denoted by $\|A\|$. We can then say that the determinant is small if

$$|A| << \|A\|$$

Several norms of a matrix have been defined in existing literature, such as

$$\|A\| = \sqrt{\sum_{i=1}^{n} \sum_{j=1}^{n} A_{ij}^2} \qquad \|A\| = \max_{1 \le i \le n} \sum_{j=1}^{n} |A_{ij}| \qquad (2.5\text{a})$$

A formal measure of conditioning is the *matrix condition number*, defined as

$$\text{cond}(A) = \|A\| \, \|A^{-1}\| \qquad (2.5\text{b})$$

If this number is close to unity, the matrix is well-conditioned. The condition number increases with the degree of ill-conditioning, reaching infinity for a singular matrix. Note that the condition number is not unique, but depends on the choice of the matrix norm. Unfortunately, the condition number is expensive to compute for large matrices. In most cases it is sufficient to gauge conditioning by comparing the determinant with the magnitudes of the elements in the matrix.

　　If the equations are ill-conditioned, small changes in the coefficient matrix result in large changes in the solution. As an illustration, consider the equations

$$2x + y = 3 \qquad 2x + 1.001y = 0$$

that have the solution $x = 1501.5$, $y = -3000$. Since $|A| = 2(1.001) - 2(1) = 0.002$ is much smaller than the coefficients, the equations are ill-conditioned. The effect of ill-conditioning can be verified by changing the second equation to $2x + 1.002y = 0$ and re-solving the equations. The result is $x = 751.5$, $y = -1500$. Note that a 0.1% change in the coefficient of y produced a 100% change in the solution.

Numerical solutions of ill-conditioned equations are not to be trusted. The reason is that the inevitable roundoff errors during the solution process are equivalent to introducing small changes into the coefficient matrix. This in turn introduces large errors into the solution, the magnitude of which depends on the severity of ill-conditioning. In suspect cases the determinant of the coefficient matrix should be computed so that the degree of ill-conditioning can be estimated. This can be done during or after the solution with only a small computational effort.

Linear Systems

Linear, algebraic equations occur in almost all branches of numerical analysis. But their most visible application in engineering is in the analysis of linear systems (any system whose response is proportional to the input is deemed to be linear). Linear systems include structures, elastic solids, heat flow, seepage of fluids, electromagnetic fields and electric circuits; i.e., most topics taught in an engineering curriculum.

If the system is discrete, such as a truss or an electric circuit, then its analysis leads directly to linear algebraic equations. In the case of a statically determinate truss, for example, the equations arise when the equilibrium conditions of the joints are written down. The unknowns $x_1, x_2, \ldots, x_n$ represent the forces in the members and the support reactions, and the constants $b_1, b_2, \ldots, b_n$ are the prescribed external loads.

The behavior of continuous systems is described by differential equations, rather than algebraic equations. However, because numerical analysis can deal only with discrete variables, it is first necessary to approximate a differential equation with a system of algebraic equations. The well-known finite difference, finite element and boundary element methods of analysis work in this manner. They use different approximations to achieve the "discretization," but in each case the final task is the same: solve a system (often a very large system) of linear, algebraic equations.

In summary, the modeling of linear systems invariably gives rise to equations of the form $\mathbf{Ax} = \mathbf{b}$, where $\mathbf{b}$ is the input and $\mathbf{x}$ represents the response of the system. The coefficient matrix $\mathbf{A}$, which reflects the characteristics of the system, is independent of the input. In other words, if the input is changed, the equations have to be solved again with a different $\mathbf{b}$, but the same $\mathbf{A}$. Therefore, it is desirable to have an equation-solving algorithm that can handle any number of constant vectors with minimal computational effort.

Methods of Solution

There are two classes of methods for solving systems of linear, algebraic equations: direct and iterative methods. The common characteristic of *direct methods* is that they transform the original equations into *equivalent equations* (equations that have the same solution) that can be solved more easily. The transformation is carried out by applying the three operations listed below. These so-called *elementary operations* do not change the solution, but they may affect the determinant of the coefficient matrix as indicated in parentheses.

- Exchanging two equations (changes sign of |A|).
- Multiplying an equation by a nonzero constant (multiplies |A| by the same constant).
- Multiplying an equation by a nonzero constant and then subtracting it from another equation (leaves |A| unchanged).

Iterative, or *indirect methods*, start with a guess of the solution **x**, and then repeatedly refine the solution until a certain convergence criterion is reached. Iterative methods are generally less efficient than their direct counterparts due to the large number of iterations required. But they do have significant computational advantages if the coefficient matrix is very large and sparsely populated (most coefficients are zero).

Overview of Direct Methods

Table 2.1 lists three popular direct methods, each of which uses elementary operations to produce its own final form of easy-to-solve equations.

Method	Initial form	Final form
Gauss elimination	$\mathbf{Ax} = \mathbf{b}$	$\mathbf{Ux} = \mathbf{c}$
LU decomposition	$\mathbf{Ax} = \mathbf{b}$	$\mathbf{LUx} = \mathbf{b}$
Gauss–Jordan elimination	$\mathbf{Ax} = \mathbf{b}$	$\mathbf{Ix} = \mathbf{c}$

Table 2.1

In the above table **U** represents an upper triangular matrix, **L** is a lower triangular matrix and **I** denotes the identity matrix. A square matrix is called *triangular* if it contains only zero elements on one side of the leading diagonal. Thus a 3×3 upper triangular matrix has the form

$$\mathbf{U} = \begin{bmatrix} U_{11} & U_{12} & U_{13} \\ 0 & U_{22} & U_{23} \\ 0 & 0 & U_{33} \end{bmatrix}$$

and a 3×3 lower triangular matrix appears as

$$\mathbf{L} = \begin{bmatrix} L_{11} & 0 & 0 \\ L_{21} & L_{22} & 0 \\ L_{31} & L_{32} & L_{33} \end{bmatrix}$$

Triangular matrices play an important role in linear algebra, since they simplify many computations. For example, consider the equations $\mathbf{Lx} = \mathbf{c}$, or

$$L_{11}x_1 = c_1$$

$$L_{21}x_1 + L_{22}x_2 = c_2$$

$$L_{31}x_1 + L_{32}x_2 + L_{33}x_3 = c_3$$

$$\vdots$$

If we solve the equations forward, starting with the first equation, the computations are very easy, since each equation would contain only one unknown at a time. The solution would thus proceed as follows:

$$x_1 = c_1/L_{11}$$

$$x_2 = (c_2 - L_{21}x_1)/L_{22}$$

$$x_3 = (c_3 - L_{31}x_1 - L_{32}x_2)/L_{33}$$

$$\vdots$$

This procedure is known as *forward substitution*. In a similar way, $\mathbf{Ux} = \mathbf{c}$, encountered in Gauss elimination, can easily be solved by *back substitution*, which starts with the last equation and proceeds backward through the equations.

The equations $\mathbf{LUx} = \mathbf{b}$, which are associated with LU decomposition, can also be solved quickly if we replace them with two sets of equivalent equations: $\mathbf{Ly} = \mathbf{b}$ and $\mathbf{Ux} = \mathbf{y}$. Now $\mathbf{Ly} = \mathbf{b}$ can be solved for $\mathbf{y}$ by forward substitution, followed by the solution of $\mathbf{Ux} = \mathbf{y}$ by means of back substitution.

The equations $\mathbf{Ix} = \mathbf{c}$, which are produced by Gauss–Jordan elimination, are equivalent to $\mathbf{x} = \mathbf{c}$ (recall the identity $\mathbf{Ix} = \mathbf{x}$), so that $\mathbf{c}$ is already the solution.

EXAMPLE 2.1

Determine whether the following matrix is singular:

$$\mathbf{A} = \begin{bmatrix} 2.1 & -0.6 & 1.1 \\ 3.2 & 4.7 & -0.8 \\ 3.1 & -6.5 & 4.1 \end{bmatrix}$$

Solution Laplace's development (see Appendix A2) of the determinant about the first row of **A** yields

$$|A| = 2.1 \begin{vmatrix} 4.7 & -0.8 \\ -6.5 & 4.1 \end{vmatrix} + 0.6 \begin{vmatrix} 3.2 & -0.8 \\ 3.1 & 4.1 \end{vmatrix} + 1.1 \begin{vmatrix} 3.2 & 4.7 \\ 3.1 & -6.5 \end{vmatrix}$$

$$= 2.1(14.07) + 0.6(15.60) + 1.1(-35.37) = 0$$

Since the determinant is zero, the matrix is singular. It can be verified that the singularity is due to the following row dependency: (row 3) = (3 × row 1) − (row 2).

EXAMPLE 2.2
Solve the equations **Ax** = **b**, where

$$A = \begin{bmatrix} 8 & -6 & 2 \\ -4 & 11 & -7 \\ 4 & -7 & 6 \end{bmatrix} \qquad b = \begin{bmatrix} 28 \\ -40 \\ 33 \end{bmatrix}$$

knowing that the LU decomposition of the coefficient matrix is (you should verify this)

$$A = LU = \begin{bmatrix} 2 & 0 & 0 \\ -1 & 2 & 0 \\ 1 & -1 & 1 \end{bmatrix} \begin{bmatrix} 4 & -3 & 1 \\ 0 & 4 & -3 \\ 0 & 0 & 2 \end{bmatrix}$$

Solution We first solve the equations **Ly** = **b** by forward substitution:

$$
\begin{aligned}
2y_1 &= 28 & y_1 &= 28/2 = 14 \\
-y_1 + 2y_2 &= -40 & y_2 &= (-40 + y_1)/2 = (-40 + 14)/2 = -13 \\
y_1 - y_2 + y_3 &= 33 & y_3 &= 33 - y_1 + y_2 = 33 - 14 - 13 = 6
\end{aligned}
$$

The solution **x** is then obtained from **Ux** = **y** by back substitution:

$$
\begin{aligned}
2x_3 &= y_3 & x_3 &= y_3/2 = 6/2 = 3 \\
4x_2 - 3x_3 &= y_2 & x_2 &= (y_2 + 3x_3)/4 = [-13 + 3(3)]/4 = -1 \\
4x_1 - 3x_2 + x_3 &= y_1 & x_1 &= (y_1 + 3x_2 - x_3)/4 = [14 + 3(-1) - 3]/4 = 2
\end{aligned}
$$

Hence the solution is $x = \begin{bmatrix} 2 & -1 & 3 \end{bmatrix}^T$

2.2   Gauss Elimination Method

Introduction

Gauss elimination is the most familiar method for solving simultaneous equations. It consists of two parts: the elimination phase and the solution phase. As indicated in Table 2.1, the function of the elimination phase is to transform the equations into the form **Ux** = **c**. The equations are then solved by back substitution. In order to illustrate the procedure, let us solve the equations

$$4x_1 - 2x_2 + x_3 = 11 \tag{a}$$

$$-2x_1 + 4x_2 - 2x_3 = -16 \tag{b}$$

$$x_1 - 2x_2 + 4x_3 = 17 \tag{c}$$

Elimination phase The elimination phase utilizes only one of the elementary operations listed in Table 2.1—multiplying one equation (say, equation j) by a constant λ and subtracting it from another equation (equation i). The symbolic representation of this operation is

$$\text{Eq. } (i) \leftarrow \text{Eq. } (i) - \lambda \times \text{Eq. } (j) \tag{2.6}$$

The equation being subtracted, namely Eq. (j), is called the *pivot equation.*

We start the elimination by taking Eq. (a) to be the pivot equation and choosing the multipliers λ so as to eliminate x_1 from Eqs. (b) and (c):

$$\text{Eq. (b)} \leftarrow \text{Eq. (b)} - (-0.5) \times \text{Eq. (a)}$$

$$\text{Eq. (c)} \leftarrow \text{Eq. (c)} - 0.25 \times \text{Eq. (a)}$$

After this transformation, the equations become

$$4x_1 - 2x_2 + x_3 = 11 \tag{a}$$

$$3x_2 - 1.5x_3 = -10.5 \tag{b}$$

$$-1.5x_2 + 3.75x_3 = 14.25 \tag{c}$$

This completes the first pass. Now we pick (b) as the pivot equation and eliminate x_2 from (c):

$$\text{Eq. (c)} \leftarrow \text{Eq. (c)} - (-0.5) \times \text{Eq. (b)}$$

which yields the equations

$$4x_1 - 2x_2 + x_3 = 11 \tag{a}$$

$$3x_2 - 1.5x_3 = -10.5 \tag{b}$$

$$3x_3 = 9 \tag{c}$$

The elimination phase is now complete. The original equations have been replaced by equivalent equations that can be easily solved by back substitution.

As pointed out before, the augmented coefficient matrix is a more convenient instrument for performing the computations. Thus the original equations would be written as

$$\begin{bmatrix} 4 & -2 & 1 & \bigm| & 11 \\ -2 & 4 & -2 & \bigm| & -16 \\ 1 & -2 & 4 & \bigm| & 17 \end{bmatrix}$$

and the equivalent equations produced by the first and the second passes of Gauss elimination would appear as

$$\begin{bmatrix} 4 & -2 & 1 & \bigm| & 11.00 \\ 0 & 3 & -1.5 & \bigm| & -10.50 \\ 0 & -1.5 & 3.75 & \bigm| & 14.25 \end{bmatrix}$$

$$\begin{bmatrix} 4 & -2 & 1 & \bigm| & 11.0 \\ 0 & 3 & -1.5 & \bigm| & -10.5 \\ 0 & 0 & 3 & \bigm| & 9.0 \end{bmatrix}$$

It is important to note that the elementary row operation in Eq. (2.6) leaves the determinant of the coefficient matrix unchanged. This is rather fortunate, since the determinant of a triangular matrix is very easy to compute—it is the product of the diagonal elements (you can verify this quite easily). In other words,

$$|\mathbf{A}| = |\mathbf{U}| = U_{11} \times U_{22} \times \cdots \times U_{nn} \qquad (2.7)$$

Back substitution phase The unknowns can now be computed by back substitution in the manner described in the previous article. Solving Eqs. (c), (b) and (a) in that order, we get

$$x_3 = 9/3 = 3$$

$$x_2 = (-10.5 + 1.5x_3)/3 = [-10.5 + 1.5(3)]/3 = -2$$

$$x_1 = (11 + 2x_2 - x_3)/4 = [11 + 2(-2) - 3]/4 = 1$$

Algorithm for Gauss Elimination Method

Elimination phase Let us look at the equations at some instant during the elimination phase. Assume that the first k rows of **A** have already been transformed to upper triangular form. Therefore, the current pivot equation is the kth equation, and all the equations below it are still to be transformed. This situation is depicted by the augmented coefficient matrix shown below. Note that the components of **A** are not the coefficients of the original equations (except for the first row), since they have been altered by the elimination procedure. The same applies to the components of the constant vector **b**.

$$
\left[\begin{array}{ccccccc|c}
A_{11} & A_{12} & A_{13} & \cdots & A_{1k} & \cdots & A_{1j} & \cdots & A_{1n} & b_1 \\
0 & A_{22} & A_{23} & \cdots & A_{2k} & \cdots & A_{2j} & \cdots & A_{2n} & b_2 \\
0 & 0 & A_{33} & \cdots & A_{3k} & \cdots & A_{3j} & \cdots & A_{3n} & b_3 \\
\vdots & \vdots & \vdots & & \vdots & & \vdots & & \vdots & \vdots \\
0 & 0 & 0 & \cdots & A_{kk} & \cdots & A_{kj} & \cdots & A_{kn} & b_k \\
\vdots & \vdots & \vdots & & \vdots & & \vdots & & \vdots & \vdots \\
0 & 0 & 0 & \cdots & A_{ik} & \cdots & A_{ij} & \cdots & A_{in} & b_i \\
\vdots & \vdots & \vdots & & \vdots & & \vdots & & \vdots & \vdots \\
0 & 0 & 0 & \cdots & A_{nk} & \cdots & A_{nj} & \cdots & A_{nn} & b_n
\end{array}\right]
\begin{array}{l}
\\ \\ \\ \\ \leftarrow \text{ pivot row} \\ \\ \leftarrow \text{ row being} \\ \quad \text{ transformed} \\ \\ \\
\end{array}
$$

Let the ith row be a typical row below the pivot equation that is to be transformed, meaning that the element A_{ik} is to be eliminated. We can achieve this by multiplying the pivot row by $\lambda = A_{ik}/A_{kk}$ and subtracting it from the ith row. The corresponding changes in the ith row are

$$A_{ij} \leftarrow A_{ij} - \lambda A_{kj}, \quad j = k, k+1, \dots, n \tag{2.8a}$$

$$b_i \leftarrow b_i - \lambda b_k \tag{2.8b}$$

To transform the entire coefficient matrix to upper triangular form, k and i in Eqs. (2.8) must have the ranges $k = 1, 2, \dots, n-1$ (chooses the pivot row), $i = k+1, k+2 \dots, n$ (chooses the row to be transformed). The algorithm for the elimination phase now almost writes itself:

```
for k = 1:n-1
    for i= k+1:n
        if A(i,k) ~= 0
            lambda = A(i,k)/A(k,k);
            A(i,k+1:n) = A(i,k+1:n) - lambda*A(k,k+1:n);
            b(i)= b(i) - lambda*b(k);
        end
    end
end
```

In order to avoid unnecessary operations, the above algorithm departs slightly from Eqs. (2.8) in the following ways:

- If A_{ik} happens to be zero, the transformation of row i is skipped.
- The index j in Eq. (2.8a) starts with $k+1$ rather than k. Therefore, A_{ik} is not replaced by zero, but retains its original value. As the solution phase never accesses

the lower triangular portion of the coefficient matrix anyway, its contents are irrelevant.

Back substitution phase   After Gauss elimination the augmented coefficient matrix has the form

$$
\begin{bmatrix} \mathbf{A} \mid \mathbf{b} \end{bmatrix} =
\begin{bmatrix}
A_{11} & A_{12} & A_{13} & \cdots & A_{1n} & b_1 \\
0 & A_{22} & A_{23} & \cdots & A_{2n} & b_2 \\
0 & 0 & A_{33} & \cdots & A_{3n} & b_3 \\
\vdots & \vdots & \vdots & & \vdots & \vdots \\
0 & 0 & 0 & \cdots & A_{nn} & b_n
\end{bmatrix}
$$

The last equation, $A_{nn}x_n = b_n$, is solved first, yielding

$$
x_n = b_n / A_{nn} \tag{2.9}
$$

Consider now the stage of back substitution where $x_n, x_{n-1}, \ldots, x_{k+1}$ have been already been computed (in that order), and we are about to determine x_k from the kth equation

$$
A_{kk}x_k + A_{k,k+1}x_{k+1} + \cdots + A_{kn}x_n = b_k
$$

The solution is

$$
x_k = \left(b_k - \sum_{j=k+1}^{n} A_{kj}x_j \right) \frac{1}{A_{kk}}, \quad k = n-1, n-2, \ldots, 1 \tag{2.10}
$$

The corresponding algorithm for back substitution is:

```
for k = n:-1:1
    b(k) = (b(k) - A(k,k+1:n)*b(k+1:n))/A(k,k);
end
```

■ gauss

The function gauss combines the elimination and the back substitution phases. During back substitution b is overwritten by the solution vector x, so that b contains the solution upon exit.

```
function [x,det] = gauss(A,b)
% Solves A*x = b by Gauss elimination and computes det(A).
% USAGE: [x,det] = gauss(A,b)

if size(b,2) > 1; b = b'; end    % b must be column vector
n = length(b);
```

```
for k = 1:n-1                        % Elimination phase
    for i= k+1:n
        if A(i,k) ~= 0
            lambda = A(i,k)/A(k,k);
            A(i,k+1:n) = A(i,k+1:n) - lambda*A(k,k+1:n);
            b(i)= b(i) - lambda*b(k);
        end
    end
end
if nargout == 2; det = prod(diag(A)); end
for k = n:-1:1                        % Back substitution phase
    b(k) = (b(k) - A(k,k+1:n)*b(k+1:n))/A(k,k);
end
x = b;
```

Multiple Sets of Equations

As mentioned before, it is frequently necessary to solve the equations $\mathbf{Ax} = \mathbf{b}$ for several constant vectors. Let there be m such constant vectors, denoted by $\mathbf{b}_1, \mathbf{b}_2, \ldots, \mathbf{b}_m$ and let the corresponding solution vectors be $\mathbf{x}_1, \mathbf{x}_2, \ldots, \mathbf{x}_m$. We denote multiple sets of equations by $\mathbf{AX} = \mathbf{B}$, where

$$\mathbf{X} = \begin{bmatrix} \mathbf{x}_1 & \mathbf{x}_2 & \cdots & \mathbf{x}_m \end{bmatrix} \qquad \mathbf{B} = \begin{bmatrix} \mathbf{b}_1 & \mathbf{b}_2 & \cdots & \mathbf{b}_m \end{bmatrix}$$

are $n \times m$ matrices whose columns consist of solution vectors and constant vectors, respectively.

An economical way to handle such equations during the elimination phase is to include all m constant vectors in the augmented coefficient matrix, so that they are transformed simultaneously with the coefficient matrix. The solutions are then obtained by back substitution in the usual manner, one vector at a time. It would quite easy to make the corresponding changes in gauss. However, the LU decomposition method, described in the next article, is more versatile in handling multiple constant vectors.

EXAMPLE 2.3
Use Gauss elimination to solve the equations $\mathbf{AX} = \mathbf{B}$, where

$$\mathbf{A} = \begin{bmatrix} 6 & -4 & 1 \\ -4 & 6 & -4 \\ 1 & -4 & 6 \end{bmatrix} \qquad \mathbf{B} = \begin{bmatrix} -14 & 22 \\ 36 & -18 \\ 6 & 7 \end{bmatrix}$$

Solution The augmented coefficient matrix is

$$\begin{bmatrix} 6 & -4 & 1 & -14 & 22 \\ -4 & 6 & -4 & 36 & -18 \\ 1 & -4 & 6 & 6 & 7 \end{bmatrix}$$

The elimination phase consists of the following two passes:

$$\text{row } 2 \leftarrow \text{row } 2 + (2/3) \times \text{row } 1$$

$$\text{row } 3 \leftarrow \text{row } 3 - (1/6) \times \text{row } 1$$

$$\begin{bmatrix} 6 & -4 & 1 & -14 & 22 \\ 0 & 10/3 & -10/3 & 80/3 & -10/3 \\ 0 & -10/3 & 35/6 & 25/3 & 10/3 \end{bmatrix}$$

and

$$\text{row } 3 \leftarrow \text{row } 3 + \text{row } 2$$

$$\begin{bmatrix} 6 & -4 & 1 & -14 & 22 \\ 0 & 10/3 & -10/3 & 80/3 & -10/3 \\ 0 & 0 & 5/2 & 35 & 0 \end{bmatrix}$$

In the solution phase, we first compute $\mathbf{x}_1$ by back substitution:

$$X_{31} = \frac{35}{5/2} = 14$$

$$X_{21} = \frac{80/3 + (10/3)X_{31}}{10/3} = \frac{80/3 + (10/3)14}{10/3} = 22$$

$$X_{11} = \frac{-14 + 4X_{21} - X_{31}}{6} = \frac{-14 + 4(22) - 14}{6} = 10$$

Thus the first solution vector is

$$\mathbf{x}_1 = \begin{bmatrix} X_{11} & X_{21} & X_{31} \end{bmatrix}^T = \begin{bmatrix} 10 & 22 & 14 \end{bmatrix}^T$$

The second solution vector is computed next, also using back substitution:

$$X_{32} = 0$$

$$X_{22} = \frac{-10/3 + (10/3)X_{32}}{10/3} = \frac{-10/3 + 0}{10/3} = -1$$

$$X_{12} = \frac{22 + 4X_{22} - X_{32}}{6} = \frac{22 + 4(-1) - 0}{6} = 3$$

Therefore,

$$\mathbf{x}_2 = \begin{bmatrix} X_{12} & X_{22} & X_{32} \end{bmatrix}^T = \begin{bmatrix} 3 & -1 & 0 \end{bmatrix}^T$$

EXAMPLE 2.4

An $n \times n$ Vandermode matrix $\mathbf{A}$ is defined by

$$A_{ij} = v_i^{n-j}, \quad i = 1, 2, \ldots, n, \quad j = 1, 2, \ldots, n$$

where $\mathbf{v}$ is a vector. In MATLAB a Vandermode matrix can be generated by the command vander(v). Use the function gauss to compute the solution of $\mathbf{Ax} = \mathbf{b}$, where $\mathbf{A}$ is the 6×6 Vandermode matrix generated from the vector

$$\mathbf{v} = \begin{bmatrix} 1.0 & 1.2 & 1.4 & 1.6 & 1.8 & 2.0 \end{bmatrix}^T$$

and

$$\mathbf{b} = \begin{bmatrix} 0 & 1 & 0 & 1 & 0 & 1 \end{bmatrix}^T$$

Also evaluate the accuracy of the solution (Vandermode matrices tend to be ill-conditioned).

Solution We used the program shown below. After constructing $\mathbf{A}$ and $\mathbf{b}$, the output format was changed to long so that the solution would be printed to 14 decimal places. Here are the results:

```
% Example 2.4 (Gauss elimination)
A = vander(1:0.2:2);
b = [0 1 0 1 0 1]';
format long
[x,det] = gauss(A,b)
x =
  1.0e+004 *
   0.04166666666701
  -0.31250000000246
   0.92500000000697
  -1.35000000000972
   0.97093333334002
  -0.27510000000181

det =
   -1.132462079991823e-006
```

As the determinant is quite small relative to the elements of $\mathbf{A}$ (you may want to print $\mathbf{A}$ to verify this), we expect detectable roundoff error. Inspection of $\mathbf{x}$ leads us to suspect that the exact solution is

$$\mathbf{x} = \begin{bmatrix} 1250/3 & -3125 & 9250 & -13500 & 29128/3 & -2751 \end{bmatrix}^T$$

in which case the numerical solution would be accurate to 9 decimal places.

Another way to gauge the accuracy of the solution is to compute $\mathbf{Ax}$ and compare the result to $\mathbf{b}$:

```
>> A*x
ans =
  -0.00000000000091
   0.99999999999909
  -0.00000000000819
   0.99999999998272
  -0.00000000005366
   0.99999999994998
```

The result seems to confirm our previous conclusion.

2.3   LU Decomposition Methods

Introduction

It is possible to show that any square matrix $\mathbf{A}$ can be expressed as a product of a lower triangular matrix $\mathbf{L}$ and an upper triangular matrix $\mathbf{U}$:

$$\mathbf{A} = \mathbf{LU} \tag{2.11}$$

The process of computing $\mathbf{L}$ and $\mathbf{U}$ for a given $\mathbf{A}$ is known as *LU decomposition* or *LU factorization*. LU decomposition is not unique (the combinations of $\mathbf{L}$ and $\mathbf{U}$ for a prescribed $\mathbf{A}$ are endless), unless certain constraints are placed on $\mathbf{L}$ or $\mathbf{U}$. These constraints distinguish one type of decomposition from another. Three commonly used decompositions are listed in Table 2.2.

Name	Constraints
Doolittle's decomposition	$L_{ii} = 1, \quad i = 1, 2, \ldots, n$
Crout's decomposition	$U_{ii} = 1, \quad i = 1, 2, \ldots, n$
Choleski's decomposition	$\mathbf{L} = \mathbf{U}^T$

Table 2.2

After decomposing $\mathbf{A}$, it is easy to solve the equations $\mathbf{Ax} = \mathbf{b}$, as pointed out in Art. 2.1. We first rewrite the equations as $\mathbf{LUx} = \mathbf{b}$. Upon using the notation $\mathbf{Ux} = \mathbf{y}$, the equations become

$$\mathbf{Ly} = \mathbf{b}$$

which can be solved for $\mathbf{y}$ by forward substitution. Then

$$\mathbf{Ux} = \mathbf{y}$$

will yield $\mathbf{x}$ by the back substitution process.

The advantage of LU decomposition over the Gauss elimination method is that once $\mathbf{A}$ is decomposed, we can solve $\mathbf{Ax} = \mathbf{b}$ for as many constant vectors $\mathbf{b}$ as we please. The cost of each additional solution is relatively small, since the forward and back substitution operations are much less time consuming than the decomposition process.

Doolittle's Decomposition Method

Decomposition phase Doolittle's decomposition is closely related to Gauss elimination. In order to illustrate the relationship, consider a 3×3 matrix $\mathbf{A}$ and assume that there exist triangular matrices

$$\mathbf{L} = \begin{bmatrix} 1 & 0 & 0 \\ L_{21} & 1 & 0 \\ L_{31} & L_{32} & 1 \end{bmatrix} \qquad \mathbf{U} = \begin{bmatrix} U_{11} & U_{12} & U_{13} \\ 0 & U_{22} & U_{23} \\ 0 & 0 & U_{33} \end{bmatrix}$$

such that $\mathbf{A} = \mathbf{LU}$. After completing the multiplication on the right hand side, we get

$$\mathbf{A} = \begin{bmatrix} U_{11} & U_{12} & U_{13} \\ U_{11}L_{21} & U_{12}L_{21} + U_{22} & U_{13}L_{21} + U_{23} \\ U_{11}L_{31} & U_{12}L_{31} + U_{22}L_{32} & U_{13}L_{31} + U_{23}L_{32} + U_{33} \end{bmatrix} \qquad (2.12)$$

Let us now apply Gauss elimination to Eq. (2.12). The first pass of the elimination procedure consists of choosing the first row as the pivot row and applying the elementary operations

$$\text{row 2} \leftarrow \text{row 2} - L_{21} \times \text{row 1 (eliminates } A_{21})$$

$$\text{row 3} \leftarrow \text{row 3} - L_{31} \times \text{row 1 (eliminates } A_{31})$$

The result is

$$\mathbf{A}' = \begin{bmatrix} U_{11} & U_{12} & U_{13} \\ 0 & U_{22} & U_{23} \\ 0 & U_{22}L_{32} & U_{23}L_{32} + U_{33} \end{bmatrix}$$

In the next pass we take the second row as the pivot row, and utilize the operation

$$\text{row 3} \leftarrow \text{row 3} - L_{32} \times \text{row 2 (eliminates } A_{32})$$

ending up with

$$\mathbf{A}'' = \mathbf{U} = \begin{bmatrix} U_{11} & U_{12} & U_{13} \\ 0 & U_{22} & U_{23} \\ 0 & 0 & U_{33} \end{bmatrix}$$

The foregoing illustration reveals two important features of Doolittle's decomposition:

- The matrix $\mathbf{U}$ is identical to the upper triangular matrix that results from Gauss elimination.
- The off-diagonal elements of $\mathbf{L}$ are the pivot equation multipliers used during Gauss elimination; that is, L_{ij} is the multiplier that eliminated A_{ij}.

It is usual practice to store the multipliers in the lower triangular portion of the coefficient matrix, replacing the coefficients as they are eliminated (L_{ij} replacing A_{ij}). The diagonal elements of $\mathbf{L}$ do not have to be stored, since it is understood that each of them is unity. The final form of the coefficient matrix would thus be the following mixture of $\mathbf{L}$ and $\mathbf{U}$:

$$[\mathbf{L} \setminus \mathbf{U}] = \begin{bmatrix} U_{11} & U_{12} & U_{13} \\ L_{21} & U_{22} & U_{23} \\ L_{31} & L_{32} & U_{33} \end{bmatrix} \tag{2.13}$$

The algorithm for Doolittle's decomposition is thus identical to the Gauss elimination procedure in `gauss`, except that each multiplier λ is now stored in the lower triangular portion of $\mathbf{A}$.

■ LUdec

In this version of LU decomposition the original $\mathbf{A}$ is destroyed and replaced by its decomposed form $[\mathbf{L} \setminus \mathbf{U}]$.

```
function A = LUdec(A)
% LU decomposition of matrix A; returns A = [L\U].
% USAGE: A = LUdec(A)

n = size(A,1);
for k = 1:n-1
    for i = k+1:n
        if A(i,k) ~= 0.0
            lambda = A(i,k)/A(k,k);
            A(i,k+1:n) = A(i,k+1:n) - lambda*A(k,k+1:n);
```

```
            A(i,k) = lambda;
        end
    end
end
```

Solution phase Consider now the procedure for solving $\mathbf{Ly} = \mathbf{b}$ by forward substitution. The scalar form of the equations is (recall that $L_{ii} = 1$)

$$y_1 = b_1$$

$$L_{21}y_1 + y_2 = b_2$$

$$\vdots$$

$$L_{k1}y_1 + L_{k2}y_2 + \cdots + L_{k,k-1}y_{k-1} + y_k = b_k$$

$$\vdots$$

Solving the kth equation for y_k yields

$$y_k = b_k - \sum_{j=1}^{k-1} L_{kj}y_j, \quad k = 2, 3, \dots, n \tag{2.14}$$

Letting $\mathbf{y}$ overwrite $\mathbf{b}$, we obtain the forward substitution algorithm:

```
for k = 2:n
    y(k)= b(k) - A(k,1:k-1)*y(1:k-1);
end
```

The back substitution phase for solving $\mathbf{Ux} = \mathbf{y}$ is identical to that used in the Gauss elimination method.

■ LUsol

This function carries out the solution phase (forward and back substitutions). It is assumed that the original coefficient matrix has been decomposed, so that the input is $\mathbf{A} = [\mathbf{L} \backslash \mathbf{U}]$. The contents of $\mathbf{b}$ are replaced by $\mathbf{y}$ during forward substitution. Similarly, back substitution overwrites $\mathbf{y}$ with the solution $\mathbf{x}$.

```
function x = LUsol(A,b)
% Solves L*U*b = x, where A contains both L and U;
% that is, A has the form [L\U].
% USAGE: x = LUsol(A,b)
```

```
if size(b,2) > 1; b = b'; end
n = length(b);
for k = 2:n
    b(k) = b(k) - A(k,1:k-1)*b(1:k-1);
end
for k = n:-1:1
    b(k) = (b(k) - A(k,k+1:n)*b(k+1:n))/A(k,k);
end
x = b;
```

Choleski's Decomposition

Choleski's decomposition $\mathbf{A} = \mathbf{LL}^T$ has two limitations:

- Since the matrix product $\mathbf{LL}^T$ is symmetric, Choleski's decomposition requires $\mathbf{A}$ to be *symmetric.*
- The decomposition process involves taking square roots of certain combinations of the elements of $\mathbf{A}$. It can be shown that square roots of negative numbers can be avoided only if $\mathbf{A}$ is *positive definite.*

Although the number of long operations in all the decomposition methods is about the same, Choleski's decomposition is not a particularly popular means of solving simultaneous equations, mainly due to the restrictions listed above. We study it here because it is invaluable in certain other applications (e.g., in the transformation of eigenvalue problems).

Let us start by looking at Choleski's decomposition

$$\mathbf{A} = \mathbf{LL}^T \tag{2.15}$$

of a 3×3 matrix:

$$
\begin{bmatrix}
A_{11} & A_{12} & A_{13} \\
A_{21} & A_{22} & A_{23} \\
A_{31} & A_{32} & A_{33}
\end{bmatrix}
=
\begin{bmatrix}
L_{11} & 0 & 0 \\
L_{21} & L_{22} & 0 \\
L_{31} & L_{32} & L_{33}
\end{bmatrix}
\begin{bmatrix}
L_{11} & L_{21} & L_{31} \\
0 & L_{22} & L_{32} \\
0 & 0 & L_{33}
\end{bmatrix}
$$

After completing the matrix multiplication on the right hand side, we get

$$
\begin{bmatrix}
A_{11} & A_{12} & A_{13} \\
A_{21} & A_{22} & A_{23} \\
A_{31} & A_{32} & A_{33}
\end{bmatrix}
=
\begin{bmatrix}
L_{11}^2 & L_{11}L_{21} & L_{11}L_{31} \\
L_{11}L_{21} & L_{21}^2 + L_{22}^2 & L_{21}L_{31} + L_{22}L_{32} \\
L_{11}L_{31} & L_{21}L_{31} + L_{22}L_{32} & L_{31}^2 + L_{32}^2 + L_{33}^2
\end{bmatrix}
\tag{2.16}
$$

Note that the right-hand-side matrix is symmetric, as pointed out before. Equating the matrices $\mathbf{A}$ and $\mathbf{LL}^T$ element-by-element, we obtain six equations (due to symmetry only lower or upper triangular elements have to be considered) in the six unknown components of **L**. By solving these equations in a certain order, it is possible to have only one unknown in each equation.

Consider the lower triangular portion of each matrix in Eq. (2.16) (the upper triangular portion would do as well). By equating the elements in the first column, starting with the first row and proceeding downward, we can compute L_{11}, L_{21}, and L_{31} in that order:

$$A_{11} = L_{11}^2 \qquad L_{11} = \sqrt{A_{11}}$$

$$A_{21} = L_{11}L_{21} \qquad L_{21} = A_{21}/L_{11}$$

$$A_{31} = L_{11}L_{31} \qquad L_{31} = A_{31}/L_{11}$$

The second column, starting with second row, yields L_{22} and L_{32}:

$$A_{22} = L_{21}^2 + L_{22}^2 \qquad L_{22} = \sqrt{A_{22} - L_{21}^2}$$

$$A_{32} = L_{21}L_{31} + L_{22}L_{32} \qquad L_{32} = (A_{32} - L_{21}L_{31})/L_{22}$$

Finally the third column, third row gives us L_{33}:

$$A_{33} = L_{31}^2 + L_{32}^2 + L_{33}^2 \qquad L_{33} = \sqrt{A_{33} - L_{31}^2 - L_{32}^2}$$

We can now extrapolate the results for an $n \times n$ matrix. We observe that a typical element in the lower triangular portion of $\mathbf{LL}^T$ is of the form

$$(\mathbf{LL}^T)_{ij} = L_{i1}L_{j1} + L_{i2}L_{j2} + \cdots + L_{ij}L_{jj} = \sum_{k=1}^{j} L_{ik}L_{jk}, \quad i \geq j$$

Equating this term to the corresponding element of **A** yields

$$A_{ij} = \sum_{k=1}^{j} L_{ik}L_{jk}, \quad i = j, j+1, \ldots, n, \quad j = 1, 2, \ldots, n \qquad (2.17)$$

The range of indices shown limits the elements to the lower triangular part. For the first column ($j = 1$), we obtain from Eq. (2.17)

$$L_{11} = \sqrt{A_{11}} \qquad L_{i1} = A_{i1}/L_{11}, \quad i = 2, 3, \ldots, n \qquad (2.18)$$

Proceeding to other columns, we observe that the unknown in Eq. (2.17) is L_{ij} (the other elements of **L** appearing in the equation have already been computed). Taking the term containing L_{ij} outside the summation in Eq. (2.17), we obtain

$$A_{ij} = \sum_{k=1}^{j-1} L_{ik}L_{jk} + L_{ij}L_{jj}$$

If $i = j$ (a diagonal term), the solution is

$$L_{jj} = \sqrt{A_{jj} - \sum_{k=1}^{j-1} L_{jk}^2}, \quad j = 2, 3, \ldots, n \tag{2.19}$$

For a nondiagonal term we get

$$L_{ij} = \left(A_{ij} - \sum_{k=1}^{j-1} L_{ik} L_{jk} \right) / L_{jj}, \quad j = 2, 3, \ldots, n-1, \quad i = j+1, j+2, \ldots, n \tag{2.20}$$

■ choleski

Note that in Eqs. (2.19) and (2.20) A_{ij} appears only in the formula for L_{ij}. Therefore, once L_{ij} has been computed, A_{ij} is no longer needed. This makes it possible to write the elements of **L** over the lower triangular portion of **A** as they are computed. The elements above the principal diagonal of **A** will remain untouched. At the conclusion of decomposition **L** is extracted with the MATLAB command tril(A). If a negative L_{jj}^2 is encountered during decomposition, an error message is printed and the program is terminated.

```
function L = choleski(A)
% Computes L in Choleski's decomposition A = LL'.
% USAGE: L = choleski(A)

n = size(A,1);
for j = 1:n
    temp = A(j,j) - dot(A(j,1:j-1),A(j,1:j-1));
    if temp < 0.0
        error('Matrix is not positive definite')
    end
    A(j,j) = sqrt(temp);
    for i = j+1:n
        A(i,j)=(A(i,j) - dot(A(i,1:j-1),A(j,1:j-1)))/A(j,j);
    end
end
L = tril(A)
```

We could also write the algorithm for forward and back substitutions that are necessary in the solution of **Ax** = **b**. But since Choleski's decomposition has no advantages over Doolittle's decomposition in the solution of simultaneous equations, we will skip that part.

EXAMPLE 2.5

Use Doolittle's decomposition method to solve the equations $\mathbf{Ax} = \mathbf{b}$, where

$$\mathbf{A} = \begin{bmatrix} 1 & 4 & 1 \\ 1 & 6 & -1 \\ 2 & -1 & 2 \end{bmatrix} \qquad \mathbf{b} = \begin{bmatrix} 7 \\ 13 \\ 5 \end{bmatrix}$$

Solution We first decompose $\mathbf{A}$ by Gauss elimination. The first pass consists of the elementary operations

$$\text{row } 2 \leftarrow \text{row } 2 - 1 \times \text{row } 1 \text{ (eliminates } A_{21})$$

$$\text{row } 3 \leftarrow \text{row } 3 - 2 \times \text{row } 1 \text{ (eliminates } A_{31})$$

Storing the multipliers $L_{21} = 1$ and $L_{31} = 2$ in place of the eliminated terms, we obtain

$$\mathbf{A}' = \begin{bmatrix} 1 & 4 & 1 \\ 1 & 2 & -2 \\ 2 & -9 & 0 \end{bmatrix}$$

The second pass of Gauss elimination uses the operation

$$\text{row } 3 \leftarrow \text{row } 3 - (-4.5) \times \text{row } 2 \text{ (eliminates } A_{32})$$

Storing the multiplier $L_{32} = -4.5$ in place of A_{32}, we get

$$\mathbf{A}'' = [\mathbf{L} \backslash \mathbf{U}] = \begin{bmatrix} 1 & 4 & 1 \\ 1 & 2 & -2 \\ 2 & -4.5 & -9 \end{bmatrix}$$

The decomposition is now complete, with

$$\mathbf{L} = \begin{bmatrix} 1 & 0 & 0 \\ 1 & 1 & 0 \\ 2 & -4.5 & 1 \end{bmatrix} \qquad \mathbf{U} = \begin{bmatrix} 1 & 4 & 1 \\ 0 & 2 & -2 \\ 0 & 0 & -9 \end{bmatrix}$$

Solution of $\mathbf{Ly} = \mathbf{b}$ by forward substitution comes next. The augmented coefficient form of the equations is

$$\begin{bmatrix} \mathbf{L} & | & \mathbf{b} \end{bmatrix} = \begin{bmatrix} 1 & 0 & 0 & | & 7 \\ 1 & 1 & 0 & | & 13 \\ 2 & -4.5 & 1 & | & 5 \end{bmatrix}$$

The solution is

$$y_1 = 7$$

$$y_2 = 13 - y_1 = 13 - 7 = 6$$

$$y_3 = 5 - 2y_1 + 4.5y_2 = 5 - 2(7) + 4.5(6) = 18$$

Finally, the equations $\mathbf{Ux} = \mathbf{y}$, or

$$\begin{bmatrix} \mathbf{U} \mid \mathbf{y} \end{bmatrix} = \begin{bmatrix} 1 & 4 & 1 & 7 \\ 0 & 2 & -2 & 6 \\ 0 & 0 & -9 & 18 \end{bmatrix}$$

are solved by back substitution. This yields

$$x_3 = \frac{18}{-9} = -2$$

$$x_2 = \frac{6 + 2x_3}{2} = \frac{6 + 2(-2)}{2} = 1$$

$$x_1 = 7 - 4x_2 - x_3 = 7 - 4(1) - (-2) = 5$$

EXAMPLE 2.6
Compute Choleski's decomposition of the matrix

$$\mathbf{A} = \begin{bmatrix} 4 & -2 & 2 \\ -2 & 2 & -4 \\ 2 & -4 & 11 \end{bmatrix}$$

Solution First we note that $\mathbf{A}$ is symmetric. Therefore, Choleski's decomposition is applicable, provided that the matrix is also positive definite. An *a priori* test for positive definiteness is not needed, since the decomposition algorithm contains its own test: if the square root of a negative number is encountered, the matrix is not positive definite and the decomposition fails.

Substituting the given matrix for $\mathbf{A}$ in Eq. (2.16), we obtain

$$\begin{bmatrix} 4 & -2 & 2 \\ -2 & 2 & -4 \\ 2 & -4 & 11 \end{bmatrix} = \begin{bmatrix} L_{11}^2 & L_{11}L_{21} & L_{11}L_{31} \\ L_{11}L_{21} & L_{21}^2 + L_{22}^2 & L_{21}L_{31} + L_{22}L_{32} \\ L_{11}L_{31} & L_{21}L_{31} + L_{22}L_{32} & L_{31}^2 + L_{32}^2 + L_{33}^2 \end{bmatrix}$$

Equating the elements in the lower (or upper) triangular portions yields

$$L_{11} = \sqrt{4} = 2$$

$$L_{21} = -2/L_{11} = -2/2 = -1$$

$$L_{31} = 2/L_{11} = 2/2 = 1$$

$$L_{22} = \sqrt{2 - L_{21}^2} = \sqrt{2 - 1^2} = 1$$

$$L_{32} = \frac{-4 - L_{21}L_{31}}{L_{22}} = \frac{-4 - (-1)(1)}{1} = -3$$

$$L_{33} = \sqrt{11 - L_{31}^2 - L_{32}^2} = \sqrt{11 - (1)^2 - (-3)^2} = 1$$

Therefore,

$$L = \begin{bmatrix} 2 & 0 & 0 \\ -1 & 1 & 0 \\ 1 & -3 & 1 \end{bmatrix}$$

The result can easily be verified by performing the multiplication LL^T.

EXAMPLE 2.7

Solve $AX = B$ with Doolittle's decomposition and compute $|A|$, where

$$A = \begin{bmatrix} 3 & -1 & 4 \\ -2 & 0 & 5 \\ 7 & 2 & -2 \end{bmatrix} \qquad B = \begin{bmatrix} 6 & -4 \\ 3 & 2 \\ 7 & -5 \end{bmatrix}$$

Solution In the program below the coefficient matrix A is first decomposed by calling LUdec. Then LUsol is used to compute the solution one vector at a time.

```
% Example 2.7 (Doolittle's decomposition)
A = [3 -1 4; -2 0 5; 7 2 -2];
B = [6 -4; 3 2; 7 -5];
A = LUdec(A);
det = prod(diag(A))
for i = 1:size(B,2)
    X(:,i) = LUsol(A,B(:,i));
end
X
```

Here are the results:

```
>> det =
   -77
X =
    1.0000   -1.0000
    1.0000    1.0000
    1.0000    0.0000
```

EXAMPLE 2.8

Test the function choleski by decomposing

$$A = \begin{bmatrix} 1.44 & -0.36 & 5.52 & 0.00 \\ -0.36 & 10.33 & -7.78 & 0.00 \\ 5.52 & -7.78 & 28.40 & 9.00 \\ 0.00 & 0.00 & 9.00 & 61.00 \end{bmatrix}$$

Solution

```
% Example 2.8 (Choleski decomposition)
A = [1.44 -0.36  5.52   0.00;
    -0.36 10.33 -7.78   0.00;
     5.52 -7.78 28.40   9.00;
     0.00  0.00  9.00  61.00];
 L = choleski(A)
 Check = L*L'   % Verify the result
```

```
>> L =
     1.2000         0         0         0
    -0.3000    3.2000         0         0
     4.6000   -2.0000    1.8000         0
          0         0    5.0000    6.0000
Check =
     1.4400   -0.3600    5.5200         0
    -0.3600   10.3300   -7.7800         0
     5.5200   -7.7800   28.4000    9.0000
          0         0    9.0000   61.0000
```

PROBLEM SET 2.1

1. By evaluating the determinant, classify the following matrices as singular, ill-conditioned or well-conditioned.

    (a) $\mathbf{A} = \begin{bmatrix} 1 & 2 & 3 \\ 2 & 3 & 4 \\ 3 & 4 & 5 \end{bmatrix}$

    (b) $\mathbf{A} = \begin{bmatrix} 2.11 & -0.80 & 1.72 \\ -1.84 & 3.03 & 1.29 \\ -1.57 & 5.25 & 4.30 \end{bmatrix}$

    (c) $\mathbf{A} = \begin{bmatrix} 2 & -1 & 0 \\ -1 & 2 & -1 \\ 0 & -1 & 2 \end{bmatrix}$

    (d) $\mathbf{A} = \begin{bmatrix} 4 & 3 & -1 \\ 7 & -2 & 3 \\ 5 & -18 & 13 \end{bmatrix}$

2. Given the LU decomposition $\mathbf{A} = \mathbf{LU}$, determine $\mathbf{A}$ and $|\mathbf{A}|$.

    (a) $\mathbf{L} = \begin{bmatrix} 1 & 0 & 0 \\ 1 & 1 & 0 \\ 1 & 5/3 & 1 \end{bmatrix}$ $\mathbf{U} = \begin{bmatrix} 1 & 2 & 4 \\ 0 & 3 & 21 \\ 0 & 0 & 0 \end{bmatrix}$

    (b) $\mathbf{L} = \begin{bmatrix} 2 & 0 & 0 \\ -1 & 1 & 0 \\ 1 & -3 & 1 \end{bmatrix}$ $\mathbf{U} = \begin{bmatrix} 2 & -1 & 1 \\ 0 & 1 & -3 \\ 0 & 0 & 1 \end{bmatrix}$

3. Utilize the results of LU decomposition

$$A = LU = \begin{bmatrix} 1 & 0 & 0 \\ 3/2 & 1 & 0 \\ 1/2 & 11/13 & 1 \end{bmatrix} \begin{bmatrix} 2 & -3 & -1 \\ 0 & 13/2 & -7/2 \\ 0 & 0 & 32/13 \end{bmatrix}$$

to solve $Ax = b$, where $b^T = \begin{bmatrix} 1 & -1 & 2 \end{bmatrix}$.

4. Use Gauss elimination to solve the equations $Ax = b$, where

$$A = \begin{bmatrix} 2 & -3 & -1 \\ 3 & 2 & -5 \\ 2 & 4 & -1 \end{bmatrix} \qquad b = \begin{bmatrix} 3 \\ -9 \\ -5 \end{bmatrix}$$

5. Solve the equations $AX = B$ by Gauss elimination, where

$$A = \begin{bmatrix} 2 & 0 & -1 & 0 \\ 0 & 1 & 2 & 0 \\ -1 & 2 & 0 & 1 \\ 0 & 0 & 1 & -2 \end{bmatrix} \qquad B = \begin{bmatrix} 1 & 0 \\ 0 & 0 \\ 0 & 1 \\ 0 & 0 \end{bmatrix}$$

6. Solve the equations $Ax = b$ by Gauss elimination, where

$$A = \begin{bmatrix} 0 & 0 & 2 & 1 & 2 \\ 0 & 1 & 0 & 2 & -1 \\ 1 & 2 & 0 & -2 & 0 \\ 0 & 0 & 0 & -1 & 1 \\ 0 & 1 & -1 & 1 & -1 \end{bmatrix} \qquad b = \begin{bmatrix} 1 \\ 1 \\ -4 \\ -2 \\ -1 \end{bmatrix}$$

Hint: reorder the equations before solving.

7. Find L and U so that

$$A = LU = \begin{bmatrix} 4 & -1 & 0 \\ -1 & 4 & -1 \\ 0 & -1 & 4 \end{bmatrix}$$

using (a) Doolittle's decomposition; (b) Choleski's decomposition.

8. Use Doolittle's decomposition method to solve $Ax = b$, where

$$A = \begin{bmatrix} -3 & 6 & -4 \\ 9 & -8 & 24 \\ -12 & 24 & -26 \end{bmatrix} \qquad b = \begin{bmatrix} -3 \\ 65 \\ -42 \end{bmatrix}$$

9. Solve the equations $Ax = b$ by Doolittle's decomposition method, where

$$A = \begin{bmatrix} 2.34 & -4.10 & 1.78 \\ -1.98 & 3.47 & -2.22 \\ 2.36 & -15.17 & 6.18 \end{bmatrix} \qquad b = \begin{bmatrix} 0.02 \\ -0.73 \\ -6.63 \end{bmatrix}$$

10. Solve the equations $\mathbf{AX} = \mathbf{B}$ by Doolittle's decomposition method, where

$$\mathbf{A} = \begin{bmatrix} 4 & -3 & 6 \\ 8 & -3 & 10 \\ -4 & 12 & -10 \end{bmatrix} \qquad \mathbf{B} = \begin{bmatrix} 1 & 0 \\ 0 & 1 \\ 0 & 0 \end{bmatrix}$$

11. Solve the equations $\mathbf{Ax} = \mathbf{b}$ by Choleski's decomposition method, where

$$\mathbf{A} = \begin{bmatrix} 1 & 1 & 1 \\ 1 & 2 & 2 \\ 1 & 2 & 3 \end{bmatrix} \qquad \mathbf{b} = \begin{bmatrix} 1 \\ 3/2 \\ 3 \end{bmatrix}$$

12. Solve the equations

$$\begin{bmatrix} 4 & -2 & -3 \\ 12 & 4 & -10 \\ -16 & 28 & 18 \end{bmatrix} \begin{bmatrix} x_1 \\ x_2 \\ x_3 \end{bmatrix} = \begin{bmatrix} 1.1 \\ 0 \\ -2.3 \end{bmatrix}$$

by Doolittle's decomposition method.

13. Determine $\mathbf{L}$ that results from Choleski's decomposition of the diagonal matrix

$$\mathbf{A} = \begin{bmatrix} \alpha_1 & 0 & 0 & \cdots \\ 0 & \alpha_2 & 0 & \cdots \\ 0 & 0 & \alpha_3 & \cdots \\ \vdots & \vdots & \vdots & \ddots \end{bmatrix}$$

14. ■ Modify the function gauss so that it will work with m constant vectors. Test the program by solving $\mathbf{AX} = \mathbf{B}$, where

$$\mathbf{A} = \begin{bmatrix} 2 & -1 & 0 \\ -1 & 2 & -1 \\ 0 & -1 & 1 \end{bmatrix} \qquad \mathbf{B} = \begin{bmatrix} 1 & 0 & 0 \\ 0 & 1 & 0 \\ 0 & 0 & 1 \end{bmatrix}$$

15. ■ A well-known example of an ill-conditioned matrix is the *Hilbert matrix*

$$\mathbf{A} = \begin{bmatrix} 1 & 1/2 & 1/3 & \cdots \\ 1/2 & 1/3 & 1/4 & \cdots \\ 1/3 & 1/4 & 1/5 & \cdots \\ \vdots & \vdots & \vdots & \ddots \end{bmatrix}$$

Write a program that specializes in solving the equations $\mathbf{Ax} = \mathbf{b}$ by Doolittle's decomposition method, where $\mathbf{A}$ is the Hilbert matrix of arbitrary size $n \times n$, and

$$b_i = \sum_{j=1}^{n} A_{ij}$$

The program should have no input apart from n. By running the program, determine the largest n for which the solution is within 6 significant figures of the exact solution

$$\mathbf{x} = \begin{bmatrix} 1 & 1 & 1 & \cdots \end{bmatrix}^T$$

(the results depend on the software and the hardware used).

16. ■ Write a function for the solution phase of Choleski's decomposition method. Test the function by solving the equations $\mathbf{Ax} = \mathbf{b}$, where

$$\mathbf{A} = \begin{bmatrix} 4 & -2 & 2 \\ -2 & 2 & -4 \\ 2 & -4 & 11 \end{bmatrix} \qquad \mathbf{b} = \begin{bmatrix} 6 \\ -10 \\ 27 \end{bmatrix}$$

Use the function `choleski` for the decomposition phase.

17. ■ Determine the coefficients of the polynomial $y = a_0 + a_1 x + a_2 x^2 + a_3 x^3$ that passes through the points $(0, 10)$, $(1, 35)$, $(3, 31)$ and $(4, 2)$.

18. ■ Determine the 4th degree polynomial $y(x)$ that passes through the points $(0, -1)$, $(1, 1)$, $(3, 3)$, $(5, 2)$ and $(6, -2)$.

19. ■ Find the 4th degree polynomial $y(x)$ that passes through the points $(0, 1)$, $(0.75, -0.25)$ and $(1, 1)$, and has zero curvature at $(0, 1)$ and $(1, 1)$.

20. ■ Solve the equations $\mathbf{Ax} = \mathbf{b}$, where

$$\mathbf{A} = \begin{bmatrix} 3.50 & 2.77 & -0.76 & 1.80 \\ -1.80 & 2.68 & 3.44 & -0.09 \\ 0.27 & 5.07 & 6.90 & 1.61 \\ 1.71 & 5.45 & 2.68 & 1.71 \end{bmatrix} \qquad \mathbf{b} = \begin{bmatrix} 7.31 \\ 4.23 \\ 13.85 \\ 11.55 \end{bmatrix}$$

By computing $|\mathbf{A}|$ and $\mathbf{Ax}$ comment on the accuracy of the solution.

2.4   Symmetric and Banded Coefficient Matrices

Introduction

Engineering problems often lead to coefficient matrices that are *sparsely populated*, meaning that most elements of the matrix are zero. If all the nonzero terms are clustered about the leading diagonal, then the matrix is said to be *banded*. An example of a banded matrix is

$$
\mathbf{A} = \begin{bmatrix}
X & X & 0 & 0 & 0 \\
X & X & X & 0 & 0 \\
0 & X & X & X & 0 \\
0 & 0 & X & X & X \\
0 & 0 & 0 & X & X
\end{bmatrix}
$$

where X's denote the nonzero elements that form the populated band (some of these elements may be zero). All the elements lying outside the band are zero. The matrix shown above has a bandwidth of three, since there are at most three nonzero elements in each row (or column). Such a matrix is called *tridiagonal*.

If a banded matrix is decomposed in the form $\mathbf{A} = \mathbf{LU}$, both $\mathbf{L}$ and $\mathbf{U}$ will retain the banded structure of $\mathbf{A}$. For example, if we decomposed the matrix shown above, we would get

$$
\mathbf{L} = \begin{bmatrix}
X & 0 & 0 & 0 & 0 \\
X & X & 0 & 0 & 0 \\
0 & X & X & 0 & 0 \\
0 & 0 & X & X & 0 \\
0 & 0 & 0 & X & X
\end{bmatrix}
\qquad
\mathbf{U} = \begin{bmatrix}
X & X & 0 & 0 & 0 \\
0 & X & X & 0 & 0 \\
0 & 0 & X & X & 0 \\
0 & 0 & 0 & X & X \\
0 & 0 & 0 & 0 & X
\end{bmatrix}
$$

The banded structure of a coefficient matrix can be exploited to save storage and computation time. If the coefficient matrix is also symmetric, further economies are possible. In this article we show how the methods of solution discussed previously can be adapted for banded and symmetric coefficient matrices.

Tridiagonal Coefficient Matrix

Consider the solution of $\mathbf{Ax} = \mathbf{b}$ by Doolittle's decomposition, where $\mathbf{A}$ is the $n \times n$ tridiagonal matrix

$$
\mathbf{A} = \begin{bmatrix}
d_1 & e_1 & 0 & 0 & \cdots & 0 \\
c_1 & d_2 & e_2 & 0 & \cdots & 0 \\
0 & c_2 & d_3 & e_3 & \cdots & 0 \\
0 & 0 & c_3 & d_4 & \cdots & 0 \\
\vdots & \vdots & \vdots & \vdots & \ddots & \vdots \\
0 & 0 & \cdots & 0 & c_{n-1} & d_n
\end{bmatrix}
$$

As the notation implies, we are storing the nonzero elements of $\mathbf{A}$ in the vectors

$$\mathbf{c} = \begin{bmatrix} c_1 \\ c_2 \\ \vdots \\ c_{n-1} \end{bmatrix} \qquad \mathbf{d} = \begin{bmatrix} d_1 \\ d_2 \\ \vdots \\ d_{n-1} \\ d_n \end{bmatrix} \qquad \mathbf{e} = \begin{bmatrix} e_1 \\ e_2 \\ \vdots \\ e_{n-1} \end{bmatrix}$$

The resulting saving of storage can be significant. For example, a 100×100 tridiagonal matrix, containing 10,000 elements, can be stored in only $99 + 100 + 99 = 298$ locations, which represents a compression ratio of about 33:1.

We now apply LU decomposition to the coefficient matrix. We reduce row k by getting rid of c_{k-1} with the elementary operation

$$\text{row } k \leftarrow \text{row } k - (c_{k-1}/d_{k-1}) \times \text{row } (k-1), \quad k = 2, 3, \ldots, n$$

The corresponding change in d_k is

$$d_k \leftarrow d_k - (c_{k-1}/d_{k-1})e_{k-1} \tag{2.21}$$

whereas e_k is not affected. In order to finish up with Doolittle's decomposition of the form $[\mathbf{L}\backslash\mathbf{U}]$, we store the multiplier $\lambda = c_{k-1}/d_{k-1}$ in the location previously occupied by c_{k-1}:

$$c_{k-1} \leftarrow c_{k-1}/d_{k-1} \tag{2.22}$$

Thus the decomposition algorithm is

```
for k = 2:n
    lambda = c(k-1)/d(k-1);
    d(k) = d(k) - lambda*e(k-1);
    c(k-1) = lambda;
end
```

Next we look at the solution phase, i.e., the solution of the $\mathbf{Ly} = \mathbf{b}$, followed by $\mathbf{Ux} = \mathbf{y}$. The equations $\mathbf{Ly} = \mathbf{b}$ can be portrayed by the augmented coefficient matrix

$$\begin{bmatrix} \mathbf{L} \mid \mathbf{b} \end{bmatrix} = \begin{bmatrix} 1 & 0 & 0 & 0 & \cdots & 0 & b_1 \\ c_1 & 1 & 0 & 0 & \cdots & 0 & b_2 \\ 0 & c_2 & 1 & 0 & \cdots & 0 & b_3 \\ 0 & 0 & c_3 & 1 & \cdots & 0 & b_4 \\ \vdots & \vdots & \vdots & \vdots & \cdots & \vdots & \vdots \\ 0 & 0 & \cdots & 0 & c_{n-1} & 1 & b_n \end{bmatrix}$$

Note that the original contents of **c** were destroyed and replaced by the multipliers during the decomposition. The solution algorithm for **y** by forward substitution is

```
y(1) = b(1)
for k = 2:n
    y(k) = b(k) - c(k-1)*y(k-1);
end
```

The augmented coefficient matrix representing $\mathbf{Ux} = \mathbf{y}$ is

$$
\begin{bmatrix} \mathbf{U} \mid \mathbf{y} \end{bmatrix} =
\begin{bmatrix}
d_1 & e_1 & 0 & \cdots & 0 & 0 & y_1 \\
0 & d_2 & e_2 & \cdots & 0 & 0 & y_2 \\
0 & 0 & d_3 & \cdots & 0 & 0 & y_3 \\
\vdots & \vdots & \vdots & & \vdots & \vdots & \vdots \\
0 & 0 & 0 & \cdots & d_{n-1} & e_{n-1} & y_{n-1} \\
0 & 0 & 0 & \cdots & 0 & d_n & y_n
\end{bmatrix}
$$

Note again that the contents of **d** were altered from the original values during the decomposition phase (but **e** was unchanged). The solution for **x** is obtained by back substitution using the algorithm

```
x(n) = y(n)/d(n);
for k = n-1:-1:1
    x(k) = (y(k) - e(k)*x(k+1))/d(k);
end
```

■ LUdec3

The function LUdec3 contains the code for the decomposition phase. The original vectors **c** and **d** are destroyed and replaced by the vectors of the decomposed matrix.

```
function [c,d,e] = LUdec3(c,d,e)
% LU decomposition of tridiagonal matrix A = [c\d\e].
% USAGE: [c,d,e] = LUdec3(c,d,e)

n = length(d);
for k = 2:n
    lambda = c(k-1)/d(k-1);
    d(k) = d(k) - lambda*e(k-1);
    c(k-1) = lambda;
end
```

■ **LUsol3**

This is the function for the solution phase. The vector $\mathbf{y}$ overwrites the constant vector $\mathbf{b}$ during the forward substitution. Similarly, the solution vector $\mathbf{x}$ replaces $\mathbf{y}$ in the back substitution process.

```
function x = LUsol3(c,d,e,b)
% Solves A*x = b where A = [c\d\e] is the LU
% decomposition of the original tridiagonal A.
% USAGE: x = LUsol3(c,d,e,b)

n = length(d);
for k = 2:n                % Forward substitution
    b(k) = b(k) - c(k-1)*b(k-1);
end
b(n) = b(n)/d(n);          % Back substitution
for k = n-1:-1:1
    b(k) = (b(k) -e(k)*b(k+1))/d(k);
end
x = b;
```

Symmetric Coefficient Matrices

More often than not, coefficient matrices that arise in engineering problems are symmetric as well as banded. Therefore, it is worthwhile to discover special properties of such matrices, and learn how to utilize them in the construction of efficient algorithms.

If the matrix $\mathbf{A}$ is symmetric, then the LU decomposition can be presented in the form

$$\mathbf{A} = \mathbf{LU} = \mathbf{LDL}^T \tag{2.23}$$

where $\mathbf{D}$ is a diagonal matrix. An example is Choleski's decomposition $\mathbf{A} = \mathbf{LL}^T$ that was discussed in the previous article (in this case $\mathbf{D} = \mathbf{I}$). For Doolittle's decomposition we have

$$\mathbf{U} = \mathbf{DL}^T = \begin{bmatrix} D_1 & 0 & 0 & \cdots & 0 \\ 0 & D_2 & 0 & \cdots & 0 \\ 0 & 0 & D_3 & \cdots & 0 \\ \vdots & \vdots & \vdots & \ddots & \vdots \\ 0 & 0 & 0 & \cdots & D_n \end{bmatrix} \begin{bmatrix} 1 & L_{21} & L_{31} & \cdots & L_{n1} \\ 0 & 1 & L_{32} & \cdots & L_{n2} \\ 0 & 0 & 1 & \cdots & L_{n3} \\ \vdots & \vdots & \vdots & \ddots & \vdots \\ 0 & 0 & 0 & \cdots & 1 \end{bmatrix}$$

which gives

$$
\mathbf{U} = \begin{bmatrix}
D_1 & D_1 L_{21} & D_1 L_{31} & \cdots & D_1 L_{n1} \\
0 & D_2 & D_2 L_{32} & \cdots & D_2 L_{n2} \\
0 & 0 & D_3 & \cdots & D_3 L_{3n} \\
\vdots & \vdots & \vdots & \ddots & \vdots \\
0 & 0 & 0 & \cdots & D_n
\end{bmatrix}
\tag{2.24}
$$

We see that during decomposition of a symmetric matrix only $\mathbf{U}$ has to be stored, since $\mathbf{D}$ and $\mathbf{L}$ can be easily recovered from $\mathbf{U}$. Thus Gauss elimination, which results in an upper triangular matrix of the form shown in Eq. (2.24), is sufficient to decompose a symmetric matrix.

There is an alternative storage scheme that can be employed during **LU** decomposition. The idea is to arrive at the matrix

$$
\mathbf{U}^* = \begin{bmatrix}
D_1 & L_{21} & L_{31} & \cdots & L_{n1} \\
0 & D_2 & L_{32} & \cdots & L_{n2} \\
0 & 0 & D_3 & \cdots & L_{n3} \\
\vdots & \vdots & \vdots & \ddots & \vdots \\
0 & 0 & 0 & \cdots & D_n
\end{bmatrix}
\tag{2.25}
$$

Here $\mathbf{U}$ can be recovered from $U_{ij} = D_i L_{ji}$. It turns out that this scheme leads to a computationally more efficient solution phase; therefore, we adopt it for symmetric, banded matrices.

Symmetric, Pentadiagonal Coefficient Matrix

We encounter pentadiagonal (bandwidth $= 5$) coefficient matrices in the solution of fourth-order, ordinary differential equations by finite differences. Often these matrices are symmetric, in which case an $n \times n$ matrix has the form

$$
\mathbf{A} = \begin{bmatrix}
d_1 & e_1 & f_1 & 0 & 0 & 0 & \cdots & 0 \\
e_1 & d_2 & e_2 & f_2 & 0 & 0 & \cdots & 0 \\
f_1 & e_2 & d_3 & e_3 & f_3 & 0 & \cdots & 0 \\
0 & f_2 & e_3 & d_4 & e_4 & f_4 & \cdots & 0 \\
\vdots & \vdots & \vdots & \vdots & \vdots & \vdots & \ddots & \vdots \\
0 & \cdots & 0 & f_{n-4} & e_{n-3} & d_{n-2} & e_{n-2} & f_{n-2} \\
0 & \cdots & 0 & 0 & f_{n-3} & e_{n-2} & d_{n-1} & e_{n-1} \\
0 & \cdots & 0 & 0 & 0 & f_{n-2} & e_{n-1} & d_n
\end{bmatrix}
\tag{2.26}
$$

As in the case of tridiagonal matrices, we store the nonzero elements in the three vectors

$$
\mathbf{d} = \begin{bmatrix} d_1 \\ d_2 \\ \vdots \\ d_{n-2} \\ d_{n-1} \\ d_n \end{bmatrix} \qquad \mathbf{e} = \begin{bmatrix} e_1 \\ e_2 \\ \vdots \\ e_{n-2} \\ e_{n-1} \end{bmatrix} \qquad \mathbf{f} = \begin{bmatrix} f_1 \\ f_2 \\ \vdots \\ f_{n-2} \end{bmatrix}
$$

Let us now look at the solution of the equations $\mathbf{Ax} = \mathbf{b}$ by Doolittle's decomposition. The first step is to transform $\mathbf{A}$ to upper triangular form by Gauss elimination. If elimination has progressed to the stage where the kth row has become the pivot row, we have the following situation:

$$
\mathbf{A} = \begin{bmatrix}
\ddots & \vdots & \vdots & \vdots & \vdots & \vdots & \vdots & \vdots & \\
\cdots & 0 & d_k & e_k & f_k & 0 & 0 & 0 & \cdots \\
\cdots & 0 & e_k & d_{k+1} & e_{k+1} & f_{k+1} & 0 & 0 & \cdots \\
\cdots & 0 & f_k & e_{k+1} & d_{k+2} & e_{k+2} & f_{k+2} & 0 & \cdots \\
\cdots & 0 & 0 & f_{k+1} & e_{k+2} & d_{k+3} & e_{k+3} & f_{k+3} & \cdots \\
& \vdots & \vdots & \vdots & \vdots & \vdots & \vdots & \vdots & \ddots
\end{bmatrix} \quad \leftarrow
$$

The elements e_k and f_k below the pivot row are eliminated by the operations

$$\text{row } (k+1) \leftarrow \text{row } (k+1) - (e_k/d_k) \times \text{row } k$$

$$\text{row } (k+2) \leftarrow \text{row } (k+2) - (f_k/d_k) \times \text{row } k$$

The only terms (other than those being eliminated) that are changed by the above operations are

$$
\begin{aligned}
d_{k+1} &\leftarrow d_{k+1} - (e_k/d_k)e_k \\
e_{k+1} &\leftarrow e_{k+1} - (e_k/d_k)f_k \\
d_{k+2} &\leftarrow d_{k+2} - (f_k/d_k)f_k
\end{aligned}
\qquad (2.27\text{a})
$$

Storage of the multipliers in the *upper* triangular portion of the matrix results in

$$
e_k \leftarrow e_k/d_k \qquad f_k \leftarrow f_k/d_k \qquad (2.27\text{b})
$$

At the conclusion of the elimination phase the matrix has the form (do not confuse $\mathbf{d}$, $\mathbf{e}$ and $\mathbf{f}$ with the original contents of $\mathbf{A}$)

$$\mathbf{U}^* = \begin{bmatrix} d_1 & e_1 & f_1 & 0 & \cdots & 0 \\ 0 & d_2 & e_2 & f_2 & \cdots & 0 \\ 0 & 0 & d_3 & e_3 & \cdots & 0 \\ \vdots & \vdots & \vdots & \vdots & \ddots & \vdots \\ 0 & 0 & \cdots & 0 & d_{n-1} & e_{n-1} \\ 0 & 0 & \cdots & 0 & 0 & d_n \end{bmatrix}$$

Next comes the solution phase. The equations $\mathbf{Ly} = \mathbf{b}$ have the augmented coefficient matrix

$$\begin{bmatrix} \mathbf{L} \mid \mathbf{b} \end{bmatrix} = \left[\begin{array}{cccccc|c} 1 & 0 & 0 & 0 & \cdots & 0 & b_1 \\ e_1 & 1 & 0 & 0 & \cdots & 0 & b_2 \\ f_1 & e_2 & 1 & 0 & \cdots & 0 & b_3 \\ 0 & f_2 & e_3 & 1 & \cdots & 0 & b_4 \\ \vdots & \vdots & \vdots & \vdots & \ddots & \vdots & \vdots \\ 0 & 0 & 0 & f_{n-2} & e_{n-1} & 1 & b_n \end{array} \right]$$

Solution by forward substitution yields

$$y_1 = b_1$$
$$y_2 = b_2 - e_1 y_1 \tag{2.28}$$
$$\vdots$$
$$y_k = b_k - f_{k-2} y_{k-2} - e_{k-1} y_{k-1}, \quad k = 3, 4, \ldots, n$$

The equations to be solved by back substitution, namely $\mathbf{Ux} = \mathbf{y}$, have the augmented coefficient matrix

$$\begin{bmatrix} \mathbf{U} \mid \mathbf{y} \end{bmatrix} = \left[\begin{array}{cccccc|c} d_1 & d_1 e_1 & d_1 f_1 & 0 & \cdots & 0 & y_1 \\ 0 & d_2 & d_2 e_2 & d_2 f_2 & \cdots & 0 & y_2 \\ 0 & 0 & d_3 & d_3 e_3 & \cdots & 0 & y_3 \\ \vdots & \vdots & \vdots & \vdots & \ddots & \vdots & \vdots \\ 0 & 0 & \cdots & 0 & d_{n-1} & d_{n-1} e_{n-1} & y_{n-1} \\ 0 & 0 & \cdots & 0 & 0 & d_n & y_n \end{array} \right]$$

the solution of which is obtained by back substitution:

$$x_n = y_n / d_n$$
$$x_{n-1} = y_{n-1} / d_{n-1} - e_{n-1} x_n$$
$$x_k = y_k / d_k - e_k x_{k+1} - f_k x_{k+2}, \quad k = n - 2, n - 3, \ldots, 1 \tag{2.29}$$

■ LUdec5

The function LUdec3 decomposes a symmetric, pentadiagonal matrix **A** stored in the form **A** = [**f****e****d****e****f**]. The original vectors **d**, **e** and **f** are destroyed and replaced by the vectors of the decomposed matrix.

```
function [d,e,f] = LUdec5(d,e,f)
% LU decomposition of pentadiagonal matrix A = [f\e\d\e\f].
% USAGE: [d,e,f] = LUdec5(d,e,f)

n = length(d);
for k = 1:n-2
    lambda = e(k)/d(k);
    d(k+1) = d(k+1) - lambda*e(k);
    e(k+1) = e(k+1) - lambda*f(k);
    e(k) = lambda;
    lambda = f(k)/d(k);
    d(k+2) = d(k+2) - lambda*f(k);
    f(k) = lambda;
end
lambda = e(n-1)/d(n-1);
d(n) = d(n) - lambda*e(n-1);
e(n-1) = lambda;
```

■ LUsol5

LUsol5 is the function for the solution phase. As in LUsol3, the vector **y** over-writes the constant vector **b** during forward substitution and **x** replaces **y** during back substitution.

```
function x = LUsol5(d,e,f,b)
% Solves A*x = b where A = [f\e\d\e\f] is the LU
% decomposition of the original pentadiagonal A.
% USAGE: x = LUsol5(d,e,f,b)

n = length(d);
b(2) = b(2) - e(1)*b(1);      % Forward substitution
for k = 3:n
    b(k) = b(k) - e(k-1)*b(k-1) - f(k-2)*b(k-2);
end
```

```
b(n) = b(n)/d(n);              % Back substitution
b(n-1) = b(n-1)/d(n-1) - e(n-1)*b(n);
for k = n-2:-1:1
    b(k) = b(k)/d(k) - e(k)*b(k+1) - f(k)*b(k+2);
end
x = b;
```

EXAMPLE 2.9

As a result of Gauss elimination, a symmetric matrix **A** was transformed to the upper triangular form

$$\mathbf{U} = \begin{bmatrix} 4 & -2 & 1 & 0 \\ 0 & 3 & -3/2 & 1 \\ 0 & 0 & 3 & -3/2 \\ 0 & 0 & 0 & 35/12 \end{bmatrix}$$

Determine the original matrix **A**.

Solution First we find **L** in the decomposition **A** = **LU**. Dividing each row of **U** by its diagonal element yields

$$\mathbf{L}^T = \begin{bmatrix} 1 & -1/2 & 1/4 & 0 \\ 0 & 1 & -1/2 & 1/3 \\ 0 & 0 & 1 & -1/2 \\ 0 & 0 & 0 & 1 \end{bmatrix}$$

Therefore, **A** = **LU** becomes

$$\mathbf{A} = \begin{bmatrix} 1 & 0 & 0 & 0 \\ -1/2 & 1 & 0 & 0 \\ 1/4 & -1/2 & 1 & 0 \\ 0 & 1/3 & -1/2 & 1 \end{bmatrix} \begin{bmatrix} 4 & -2 & 1 & 0 \\ 0 & 3 & -3/2 & 1 \\ 0 & 0 & 3 & -3/2 \\ 0 & 0 & 0 & 35/12 \end{bmatrix}$$

$$= \begin{bmatrix} 4 & -2 & 1 & 0 \\ -2 & 4 & -2 & 1 \\ 1 & -2 & 4 & -2 \\ 0 & 1 & -2 & 4 \end{bmatrix}$$

EXAMPLE 2.10

Determine **L** and **D** that result from Doolittle's decomposition **A** = **LDL**^{T} of the symmetric matrix

$$\mathbf{A} = \begin{bmatrix} 3 & -3 & 3 \\ -3 & 5 & 1 \\ 3 & 1 & 10 \end{bmatrix}$$

Solution We use Gauss elimination, storing the multipliers in the *upper* triangular portion of **A**. At the completion of elimination, the matrix will have the form of $\mathbf{U}^*$ in Eq. (2.25).

The terms to be eliminated in the first pass are A_{21} and A_{31} using the elementary operations

$$\text{row } 2 \leftarrow \text{row } 2 - (-1) \times \text{row } 1$$

$$\text{row } 3 \leftarrow \text{row } 3 - (1) \times \text{row } 1$$

Storing the multipliers (-1 and 1) in the locations occupied by A_{12} and A_{13}, we get

$$\mathbf{A}' = \begin{bmatrix} 3 & -1 & 1 \\ 0 & 2 & 4 \\ 0 & 4 & 7 \end{bmatrix}$$

The second pass is the operation

$$\text{row } 3 \leftarrow \text{row } 3 - 2 \times \text{row } 2$$

which yields after overwriting A_{23} with the multiplier 2

$$\mathbf{A}'' = \begin{bmatrix} 0\backslash\mathbf{D}\backslash\mathbf{L}^T \end{bmatrix} = \begin{bmatrix} 3 & -1 & 1 \\ 0 & 2 & 2 \\ 0 & 0 & -1 \end{bmatrix}$$

Hence

$$\mathbf{L} = \begin{bmatrix} 1 & 0 & 0 \\ -1 & 1 & 0 \\ 1 & 2 & 1 \end{bmatrix} \qquad \mathbf{D} = \begin{bmatrix} 3 & 0 & 0 \\ 0 & 2 & 0 \\ 0 & 0 & -1 \end{bmatrix}$$

EXAMPLE 2.11
Solve $\mathbf{Ax} = \mathbf{b}$, where

$$\mathbf{A} = \begin{bmatrix} 6 & -4 & 1 & 0 & 0 & \cdots \\ -4 & 6 & -4 & 1 & 0 & \cdots \\ 1 & -4 & 6 & -4 & 1 & \cdots \\ & \ddots & \ddots & \ddots & \ddots & \\ \cdots & 0 & 1 & -4 & 6 & -4 \\ \cdots & 0 & 0 & 1 & -4 & 7 \end{bmatrix} \begin{bmatrix} x_1 \\ x_2 \\ x_3 \\ \vdots \\ x_9 \\ x_{10} \end{bmatrix} = \begin{bmatrix} 3 \\ 0 \\ 0 \\ \vdots \\ 0 \\ 4 \end{bmatrix}$$

Solution As the coefficient matrix is symmetric and pentadiagonal, we utilize the functions LUdec5 and LUsol5:

```
% Example 2.11 (Solution of pentadiagonal eqs.)
n = 10;
```

```
d = 6*ones(n,1); d(n) = 7;
e = -4*ones(n-1,1);
f = ones(n-2,1);
b = zeros(n,1); b(1) = 3; b(n) = 4;
[d,e,f] = LUdec5(d,e,f);
x = LUsol5(d,e,f,b)
```

The output from the program is

```
>> x =
    2.3872
    4.1955
    5.4586
    6.2105
    6.4850
    6.3158
    5.7368
    4.7820
    3.4850
    1.8797
```

2.5   Pivoting

Introduction

Sometimes the order in which the equations are presented to the solution algorithm has a significant effect on the results. For example, consider the equations

$$2x_1 - x_2 = 1$$

$$-x_1 + 2x_2 - x_3 = 0$$

$$-x_2 + x_3 = 0$$

The corresponding augmented coefficient matrix is

$$\begin{bmatrix} \mathbf{A} \mid \mathbf{b} \end{bmatrix} = \begin{bmatrix} 2 & -1 & 0 & 1 \\ -1 & 2 & -1 & 0 \\ 0 & -1 & 1 & 0 \end{bmatrix} \tag{a}$$

Equations (a) are in the "right order" in the sense that we would have no trouble obtaining the correct solution $x_1 = x_2 = x_3 = 1$ by Gauss elimination or LU decomposition. Now suppose that we exchange the first and third equations, so that the augmented coefficient matrix becomes

$$\left[\mathbf{A} \mid \mathbf{b}\right] = \begin{bmatrix} 0 & -1 & 1 & 0 \\ -1 & 2 & -1 & 0 \\ 2 & -1 & 0 & 1 \end{bmatrix} \tag{b}$$

Since we did not change the equations (only their order was altered), the solution is still $x_1 = x_2 = x_3 = 1$. However, Gauss elimination fails immediately due to the presence of the zero pivot element (the element A_{11}).

The above example demonstrates that it is sometimes essential to reorder the equations during the elimination phase. The reordering, or *row pivoting*, is also required if the pivot element is not zero, but very small in comparison to other elements in the pivot row, as demonstrated by the following set of equations:

$$\left[\mathbf{A} \mid \mathbf{b}\right] = \begin{bmatrix} \varepsilon & -1 & 1 & 0 \\ -1 & 2 & -1 & 0 \\ 2 & -1 & 0 & 1 \end{bmatrix} \tag{c}$$

These equations are the same as Eqs. (b), except that the small number ε replaces the zero element A_{11} in Eq. (b). Therefore, if we let $\varepsilon \to 0$, the solutions of Eqs. (b) and (c) should become identical. After the first phase of Gauss elimination, the augmented coefficient matrix becomes

$$\left[\mathbf{A}' \mid \mathbf{b}'\right] = \begin{bmatrix} \varepsilon & -1 & 1 & 0 \\ 0 & 2 - 1/\varepsilon & -1 + 1/\varepsilon & 0 \\ 0 & -1 + 2/\varepsilon & -2/\varepsilon & 1 \end{bmatrix} \tag{d}$$

Because the computer works with a fixed word length, all numbers are rounded off to a finite number of significant figures. If ε is very small, then $1/\varepsilon$ is huge, and an element such as $2 - 1/\varepsilon$ is rounded to $-1/\varepsilon$. Therefore, for sufficiently small ε, the Eqs. (d) are actually stored as

$$\left[\mathbf{A}' \mid \mathbf{b}'\right] = \begin{bmatrix} \varepsilon & -1 & 1 & 0 \\ 0 & -1/\varepsilon & 1/\varepsilon & 0 \\ 0 & 2/\varepsilon & -2/\varepsilon & 1 \end{bmatrix}$$

Because the second and third equations obviously contradict each other, the solution process fails again. This problem would not arise if the first and second, or the first and the third, equations were interchanged in Eqs. (c) before the elimination.

The last example illustrates the extreme case where ε was so small that roundoff errors resulted in total failure of the solution. If we were to make ε somewhat bigger so that the solution would not "bomb" any more, the roundoff errors might still be large enough to render the solution unreliable. Again, this difficulty could be avoided by pivoting.

Diagonal Dominance

An $n \times n$ matrix $\mathbf{A}$ is said to be *diagonally dominant* if each diagonal element is larger than the sum of the other elements in the same row (we are talking here about absolute values). Thus diagonal dominance requires that

$$|A_{ii}| > \sum_{\substack{j=1 \\ j \neq i}}^{n} |A_{ij}| \ (i = 1, 2, \dots, n) \tag{2.30}$$

For example, the matrix

$$\begin{bmatrix} -2 & 4 & -1 \\ 1 & -1 & 3 \\ 4 & -2 & 1 \end{bmatrix}$$

is not diagonally dominant, but if we rearrange the rows in the following manner

$$\begin{bmatrix} 4 & -2 & 1 \\ -2 & 4 & -1 \\ 1 & -1 & 3 \end{bmatrix}$$

then we have diagonal dominance.

It can be shown that if the coefficient matrix $\mathbf{A}$ of the equations $\mathbf{Ax} = \mathbf{b}$ is diagonally dominant, then the solution does not benefit from pivoting; that is, the equations are already arranged in the optimal order. It follows that the strategy of pivoting should be to reorder the equations so that the coefficient matrix is as close to diagonal dominance as possible. This is the principle behind scaled row pivoting, discussed next.

Gauss Elimination with Scaled Row Pivoting

Consider the solution of $\mathbf{Ax} = \mathbf{b}$ by Gauss elimination with row pivoting. Recall that pivoting aims at improving diagonal dominance of the coefficient matrix, i.e., making the pivot element as large as possible in comparison to other elements in the pivot row. The comparison is made easier if we establish an array $\mathbf{s}$, with the elements

$$s_i = \max_j |A_{ij}|, \quad i = 1, 2, \dots, n \tag{2.31}$$

Thus s_i, called the *scale factor* of row i, contains the absolute value of the largest element in the ith row of $\mathbf{A}$. The vector $\mathbf{s}$ can be obtained with the following algorithm:

```
for i = 1:n
    s(i) = max (abs(A(i,1:n)))
end
```

The *relative size* of any element A_{ij} (i.e., relative to the largest element in the ith row) is defined as the ratio

$$r_{ij} = \frac{|A_{ij}|}{s_i} \tag{2.32}$$

Suppose that the elimination phase has reached the stage where the kth row has become the pivot row. The augmented coefficient matrix at this point is shown below.

$$
\begin{bmatrix}
A_{11} & A_{12} & A_{13} & A_{14} & \cdots & A_{1n} & b_1 \\
0 & A_{22} & A_{23} & A_{24} & \cdots & A_{2n} & b_2 \\
0 & 0 & A_{33} & A_{34} & \cdots & A_{3n} & b_3 \\
\vdots & \vdots & \vdots & \vdots & \cdots & \vdots & \vdots \\
0 & \cdots & 0 & A_{kk} & \cdots & A_{kn} & b_k \\
\vdots & \cdots & \vdots & \vdots & \cdots & \vdots & \vdots \\
0 & \cdots & 0 & A_{nk} & \cdots & A_{nn} & b_n
\end{bmatrix} \leftarrow
$$

We don't automatically accept A_{kk} as the pivot element, but look in the kth column below A_{kk} for a "better" pivot. The best choice is the element A_{pk} that has the largest relative size; that is, we choose p such that

$$r_{pk} = \max_{j \geq k} r_{jk}$$

If we find such an element, then we interchange the rows k and p, and proceed with the elimination pass as usual. Note that the corresponding row interchange must also be carried out in the scale factor array **s**. The algorithm that does all this is

```
for k = 1:n-1
    % Find element with largest relative size
    % and the corresponding row number p
    [Amax,p] = max(abs(A(k:n,k))./s(k:n));
    p = p + k - 1;
    % If this element is very small, matrix is singular
    if Amax < eps
        error('Matrix is singular')
    end
    % Interchange rows k and p if needed
    if p ~= k
        b = swapRows(b,k,p);
        s = swapRows(s,k,p);
        A = swapRows(A,k,p);
    end
```

```
    % Elimination pass
    ⋮

end
```

■ swapRows

The function swapRows interchanges rows *i* and *j* of a matrix or vector **v**:

```
function v = swapRows(v,i,j)
% Swap rows i and j of vector or matrix v.
% USAGE: v = swapRows(v,i,j)

temp = v(i,:);
v(i,:) = v(j,:);
v(j,:) = temp;
```

■ gaussPiv

The function gaussPiv performs Gauss elimination with row pivoting. Apart from row swapping, the elimination and solution phases are identical to those of function gauss in Art. 2.2.

```
function x = gaussPiv(A,b)
% Solves A*x = b by Gauss elimination with row pivoting.
% USAGE: x = gaussPiv(A,b)

if size(b,2) > 1; b = b'; end
n = length(b); s = zeros(n,1);
%----------Set up scale factor array----------
for i = 1:n; s(i) = max(abs(A(i,1:n))); end
%--------Exchange rows if necessary----------
for k = 1:n-1
    [Amax,p] = max(abs(A(k:n,k))./s(k:n));
    p = p + k - 1;
    if Amax < eps; error('Matrix is singular'); end
    if p ~= k
        b = swapRows(b,k,p);
        s = swapRows(s,k,p);
        A = swapRows(A,k,p);
    end
```

```
%-------------Elimination pass--------------
    for i = k+1:n
        if A(i,k) ~= 0
            lambda = A(i,k)/A(k,k);
            A(i,k+1:n) = A(i,k+1:n) - lambda*A(k,k+1:n);
            b(i) = b(i) - lambda*b(k);
        end
    end
end
%-----------Back substitution phase----------
for k = n:-1:1
    b(k) = (b(k) - A(k,k+1:n)*b(k+1:n))/A(k,k);
end
x = b;
```

■ LUdecPiv

The Gauss elimination algorithm can be changed to Doolittle's decomposition with minor changes. The most important of these is keeping a record of the row interchanges during the decomposition phase. In LUdecPiv this record is kept in the permutation array perm, initially set to $[1, 2, \ldots, n]^T$. Whenever two rows are interchanged, the corresponding interchange is also carried out in perm. Thus perm shows how the original rows were permuted. This information is then passed to the function LUsolPiv, which rearranges the elements of the constant vector in the same order before carrying out forward and back substitutions.

```
function [A,perm] = LUdecPiv(A)
% LU decomposition of matrix A; returns A = [L\U]
% and the row permutation vector 'perm'.
% USAGE: [A,perm] = LUdecPiv(A)

n = size(A,1); s = zeros(n,1);
perm = (1:n)';
%----------Set up scale factor array----------
for i = 1:n; s(i) = max(abs(A(i,1:n))); end
%--------Exchange rows if necessary----------
for k = 1:n-1
    [Amax,p] = max(abs(A(k:n,k))./s(k:n));
    p = p + k - 1;
    if Amax < eps
        error('Matrix is singular')
    end
```

```
    if p ~= k
        s = swapRows(s,k,p);
        A = swapRows(A,k,p);
        perm = swapRows(perm,k,p);
    end
%-------------Elimination pass--------------
    for i = k+1:n
        if A(i,k) ~= 0
            lambda = A(i,k)/A(k,k);
            A(i,k+1:n) = A(i,k+1:n) - lambda*A(k,k+1:n);
            A(i,k) = lambda;
        end
    end
end
```

■ LUsolPiv

```
function x = LUsolPiv(A,b,perm)
% Solves L*U*b = x, where A contains row-wise
% permutation of L and U in the form A = [L\U].
% Vector 'perm' holds the row permutation data.
% USAGE:  x = LUsolPiv(A,b,perm)

%----------Rearrange b, store it in x--------
if size(b) > 1; b = b'; end
n = size(A,1);
x = b;
for i = 1:n; x(i) = b(perm(i)); end
%-------Forward and back substitution--------
for k = 2:n
    x(k) = x(k) - A(k,1:k-1)*x(1:k-1);
end
for k = n:-1:1
    x(k) = (x(k) - A(k,k+1:n)*x(k+1:n))/A(k,k);
end
```

When to Pivot

Pivoting has a couple of drawbacks. One of these is the increased cost of computation; the other is the destruction of the symmetry and banded structure of the coefficient matrix. The latter is of particular concern in engineering computing, where the co-efficient matrices are frequently banded and symmetric, a property that is utilized in the solution, as seen in the previous article. Fortunately, these matrices are often diagonally dominant as well, so that they would not benefit from pivoting anyway.

There are no infallible rules for determining when pivoting should be used. Experience indicates that pivoting is likely to be counterproductive if the coefficient matrix is banded. Positive definite and, to a lesser degree, symmetric matrices also seldom gain from pivoting. And we should not forget that pivoting is not the only means of controlling roundoff errors—there is also double precision arithmetic.

It should be strongly emphasized that the above rules of thumb are only meant for equations that stem from real engineering problems. It is not difficult to concoct "textbook" examples that do not conform to these rules.

EXAMPLE 2.12
Employ Gauss elimination with scaled row pivoting to solve the equations $\mathbf{Ax} = \mathbf{b}$, where

$$\mathbf{A} = \begin{bmatrix} 2 & -2 & 6 \\ -2 & 4 & 3 \\ -1 & 8 & 4 \end{bmatrix} \qquad \mathbf{b} = \begin{bmatrix} 16 \\ 0 \\ -1 \end{bmatrix}$$

Solution The augmented coefficient matrix and the scale factor array are

$$\begin{bmatrix} \mathbf{A} \mid \mathbf{b} \end{bmatrix} = \begin{bmatrix} 2 & -2 & 6 & | & 16 \\ -2 & 4 & 3 & | & 0 \\ -1 & 8 & 4 & | & -1 \end{bmatrix} \qquad \mathbf{s} = \begin{bmatrix} 6 \\ 4 \\ 8 \end{bmatrix}$$

Note that $\mathbf{s}$ contains the absolute value of the largest element in each row of $\mathbf{A}$. At this stage, all the elements in the first column of $\mathbf{A}$ are potential pivots. To determine the best pivot element, we calculate the relative sizes of the elements in the first column:

$$\begin{bmatrix} r_{11} \\ r_{21} \\ r_{31} \end{bmatrix} = \begin{bmatrix} |A_{11}|/s_1 \\ |A_{21}|/s_2 \\ |A_{31}|/s_3 \end{bmatrix} = \begin{bmatrix} 1/3 \\ 1/2 \\ 1/8 \end{bmatrix}$$

Since r_{21} is the biggest element, we conclude that A_{21} makes the best pivot element. Therefore, we exchange rows 1 and 2 of the augmented coefficient matrix and the scale factor array, obtaining

$$\begin{bmatrix} \mathbf{A} \mid \mathbf{b} \end{bmatrix} = \begin{bmatrix} -2 & 4 & 3 & | & 0 \\ 2 & -2 & 6 & | & 16 \\ -1 & 8 & 4 & | & -1 \end{bmatrix} \leftarrow \qquad \mathbf{s} = \begin{bmatrix} 4 \\ 6 \\ 8 \end{bmatrix}$$

Now the first pass of Gauss elimination is carried out (the arrow points to the pivot row), yielding

$$\left[\mathbf{A'} \mid \mathbf{b'}\right] = \left[\begin{array}{ccc|c} -2 & 4 & 3 & 0 \\ 0 & 2 & 9 & 16 \\ 0 & 6 & 5/2 & -1 \end{array}\right] \qquad \mathbf{s} = \left[\begin{array}{c} 4 \\ 6 \\ 8 \end{array}\right]$$

The potential pivot elements for the next elimination pass are A_{22} and A_{32}. We determine the "winner" from

$$\left[\begin{array}{c} * \\ r_{22} \\ r_{32} \end{array}\right] = \left[\begin{array}{c} * \\ |A_{22}|/s_2 \\ |A_{32}|/s_3 \end{array}\right] = \left[\begin{array}{c} * \\ 1/3 \\ 3/4 \end{array}\right]$$

Note that r_{12} is irrelevant, since row 1 already acted as the pivot row. Therefore, it is excluded from further consideration. As r_{32} is larger than r_{22}, the third row is the better pivot row. After interchanging rows 2 and 3, we have

$$\left[\mathbf{A'} \mid \mathbf{b'}\right] = \left[\begin{array}{ccc|c} -2 & 4 & 3 & 0 \\ 0 & 6 & 5/2 & -1 \\ 0 & 2 & 9 & 16 \end{array}\right] \leftarrow \qquad \mathbf{s} = \left[\begin{array}{c} 4 \\ 8 \\ 6 \end{array}\right]$$

The second elimination pass now yields

$$\left[\mathbf{A''} \mid \mathbf{b''}\right] = \left[\mathbf{U} \mid \mathbf{c}\right] = \left[\begin{array}{ccc|c} -2 & 4 & 3 & 0 \\ 0 & 6 & 5/2 & -1 \\ 0 & 0 & 49/6 & 49/3 \end{array}\right]$$

This completes the elimination phase. It should be noted that $\mathbf{U}$ is the matrix that would result in the LU decomposition of the following row-wise permutation of $\mathbf{A}$ (the ordering of rows is the same as achieved by pivoting):

$$\left[\begin{array}{ccc} -2 & 4 & 3 \\ -1 & 8 & 4 \\ 2 & -2 & 6 \end{array}\right]$$

Since the solution of $\mathbf{Ux} = \mathbf{c}$ by back substitution is not affected by pivoting, we skip the detailed computation. The result is $\mathbf{x}^T = \left[\begin{array}{ccc} 1 & -1 & 2 \end{array}\right]$.

Alternate Solution It it not necessary to physically exchange equations during pivoting. We could accomplish Gauss elimination just as well by keeping the equations in place. The elimination would then proceed as follows (for the sake of brevity, we skip repeating the details of choosing the pivot equation):

$$\begin{bmatrix} \mathbf{A} \mid \mathbf{b} \end{bmatrix} = \begin{bmatrix} 2 & -2 & 6 & \mid & 16 \\ -2 & 4 & 3 & \mid & 0 \\ -1 & 8 & 4 & \mid & -1 \end{bmatrix} \leftarrow$$

$$\begin{bmatrix} \mathbf{A}' \mid \mathbf{b}' \end{bmatrix} = \begin{bmatrix} 0 & 2 & 9 & \mid & 16 \\ -2 & 4 & 3 & \mid & 0 \\ 0 & 6 & 5/2 & \mid & -1 \end{bmatrix} \leftarrow$$

$$\begin{bmatrix} \mathbf{A}'' \mid \mathbf{b}'' \end{bmatrix} = \begin{bmatrix} 0 & 0 & 49/6 & \mid & 49/3 \\ -2 & 4 & 3 & \mid & 0 \\ 0 & 6 & 5/2 & \mid & -1 \end{bmatrix}$$

But now the back substitution phase is a little more involved, since the order in which the equations must be solved has become scrambled. In hand computations this is not a problem, because we can determine the order by inspection. Unfortunately, "by inspection" does not work on a computer. To overcome this difficulty, we have to maintain an integer array **p** that keeps track of the row permutations during the elimination phase. The contents of **p** indicate the order in which the pivot rows were chosen. In this example, we would have at the end of Gauss elimination

$$\mathbf{p} = \begin{bmatrix} 2 \\ 3 \\ 1 \end{bmatrix}$$

showing that row 2 was the pivot row in the first elimination pass, followed by row 3 in the second pass. The equations are solved by back substitution in the reverse order: equation 1 is solved first for x_3, then equation 3 is solved for x_2, and finally equation 2 yields x_1.

By dispensing with swapping of equations, the scheme outlined above would probably result in a faster (and more complex) algorithm than `gaussPiv`, but the number of equations would have to be quite large before the difference becomes noticeable.

PROBLEM SET 2.2

1. Solve the equations $\mathbf{Ax} = \mathbf{b}$ by utilizing Doolittle's decomposition, where

$$\mathbf{A} = \begin{bmatrix} 3 & -3 & 3 \\ -3 & 5 & 1 \\ 3 & 1 & 5 \end{bmatrix} \qquad \mathbf{b} = \begin{bmatrix} 9 \\ -7 \\ 12 \end{bmatrix}$$

2. Use Doolittle's decomposition to solve $Ax = b$, where

$$A = \begin{bmatrix} 4 & 8 & 20 \\ 8 & 13 & 16 \\ 20 & 16 & -91 \end{bmatrix} \qquad b = \begin{bmatrix} 24 \\ 18 \\ -119 \end{bmatrix}$$

3. Determine L and D that result from Doolittle's decomposition of the matrix

$$A = \begin{bmatrix} 2 & -2 & 0 & 0 & 0 \\ -2 & 5 & -6 & 0 & 0 \\ 0 & -6 & 16 & 12 & 0 \\ 0 & 0 & 12 & 39 & -6 \\ 0 & 0 & 0 & -6 & 14 \end{bmatrix}$$

4. Solve the tridiagonal equations $Ax = b$ by Doolittle's decomposition method, where

$$A = \begin{bmatrix} 6 & 2 & 0 & 0 & 0 \\ -1 & 7 & 2 & 0 & 0 \\ 0 & -2 & 8 & 2 & 0 \\ 0 & 0 & 3 & 7 & -2 \\ 0 & 0 & 0 & 3 & 5 \end{bmatrix} \qquad b = \begin{bmatrix} 2 \\ -3 \\ 4 \\ -3 \\ 1 \end{bmatrix}$$

5. Use Gauss elimination with scaled row pivoting to solve

$$\begin{bmatrix} 4 & -2 & 1 \\ -2 & 1 & -1 \\ -2 & 3 & 6 \end{bmatrix} \begin{bmatrix} x_1 \\ x_2 \\ x_3 \end{bmatrix} = \begin{bmatrix} 2 \\ -1 \\ 0 \end{bmatrix}$$

6. Solve $Ax = b$ by Gauss elimination with scaled row pivoting, where

$$A = \begin{bmatrix} 2.34 & -4.10 & 1.78 \\ -1.98 & 3.47 & -2.22 \\ 2.36 & -15.17 & 6.81 \end{bmatrix} \qquad b = \begin{bmatrix} 0.02 \\ -0.73 \\ -6.63 \end{bmatrix}$$

7. Solve the equations

$$\begin{bmatrix} 2 & -1 & 0 & 0 \\ 0 & 0 & -1 & 1 \\ 0 & -1 & 2 & -1 \\ -1 & 2 & -1 & 0 \end{bmatrix} \begin{bmatrix} x_1 \\ x_2 \\ x_3 \\ x_4 \end{bmatrix} = \begin{bmatrix} 1 \\ 0 \\ 0 \\ 0 \end{bmatrix}$$

by Gauss elimination with scaled row pivoting.

8. ■ Solve the equations

$$\begin{bmatrix} 0 & 2 & 5 & -1 \\ 2 & 1 & 3 & 0 \\ -2 & -1 & 3 & 1 \\ 3 & 3 & -1 & 2 \end{bmatrix} \begin{bmatrix} x_1 \\ x_2 \\ x_3 \\ x_4 \end{bmatrix} = \begin{bmatrix} -3 \\ 3 \\ -2 \\ 5 \end{bmatrix}$$

9. ■ Solve the symmetric, tridiagonal equations

$$4x_1 - x_2 = 9$$

$$-x_{i-1} + 4x_i - x_{i+1} = 5, \quad i = 2, \ldots, n-1$$

$$-x_{n-1} + 4x_n = 5$$

with $n = 10$.

10. ■ Solve the equations $\mathbf{Ax} = \mathbf{b}$, where

$$\mathbf{A} = \begin{bmatrix} 1.3174 & 2.7250 & 2.7250 & 1.7181 \\ 0.4002 & 0.8278 & 1.2272 & 2.5322 \\ 0.8218 & 1.5608 & 0.3629 & 2.9210 \\ 1.9664 & 2.0011 & 0.6532 & 1.9945 \end{bmatrix} \quad \mathbf{b} = \begin{bmatrix} 8.4855 \\ 4.9874 \\ 5.6665 \\ 6.6152 \end{bmatrix}$$

11. ■ Solve the equations

$$\begin{bmatrix} 10 & -2 & -1 & 2 & 3 & 1 & -4 & 7 \\ 5 & 11 & 3 & 10 & -3 & 3 & 3 & -4 \\ 7 & 12 & 1 & 5 & 3 & -12 & 2 & 3 \\ 8 & 7 & -2 & 1 & 3 & 2 & 2 & 4 \\ 2 & -15 & -1 & 1 & 4 & -1 & 8 & 3 \\ 4 & 2 & 9 & 1 & 12 & -1 & 4 & 1 \\ -1 & 4 & -7 & -1 & 1 & 1 & -1 & -3 \\ -1 & 3 & 4 & 1 & 3 & -4 & 7 & 6 \end{bmatrix} \begin{bmatrix} x_1 \\ x_2 \\ x_3 \\ x_4 \\ x_5 \\ x_6 \\ x_7 \\ x_8 \end{bmatrix} = \begin{bmatrix} 0 \\ 12 \\ -5 \\ 3 \\ -25 \\ -26 \\ 9 \\ -7 \end{bmatrix}$$

12. ■ The system shown in Fig. (a) consists of n linear springs that support n masses. The spring stiffnesses are denoted by k_i, the weights of the masses are W_i, and x_i are the displacements of the masses (measured from the positions where the springs are undeformed). The so-called *displacement formulation* is obtained by writing the equilibrium equation of each mass and substituting $F_i = k_i(x_{i+1} - x_i)$ for the spring forces. The result is the symmetric, tridiagonal set of equations

$$(k_1 + k_2)x_1 - k_2 x_2 = W_1$$

$$-k_i x_{i-1} + (k_i + k_{i+1})x_i - k_{i+1} x_{i+1} = W_i, \quad i = 2, 3, \ldots, n-1$$

$$-k_n x_{n-1} + k_n x_n = W_n$$

Write a program that solves these equations for given values of n, **k** and **W**. Run the program with $n = 5$ and

$$k_1 = k_2 = k_3 = 10 \text{ N/mm} \quad k_4 = k_5 = 5 \text{ N/mm}$$
$$W_1 = W_3 = W_5 = 100 \text{ N} \quad W_2 = W_4 = 50 \text{ N}$$

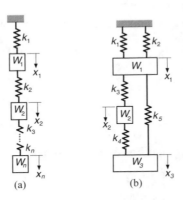

(a)   (b)

13. ■ The displacement formulation for the mass–spring system shown in Fig. (b) results in the following equilibrium equations of the masses:

$$\begin{bmatrix} k_1 + k_2 + k_3 + k_5 & -k_3 & -k_5 \\ -k_3 & k_3 + k_4 & -k_4 \\ -k_5 & -k_4 & k_4 + k_5 \end{bmatrix} \begin{bmatrix} x_1 \\ x_2 \\ x_3 \end{bmatrix} = \begin{bmatrix} W_1 \\ W_2 \\ W_3 \end{bmatrix}$$

where k_i are the spring stiffnesses, W_i represent the weights of the masses, and x_i are the displacements of the masses from the undeformed configuration of the system. Write a program that solves these equations, given **k** and **W**. Use the program to find the displacements if

$$k_1 = k_3 = k_4 = k \quad k_2 = k_5 = 2k$$
$$W_1 = W_3 = 2W \quad W_2 = W$$

14. ■

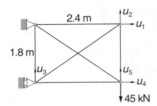

The displacement formulation for a plane truss is similar to that of a mass–spring system. The differences are: (1) the stiffnesses of the members are

$k_i = (EA/L)_i$, where E is the modulus of elasticity, A represents the cross-sectional area and L is the length of the member; (2) there are two components of displacement at each joint. For the statically indeterminate truss shown the displacement formulation yields the symmetric equations $\mathbf{Ku} = \mathbf{p}$, where

$$\mathbf{K} = \begin{bmatrix} 27.58 & 7.004 & -7.004 & 0.0000 & 0.0000 \\ 7.004 & 29.57 & -5.253 & 0.0000 & -24.32 \\ -7.004 & -5.253 & 29.57 & 0.0000 & 0.0000 \\ 0.0000 & 0.0000 & 0.0000 & 27.58 & -7.004 \\ 0.0000 & -24.32 & 0.0000 & -7.004 & 29.57 \end{bmatrix} \text{MN/m}$$

$$\mathbf{p} = \begin{bmatrix} 0 & 0 & 0 & 0 & -45 \end{bmatrix}^T \text{kN}$$

Determine the displacements u_i of the joints.

15. ■

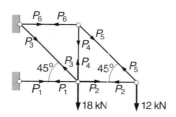

In the *force formulation* of a truss, the unknowns are the member forces P_i. For the statically determinate truss shown, the equilibrium equations of the joints are:

$$\begin{bmatrix} -1 & 1 & -1/\sqrt{2} & 0 & 0 & 0 \\ 0 & 0 & 1/\sqrt{2} & 1 & 0 & 0 \\ 0 & -1 & 0 & 0 & -1/\sqrt{2} & 0 \\ 0 & 0 & 0 & 0 & 1/\sqrt{2} & 0 \\ 0 & 0 & 0 & 0 & 1/\sqrt{2} & 1 \\ 0 & 0 & 0 & -1 & -1/\sqrt{2} & 0 \end{bmatrix} \begin{bmatrix} P_1 \\ P_2 \\ P_3 \\ P_4 \\ P_5 \\ P_6 \end{bmatrix} = \begin{bmatrix} 0 \\ 18 \\ 0 \\ 12 \\ 0 \\ 0 \end{bmatrix}$$

where the units of P_i are kN. (a) Solve the equations as they are with a computer program. (b) Rearrange the rows and columns so as to obtain a lower triangular coefficient matrix, and then solve the equations by back substitution using a calculator.

16. ■

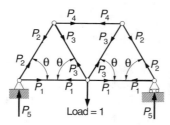

The force formulation of the symmetric truss shown results in the joint equilibrium equations

$$
\begin{bmatrix}
c & 1 & 0 & 0 & 0 \\
0 & s & 0 & 0 & 1 \\
0 & 0 & 2s & 0 & 0 \\
0 & -c & c & 1 & 0 \\
0 & s & s & 0 & 0
\end{bmatrix}
\begin{bmatrix}
P_1 \\
P_2 \\
P_3 \\
P_4 \\
P_5
\end{bmatrix}
=
\begin{bmatrix}
0 \\
0 \\
1 \\
0 \\
0
\end{bmatrix}
$$

where $s = \sin\theta$, $c = \cos\theta$ and P_i are the unknown forces. Write a program that computes the forces, given the angle θ. Run the program with $\theta = 53°$.

17. ■

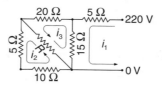

The electrical network shown can be viewed as consisting of three loops. Applying Kirhoff's law ($\sum$voltage drops $= \sum$voltage sources) to each loop yields the following equations for the loop currents i_1, i_2 and i_3:

$$
5i_1 + 15(i_1 - i_3) = 220 \text{ V}
$$

$$
R(i_2 - i_3) + 5i_2 + 10i_2 = 0
$$

$$
20i_3 + R(i_3 - i_2) + 15(i_3 - i_1) = 0
$$

Compute the three loop currents for $R = 5$, 10 and 20 Ω.

18. ■

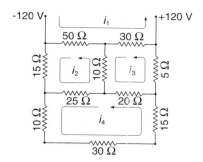

Determine the loop currents i_1 to i_4 in the electrical network shown.

19. ■ Consider the n simultaneous equations $\mathbf{A}\mathbf{x} = \mathbf{b}$, where

$$A_{ij} = (i + j)^2 \qquad b_i = \sum_{j=0}^{n-1} A_{ij}, \quad i = 0, 1, \ldots, n-1, \quad j = 0, 1, \ldots, n-1$$

The solution is $\mathbf{x} = \begin{bmatrix} 1 & 1 & \cdots & 1 \end{bmatrix}^T$. Write a program that solves these equations for any given n (pivoting is recommended). Run the program with $n = 2, 3$ and 4, and comment on the results.

*2.6 Matrix Inversion

Computing the inverse of a matrix and solving simultaneous equations are related tasks. The most economical way to invert an $n \times n$ matrix $\mathbf{A}$ is to solve the equations

$$\mathbf{A}\mathbf{X} = \mathbf{I} \tag{2.33}$$

where $\mathbf{I}$ is the $n \times n$ identity matrix. The solution $\mathbf{X}$, also of size $n \times n$, will be the inverse of $\mathbf{A}$. The proof is simple: after we premultiply both sides of Eq. (2.33) by $\mathbf{A}^{-1}$ we have $\mathbf{A}^{-1}\mathbf{A}\mathbf{X} = \mathbf{A}^{-1}\mathbf{I}$, which reduces to $\mathbf{X} = \mathbf{A}^{-1}$.

Inversion of large matrices should be avoided whenever possible due its high cost. As seen from Eq. (2.33), inversion of $\mathbf{A}$ is equivalent to solving $\mathbf{A}\mathbf{x}_i = \mathbf{b}_i, i = 1, 2, \ldots, n$, where $\mathbf{b}_i$ is the ith column of $\mathbf{I}$. If LU decomposition is employed in the solution, the solution phase (forward and back substitution) must be repeated n times, once for each $\mathbf{b}_i$. Since the cost of computation is proportional to n^3 for the decomposition phase and n^2 for each vector of the solution phase, the cost of inversion is considerably more expensive than the solution of $\mathbf{A}\mathbf{x} = \mathbf{b}$ (single constant vector $\mathbf{b}$).

Matrix inversion has another serious drawback—a banded matrix loses its structure during inversion. In other words, if A is banded or otherwise sparse, then A^{-1} is fully populated. However, the inverse of a triangular matrix remains triangular.

EXAMPLE 2.13
Write a function that inverts a matrix using LU decomposition with pivoting. Test the function by inverting

$$A = \begin{bmatrix} 0.6 & -0.4 & 1.0 \\ -0.3 & 0.2 & 0.5 \\ 0.6 & -1.0 & 0.5 \end{bmatrix}$$

Solution The function matInv listed below inverts any martix A.

```
function Ainv = matInv(A)
% Inverts martix A with LU decomposition.
% USAGE: Ainv = matInv(A)

n = size(A,1);
Ainv = eye(n);              % Store RHS vectors in Ainv.
[A,perm] = LUdecPiv(A);     % Decompose A.
% Solve for each RHS vector and store results in Ainv
% replacing the corresponding RHS vector.
for i = 1:n
    Ainv(:,i) = LUsolPiv(A,Ainv(:,i),perm);
end
```

The following test program computes the inverse of the given matrix and checks whether $AA^{-1} = I$:

```
% Example 2.13 (Matrix inversion)
A = [0.6 -0.4 1.0
    -0.3  0.2 0.5
     0.6 -1.0 0.5];
Ainv = matInv(A)
check = A*Ainv
```

Here are the results:

```
>> Ainv =
    1.6667   -2.2222   -1.1111
    1.2500   -0.8333   -1.6667
    0.5000    1.0000         0
```

```
check =
    1.0000   -0.0000   -0.0000
         0    1.0000    0.0000
         0   -0.0000    1.0000
```

EXAMPLE 2.14
Invert the matrix

$$A = \begin{bmatrix} 2 & -1 & 0 & 0 & 0 & 0 \\ -1 & 2 & -1 & 0 & 0 & 0 \\ 0 & -1 & 2 & -1 & 0 & 0 \\ 0 & 0 & -1 & 2 & -1 & 0 \\ 0 & 0 & 0 & -1 & 2 & -1 \\ 0 & 0 & 0 & 0 & -1 & 5 \end{bmatrix}$$

Solution Since the matrix is tridiagonal, we solve $AX = I$ using the functions LUdec3 and LUsol3 (LU decomposition for tridiagonal matrices):

```
% Example 2.14 (Matrix inversion)
n = 6;
d = ones(n,1)*2;
e = -ones(n-1,1);
c = e;
d(n) = 5;
[c,d,e] = LUdec3(c,d,e);
for i = 1:n
    b = zeros(n,1);
    b(i) = 1;
    Ainv(:,i) = LUsol3(c,d,e,b);
end
Ainv
```

The result is

```
>> Ainv =
    0.8400    0.6800    0.5200    0.3600    0.2000    0.0400
    0.6800    1.3600    1.0400    0.7200    0.4000    0.0800
    0.5200    1.0400    1.5600    1.0800    0.6000    0.1200
    0.3600    0.7200    1.0800    1.4400    0.8000    0.1600
    0.2000    0.4000    0.6000    0.8000    1.0000    0.2000
    0.0400    0.0800    0.1200    0.1600    0.2000    0.2400
```

Note that although A is tridiagonal, A^{-1} is fully populated.

*2.7   Iterative Methods

Introduction

So far, we have discussed only direct methods of solution. The common characteristic of these methods is that they compute the solution with a finite number of operations. Moreover, if the computer were capable of infinite precision (no roundoff errors), the solution would be exact.

Iterative, or *indirect methods*, start with an initial guess of the solution **x** and then repeatedly improve the solution until the change in **x** becomes negligible. Since the required number of iterations can be very large, the indirect methods are, in general, slower than their direct counterparts. However, iterative methods do have the following advantages that make them attractive for certain problems:

1. It is feasible to store only the nonzero elements of the coefficient matrix. This makes it possible to deal with very large matrices that are sparse, but not necessarily banded. In many problems, there is no need to store the coefficient matrix at all.
2. Iterative procedures are self-correcting, meaning that roundoff errors (or even arithmetic mistakes) in one iterative cycle are corrected in subsequent cycles.

A serious drawback of iterative methods is that they do not always converge to the solution. It can be shown that convergence is guaranteed only if the coefficient matrix is diagonally dominant. The initial guess for **x** plays no role in determining whether convergence takes place—if the procedure converges for one starting vector, it would do so for any starting vector. The initial guess affects only the number of iterations that are required for convergence.

Gauss–Seidel Method

The equations **Ax** = **b** are in scalar notation

$$\sum_{j=1}^{n} A_{ij}x_j = b_i, \quad i = 1, 2, \ldots, n$$

Extracting the term containing x_i from the summation sign yields

$$A_{ii}x_i + \sum_{\substack{j=1 \\ j \neq i}}^{n} A_{ij}x_j = b_i, \quad i = 1, 2, \ldots, n$$

Solving for x_i, we get

$$x_i = \frac{1}{A_{ii}} \left(b_i - \sum_{\substack{j=1 \\ j \neq i}}^{n} A_{ij}x_j \right), \quad i = 1, 2, \ldots, n$$

The last equation suggests the following iterative scheme

$$x_i \leftarrow \frac{1}{A_{ii}} \left(b_i - \sum_{\substack{j=1 \\ j \neq i}}^{n} A_{ij}x_j \right), \quad i = 1, 2, \ldots, n \tag{2.34}$$

We start by choosing the starting vector **x**. If a good guess for the solution is not available, **x** can be chosen randomly. Equation (2.34) is then used to recompute each element of **x**, always using the latest available values of x_j. This completes one iteration cycle. The procedure is repeated until the changes in **x** between successive iteration cycles become sufficiently small.

Convergence of the Gauss–Seidel method can be improved by a technique known as *relaxation*. The idea is to take the new value of x_i as a weighted average of its previous value and the value predicted by Eq. (2.34). The corresponding iterative formula is

$$x_i \leftarrow \frac{\omega}{A_{ii}} \left(b_i - \sum_{\substack{j=1 \\ j \neq i}}^{n} A_{ij}x_j \right) + (1 - \omega)x_i, \quad i = 1, 2, \ldots, n \tag{2.35}$$

where the weight ω is called the *relaxation factor*. It can be seen that if $\omega = 1$, no relaxation takes place, since Eqs. (2.34) and (2.35) produce the same result. If $\omega < 1$, Eq. (2.35) represents interpolation between the old x_i and the value given by Eq. (2.34). This is called *underrelaxation*. In cases where $\omega > 1$, we have extrapolation, or *overrelaxation*.

There is no practical method of determining the optimal value of ω beforehand; however, a good estimate can be computed during run time. Let $\Delta x^{(k)} = \left| \mathbf{x}^{(k-1)} - \mathbf{x}^{(k)} \right|$ be the magnitude of the change in **x** during the kth iteration (carried out without relaxation; i.e., with $\omega = 1$). If k is sufficiently large (say $k \geq 5$), it can be shown[2] that an approximation of the optimal value of ω is

$$\omega_{\text{opt}} \approx \frac{2}{1 + \sqrt{1 - \left(\Delta x^{(k+p)} / \Delta x^{(k)} \right)^{1/p}}} \tag{2.36}$$

where p is a positive integer.

---

[2] See, for example, Terrence J. Akai, *Applied Numerical Methods for Engineers*, John Wiley & Sons (1994), p. 100.

The essential elements of a Gauss–Seidel algorithm with relaxation are:

1. Carry out k iterations with $\omega = 1$ ($k = 10$ is reasonable). After the kth iteration record $\Delta x^{(k)}$.
2. Perform an additional p iterations ($p \geq 1$) and record $\Delta x^{(k+p)}$ after the last iteration.
3. Perform all subsequent iterations with $\omega = \omega_{\text{opt}}$, where ω_{opt} is computed from Eq. (2.36).

■ gaussSeidel

The function gaussSeidel is an implementation of the Gauss–Seidel method with relaxation. It automatically computes ω_{opt} from Eq. (2.36) using $k = 10$ and $p = 1$. The user must provide the function iterEqs that computes the improved **x** from the iterative formulas in Eq. (2.35)—see Example 2.17.

```
function [x,numIter,omega] = gaussSeidel(func,x,maxIter,epsilon)
% Solves Ax = b by Gauss-Seidel method with relaxation.
% USAGE: [x,numIter,omega] = gaussSeidel(func,x,maxIter,epsilon)
% INPUT:
% func    = handle of function that returns improved x using
%           the iterative formulas in Eq. (2.35).
% x       = starting solution vector
% maxIter = allowable number of iterations (default is 500)
% epsilon = error tolerance (default is 1.0e-9)
% OUTPUT:
% x       = solution vector
% numIter = number of iterations carried out
% omega   = computed relaxation factor

if nargin < 4; epsilon = 1.0e-9; end
if nargin < 3; maxIter = 500; end
k = 10; p = 1; omega = 1;
for numIter = 1:maxIter
    xOld = x;
    x = feval(func,x,omega);
    dx = sqrt(dot(x - xOld,x - xOld));
    if dx < epsilon; return; end
    if numIter == k; dx1 = dx; end
```

```
      if numIter == k + p
          omega = 2/(1 + sqrt(1 - (dx/dx1)^(1/p)));
      end
end
error('Too many iterations')
```

Conjugate Gradient Method

Consider the problem of finding the vector $\mathbf{x}$ that minimizes the scalar function

$$f(\mathbf{x}) = \frac{1}{2}\mathbf{x}^T\mathbf{A}\mathbf{x} - \mathbf{b}^T\mathbf{x} \tag{2.37}$$

where the matrix $\mathbf{A}$ is *symmetric* and *positive definite*. Because $f(\mathbf{x})$ is minimized when its gradient $\nabla f = \mathbf{A}\mathbf{x} - \mathbf{b}$ is zero, we see that minimization is equivalent to solving

$$\mathbf{A}\mathbf{x} = \mathbf{b} \tag{2.38}$$

Gradient methods accomplish the minimization by iteration, starting with an initial vector $\mathbf{x}_0$. Each iterative cycle k computes a refined solution

$$\mathbf{x}_{k+1} = \mathbf{x}_k + \alpha_k\mathbf{s}_k \tag{2.39}$$

The *step length* α_k is chosen so that $\mathbf{x}_{k+1}$ minimizes $f(\mathbf{x}_{k+1})$ in the *search direction* $\mathbf{s}_k$. That is, $\mathbf{x}_{k+1}$ must satisfy Eq. (2.38):

$$\mathbf{A}(\mathbf{x}_k + \alpha_k\mathbf{s}_k) = \mathbf{b} \tag{a}$$

Introducing the *residual*

$$\mathbf{r}_k = \mathbf{b} - \mathbf{A}\mathbf{x}_k \tag{2.40}$$

Eq. (a) becomes $\alpha\mathbf{A}\mathbf{s}_k = \mathbf{r}_k$. Premultiplying both sides by $\mathbf{s}_k^T$ and solving for α_k, we obtain

$$\alpha_k = \frac{\mathbf{s}_k^T\mathbf{r}_k}{\mathbf{s}_k^T\mathbf{A}\mathbf{s}_k} \tag{2.41}$$

We are still left with the problem of determining the search direction $\mathbf{s}_k$. Intuition tells us to choose $\mathbf{s}_k = -\nabla f = \mathbf{r}_k$, since this is the direction of the largest negative change in $f(\mathbf{x})$. The resulting procedure is known as the *method of steepest descent*. It is not a popular algorithm due to slow convergence. The more efficient conjugate gradient method uses the search direction

$$\mathbf{s}_{k+1} = \mathbf{r}_{k+1} + \beta_k\mathbf{s}_k \tag{2.42}$$

The constant β_k is chosen so that the two successive search directions are *conjugate* (noninterfering) to each other, meaning $\mathbf{s}_{k+1}^T\mathbf{A}\mathbf{s}_k = 0$. Substituting for $\mathbf{s}_{k+1}$ from

Eq. (2.42), we get $\left(\mathbf{r}_{k+1}^T + \beta_k \mathbf{s}_k^T\right) \mathbf{A}\mathbf{s}_k = 0$, which yields

$$\beta_k = -\frac{\mathbf{r}_{k+1}^T \mathbf{A}\mathbf{s}_k}{\mathbf{s}_k^T \mathbf{A}\mathbf{s}_k} \tag{2.43}$$

Here is the outline of the conjugate gradient algorithm:

- Choose $\mathbf{x}_0$ (any vector will do, but one close to solution results in fewer iterations)
- $\mathbf{r}_0 \leftarrow \mathbf{b} - \mathbf{A}\mathbf{x}_0$
- $\mathbf{s}_0 \leftarrow \mathbf{r}_0$ (lacking a previous search direction, choose the direction of steepest descent)
- do with $k = 0, 1, 2, \ldots$

$$\alpha_k \leftarrow \frac{\mathbf{s}_k^T \mathbf{r}_k}{\mathbf{s}_k^T \mathbf{A}\mathbf{s}_k}$$

$$\mathbf{x}_{k+1} \leftarrow \mathbf{x}_k + \alpha_k \mathbf{s}_k$$

$$\mathbf{r}_{k+1} \leftarrow \mathbf{b} - \mathbf{A}\mathbf{x}_{k+1}$$

if $|\mathbf{r}_{k+1}| \leq \varepsilon$ exit loop (convergence criterion; ε is the error tolerance)

$$\beta_k \leftarrow -\frac{\mathbf{r}_{k+1}^T \mathbf{A}\mathbf{s}_k}{\mathbf{s}_k^T \mathbf{A}\mathbf{s}_k}$$

$$\mathbf{s}_{k+1} \leftarrow \mathbf{r}_{k+1} + \beta_k \mathbf{s}_k$$

- end do

It can be shown that the residual vectors $\mathbf{r}_1, \mathbf{r}_2, \mathbf{r}_3, \ldots$ produced by the algorithm are mutually orthogonal; i.e., $\mathbf{r}_i \cdot \mathbf{r}_j = 0$, $i \neq j$. Now suppose that we have carried out enough iterations to have computed the whole set of n residual vectors. The residual resulting from the next iteration must be a null vector ($\mathbf{r}_{n+1} = \mathbf{0}$), indicating that the solution has been obtained. It thus appears that the conjugate gradient algorithm is not an iterative method at all, since it reaches the exact solution after n computational cycles. In practice, however, convergence is usually achieved in less than n iterations.

The conjugate gradient method is not competitive with direct methods in the solution of small sets of equations. Its strength lies in the handling of large, sparse systems (where most elements of $\mathbf{A}$ are zero). It is important to note that $\mathbf{A}$ enters the algorithm only through its multiplication by a vector; i.e., in the form $\mathbf{A}\mathbf{v}$, where $\mathbf{v}$ is a vector (either $\mathbf{x}_{k+1}$ or $\mathbf{s}_k$). If $\mathbf{A}$ is sparse, it is possible to write an efficient subroutine for the multiplication and pass it on to the conjugate gradient algorithm.

■ conjGrad

The function `conjGrad` shown below implements the conjugate gradient algorithm. The maximum allowable number of iterations is set to n. Note that `conjGrad` calls the function `Av(v)` which returns the product **Av**. This function must be supplied by the user (see Example 2.18). We must also supply the starting vector **x** and the constant (right-hand-side) vector **b**.

```
function [x,numIter] = conjGrad(func,x,b,epsilon)
% Solves Ax = b by conjugate gradient method.
% USAGE: [x,numIter] = conjGrad(func,x,b,epsilon)
% INPUT:
% func    = handle of function that returns the vector A*v
% x       = starting solution vector
% b       = constant vector in A*x = b
% epsilon = error tolerance (default = 1.0e-9)
% OUTPUT:
% x       = solution vector
% numIter = number of iterations carried out

if nargin == 3; epsilon = 1.0e-9; end
n = length(b);
r = b - feval(func,x); s = r;
for numIter = 1:n
    u = feval(func,s);
    alpha = dot(s,r)/dot(s,u);
    x = x + alpha*s;
    r = b - feval(func,x);
    if sqrt(dot(r,r)) < epsilon
        return
    else
        beta = -dot(r,u)/dot(s,u);
        s = r + beta*s;
    end
end
error('Too many iterations')
```

EXAMPLE 2.15

Solve the equations

$$\begin{bmatrix} 4 & -1 & 1 \\ -1 & 4 & -2 \\ 1 & -2 & 4 \end{bmatrix} \begin{bmatrix} x_1 \\ x_2 \\ x_3 \end{bmatrix} = \begin{bmatrix} 12 \\ -1 \\ 5 \end{bmatrix}$$

by the Gauss–Seidel method without relaxation.

Solution With the given data, the iteration formulas in Eq. (2.34) become

$$x_1 = \frac{1}{4}(12 + x_2 - x_3)$$

$$x_2 = \frac{1}{4}(-1 + x_1 + 2x_3)$$

$$x_3 = \frac{1}{4}(5 - x_1 + 2x_2)$$

Choosing the starting values $x_1 = x_2 = x_3 = 0$, we have for the first iteration

$$x_1 = \frac{1}{4}(12 + 0 - 0) = 3$$

$$x_2 = \frac{1}{4}[-1 + 3 + 2(0)] = 0.5$$

$$x_3 = \frac{1}{4}[5 - 3 + 2(0.5)] = 0.75$$

The second iteration yields

$$x_1 = \frac{1}{4}(12 + 0.5 - 0.75) = 2.9375$$

$$x_2 = \frac{1}{4}[-1 + 2.9375 + 2(0.75)] = 0.859\,38$$

$$x_3 = \frac{1}{4}[5 - 2.9375 + 2(0.85938)] = 0\,.945\,31$$

and the third iteration results in

$$x_1 = \frac{1}{4}(12 + 0.85938 - 0\,.94531) = 2.978\,52$$

$$x_2 = \frac{1}{4}[-1 + 2.97852 + 2(0\,.94531)] = 0.967\,29$$

$$x_3 = \frac{1}{4}[5 - 2.97852 + 2(0.96729)] = 0.989\,02$$

After five more iterations the results would agree with the exact solution $x_1 = 3$, $x_2 = x_3 = 1$ within five decimal places.

EXAMPLE 2.16
Solve the equations in Example 2.15 by the conjugate gradient method.

Solution The conjugate gradient method should converge after three iterations. Choosing again for the starting vector

$$\mathbf{x}_0 = \begin{bmatrix} 0 & 0 & 0 \end{bmatrix}^T$$

the computations outlined in the text proceed as follows:

$$\mathbf{r}_0 = \mathbf{b} - A\mathbf{x}_0 = \begin{bmatrix} 12 \\ -1 \\ 5 \end{bmatrix} - \begin{bmatrix} 4 & -1 & 1 \\ -1 & 4 & -2 \\ 1 & -2 & 4 \end{bmatrix} \begin{bmatrix} 0 \\ 0 \\ 0 \end{bmatrix} = \begin{bmatrix} 12 \\ -1 \\ 5 \end{bmatrix}$$

$$\mathbf{s}_0 = \mathbf{r}_0 = \begin{bmatrix} 12 \\ -1 \\ 5 \end{bmatrix}$$

$$A\mathbf{s}_0 = \begin{bmatrix} 4 & -1 & 1 \\ -1 & 4 & -2 \\ 1 & -2 & 4 \end{bmatrix} \begin{bmatrix} 12 \\ -1 \\ 5 \end{bmatrix} = \begin{bmatrix} 54 \\ -26 \\ 34 \end{bmatrix}$$

$$\alpha_0 = \frac{\mathbf{s}_0^T \mathbf{r}_0}{\mathbf{s}_0^T A\mathbf{s}_0} = \frac{12^2 + (-1)^2 + 5^2}{12(54) + (-1)(-26) + 5(34)} = 0.201\,42$$

$$\mathbf{x}_1 = \mathbf{x}_0 + \alpha_0 \mathbf{s}_0 = \begin{bmatrix} 0 \\ 0 \\ 0 \end{bmatrix} + 0.201\,42 \begin{bmatrix} 12 \\ -1 \\ 5 \end{bmatrix} = \begin{bmatrix} 2.41\,704 \\ -0.201\,42 \\ 1.007\,10 \end{bmatrix}$$

$$\mathbf{r}_1 = \mathbf{b} - A\mathbf{x}_1 = \begin{bmatrix} 12 \\ -1 \\ 5 \end{bmatrix} - \begin{bmatrix} 4 & -1 & 1 \\ -1 & 4 & -2 \\ 1 & -2 & 4 \end{bmatrix} \begin{bmatrix} 2.417\,04 \\ -0.201\,42 \\ 1.007\,10 \end{bmatrix} = \begin{bmatrix} 1.123\,32 \\ 4.236\,92 \\ -1.848\,28 \end{bmatrix}$$

$$\beta_0 = -\frac{\mathbf{r}_1^T A\mathbf{s}_0}{\mathbf{s}_0^T A\mathbf{s}_0} = -\frac{1.123\,32(54) + 4.236\,92(-26) - 1.848\,28(34)}{12(54) + (-1)(-26) + 5(34)} = 0.133\,107$$

$$\mathbf{s}_1 = \mathbf{r}_1 + \beta_0 \mathbf{s}_0 = \begin{bmatrix} 1.123\,32 \\ 4.236\,92 \\ -1.848\,28 \end{bmatrix} + 0.133\,107 \begin{bmatrix} 12 \\ -1 \\ 5 \end{bmatrix} = \begin{bmatrix} 2.720\,76 \\ 4.103\,80 \\ -1.182\,68 \end{bmatrix}$$

$$A\mathbf{s}_1 = \begin{bmatrix} 4 & -1 & 1 \\ -1 & 4 & -2 \\ 1 & -2 & 4 \end{bmatrix} \begin{bmatrix} 2.720\,76 \\ 4.103\,80 \\ -1.182\,68 \end{bmatrix} = \begin{bmatrix} 5.596\,56 \\ 16.059\,80 \\ -10.217\,60 \end{bmatrix}$$

$$\alpha_1 = \frac{\mathbf{s}_1^T \mathbf{r}_1}{\mathbf{s}_1^T A\mathbf{s}_1}$$

$$= \frac{2.720\,76(1.123\,32) + 4.103\,80(4.236\,92) + (-1.182\,68)(-1.848\,28)}{2.720\,76(5.596\,56) + 4.103\,80(16.059\,80) + (-1.182\,68)(-10.217\,60)}$$

$$= 0.24276$$

$$\mathbf{x}_2 = \mathbf{x}_1 + \alpha_1 \mathbf{s}_1 = \begin{bmatrix} 2.417\,04 \\ -0.201\,42 \\ 1.007\,10 \end{bmatrix} + 0.24276 \begin{bmatrix} 2.720\,76 \\ 4.103\,80 \\ -1.182\,68 \end{bmatrix} = \begin{bmatrix} 3.07753 \\ 0.79482 \\ 0.71999 \end{bmatrix}$$

$$\mathbf{r}_2 = \mathbf{b} - \mathbf{A}\mathbf{x}_2 = \begin{bmatrix} 12 \\ -1 \\ 5 \end{bmatrix} - \begin{bmatrix} 4 & -1 & 1 \\ -1 & 4 & -2 \\ 1 & -2 & 4 \end{bmatrix} \begin{bmatrix} 3.07753 \\ 0.79482 \\ 0.71999 \end{bmatrix} = \begin{bmatrix} -0.23529 \\ 0.33823 \\ 0.63215 \end{bmatrix}$$

$$\beta_1 = -\frac{\mathbf{r}_2^T \mathbf{A}\mathbf{s}_1}{\mathbf{s}_1^T \mathbf{A}\mathbf{s}_1}$$

$$= -\frac{(-0.23529)(5.59656) + 0.33823(16.05980) + 0.63215(-10.21760)}{2.72076(5.59656) + 4.10380(16.05980) + (-1.18268)(-10.21760)}$$

$$= 0.0251452$$

$$\mathbf{s}_2 = \mathbf{r}_2 + \beta_1 \mathbf{s}_1 = \begin{bmatrix} -0.23529 \\ 0.33823 \\ 0.63215 \end{bmatrix} + 0.0251452 \begin{bmatrix} 2.72076 \\ 4.10380 \\ -1.18268 \end{bmatrix} = \begin{bmatrix} -0.166876 \\ 0.441421 \\ 0.602411 \end{bmatrix}$$

$$\mathbf{A}\mathbf{s}_2 = \begin{bmatrix} 4 & -1 & 1 \\ -1 & 4 & -2 \\ 1 & -2 & 4 \end{bmatrix} \begin{bmatrix} -0.166876 \\ 0.441421 \\ 0.602411 \end{bmatrix} = \begin{bmatrix} -0.506514 \\ 0.727738 \\ 1.359930 \end{bmatrix}$$

$$\alpha_2 = \frac{\mathbf{r}_2^T \mathbf{s}_2}{\mathbf{s}_2^T \mathbf{A}\mathbf{s}_2}$$

$$= \frac{(-0.23529)(-0.166876) + 0.33823(0.441421) + 0.63215(0.602411)}{(-0.166876)(-0.506514) + 0.441421(0.727738) + 0.602411(1.359930)}$$

$$= 0.46480$$

$$\mathbf{x}_3 = \mathbf{x}_2 + \alpha_2 \mathbf{s}_2 = \begin{bmatrix} 3.07753 \\ 0.79482 \\ 0.71999 \end{bmatrix} + 0.46480 \begin{bmatrix} -0.166876 \\ 0.441421 \\ 0.602411 \end{bmatrix} = \begin{bmatrix} 2.99997 \\ 0.99999 \\ 0.99999 \end{bmatrix}$$

The solution $\mathbf{x}_3$ is correct to almost five decimal places. The small discrepancy is caused by roundoff errors in the computations.

EXAMPLE 2.17

Write a computer program to solve the following n simultaneous equations[3] by the Gauss–Seidel method with relaxation (the program should work with any

---

[3] Equations of this form are called *cyclic* tridiagonal. They occur in the finite difference formulation of second-order differential equations with periodic boundary conditions.

value of n):

$$
\begin{bmatrix}
2 & -1 & 0 & 0 & \cdots & 0 & 0 & 0 & 1 \\
-1 & 2 & -1 & 0 & \cdots & 0 & 0 & 0 & 0 \\
0 & -1 & 2 & -1 & \cdots & 0 & 0 & 0 & 0 \\
\vdots & \vdots & \vdots & \vdots & & \vdots & \vdots & \vdots & \vdots \\
0 & 0 & 0 & 0 & \cdots & -1 & 2 & -1 & 0 \\
0 & 0 & 0 & 0 & \cdots & 0 & -1 & 2 & -1 \\
1 & 0 & 0 & 0 & \cdots & 0 & 0 & -1 & 2
\end{bmatrix}
\begin{bmatrix}
x_1 \\ x_2 \\ x_3 \\ \vdots \\ x_{n-2} \\ x_{n-1} \\ x_n
\end{bmatrix}
=
\begin{bmatrix}
0 \\ 0 \\ 0 \\ \vdots \\ 0 \\ 0 \\ 1
\end{bmatrix}
$$

Run the program with $n = 20$. The exact solution can be shown to be $x_i = -n/4 + i/2$, $i = 1, 2, \ldots, n$.

Solution In this case the iterative formulas in Eq. (2.35) are

$$x_1 = \omega(x_2 - x_n)/2 + (1 - \omega)x_1$$

$$x_i = \omega(x_{i-1} + x_{i+1})/2 + (1 - \omega)x_i, \quad i = 2, 3, \ldots, n - 1 \tag{a}$$

$$x_n = \omega(1 - x_1 + x_{n-1})/2 + (1 - \omega)x_n$$

which are evaluated by the following function:

```
function x = fex2_17(x,omega)
% Iteration formula Eq. (2.35) for Example 2.17.

n = length(x);
x(1) = omega*(x(2) - x(n))/2 + (1-omega)*x(1);
for i = 2:n-1
    x(i) = omega*(x(i-1) + x(i+1))/2 + (1-omega)*x(i);
end
x(n) = omega *(1 - x(1) + x(n-1))/2 + (1-omega)*x(n);
```

The solution can be obtained with a single command (note that $\mathbf{x} = \mathbf{0}$ is the starting vector):

```
>> [x,numIter,omega] = gaussSeidel(@fex2_17,zeros(20,1))
```

resulting in

```
x =
   -4.5000
   -4.0000
   -3.5000
   -3.0000
```

```
    -2.5000
    -2.0000
    -1.5000
    -1.0000
    -0.5000
     0.0000
     0.5000
     1.0000
     1.5000
     2.0000
     2.5000
     3.0000
     3.5000
     4.0000
     4.5000
     5.0000
numIter =
    259
omega =
     1.7055
```

The convergence is very slow, because the coefficient matrix lacks diagonal dominance—substituting the elements of **A** in Eq. (2.30) produces an equality rather than the desired inequality. If we were to change each diagonal term of the coefficient matrix from 2 to 4, **A** would be diagonally dominant and the solution would converge in only 22 iterations.

EXAMPLE 2.18
Solve Example 2.17 with the conjugate gradient method, also using $n = 20$.

Solution For the given **A**, the components of the vector **Av** are

$$(\mathbf{Av})_1 = 2v_1 - v_2 + v_n$$

$$(\mathbf{Av})_i = -v_{i-1} + 2v_i - v_{i+1}, \quad i = 2, 3, \ldots, n-1$$

$$(\mathbf{Av})_n = -v_{n-1} + 2v_n + v_1$$

which are evaluated by the following function:

```
function Av = fex2_18(v)
% Computes the product A*v in Example 2.18
```

```
n = length(v);
Av = zeros(n,1);
Av(1) = 2*v(1) - v(2) + v(n);
Av(2:n-1) = -v(1:n-2) + 2*v(2:n-1) - v(3:n);
Av(n) = -v(n-1) + 2*v(n) + v(1);
```

The program shown below utilizes the function `conjGrad`. The solution vector **x** is initialized to zero in the program, which also sets up the constant vector **b**.

```
% Example 2.18 (Conjugate gradient method)
n = 20;
x = zeros(n,1);
b = zeros(n,1); b(n) = 1;
[x,numIter] = conjGrad(@fex2_18,x,b)
```

Running the program results in

```
x =
    -4.5000
    -4.0000
    -3.5000
    -3.0000
    -2.5000
    -2.0000
    -1.5000
    -1.0000
    -0.5000
         0
     0.5000
     1.0000
     1.5000
     2.0000
     2.5000
     3.0000
     3.5000
     4.0000
     4.5000
     5.0000
numIter =
    10
```

PROBLEM SET 2.3

1. Let

$$\mathbf{A} = \begin{bmatrix} 3 & -1 & 2 \\ 0 & 1 & 3 \\ -2 & 2 & -4 \end{bmatrix} \qquad \mathbf{B} = \begin{bmatrix} 0 & 1 & 3 \\ 3 & -1 & 2 \\ -2 & 2 & -4 \end{bmatrix}$$

(note that $\mathbf{B}$ is obtained by interchanging the first two rows of $\mathbf{A}$). Knowing that

$$\mathbf{A}^{-1} = \begin{bmatrix} 0.5 & 0 & 0.25 \\ 0.3 & 0.4 & 0.45 \\ -0.1 & 0.2 & -0.15 \end{bmatrix}$$

determine $\mathbf{B}^{-1}$.

2. Invert the triangular matrices

$$\mathbf{A} = \begin{bmatrix} 2 & 4 & 3 \\ 0 & 6 & 5 \\ 0 & 0 & 2 \end{bmatrix} \qquad \mathbf{B} = \begin{bmatrix} 2 & 0 & 0 \\ 3 & 4 & 0 \\ 4 & 5 & 6 \end{bmatrix}$$

3. Invert the triangular matrix

$$\mathbf{A} = \begin{bmatrix} 1 & 1/2 & 1/4 & 1/8 \\ 0 & 1 & 1/3 & 1/9 \\ 0 & 0 & 1 & 1/4 \\ 0 & 0 & 0 & 1 \end{bmatrix}$$

4. Invert the following matrices:

$$\text{(a) } \mathbf{A} = \begin{bmatrix} 1 & 2 & 4 \\ 1 & 3 & 9 \\ 1 & 4 & 16 \end{bmatrix} \qquad \text{(b) } \mathbf{B} = \begin{bmatrix} 4 & -1 & 0 \\ -1 & 4 & -1 \\ 0 & -1 & 4 \end{bmatrix}$$

5. Invert the matrix

$$\mathbf{A} = \begin{bmatrix} 4 & -2 & 1 \\ -2 & 1 & -1 \\ 1 & -2 & 4 \end{bmatrix}$$

6. ■ Invert the following matrices with any method:

$$\mathbf{A} = \begin{bmatrix} 5 & -3 & -1 & 0 \\ -2 & 1 & 1 & 1 \\ 3 & -5 & 1 & 2 \\ 0 & 8 & -4 & -3 \end{bmatrix} \qquad \mathbf{B} = \begin{bmatrix} 4 & -1 & 0 & 0 \\ -1 & 4 & -1 & 0 \\ 0 & -1 & 4 & -1 \\ 0 & 0 & -1 & 4 \end{bmatrix}$$

7. ■ Invert the matrix with any method;

$$A = \begin{bmatrix} 1 & 3 & -9 & 6 & 4 \\ 2 & -1 & 6 & 7 & 1 \\ 3 & 2 & -3 & 15 & 5 \\ 8 & -1 & 1 & 4 & 2 \\ 11 & 1 & -2 & 18 & 7 \end{bmatrix}$$

and comment on the reliability of the result.

8. ■ The joint displacements $\mathbf{u}$ of the plane truss in Prob. 14, Problem Set 2.2 are related to the applied joint forces $\mathbf{p}$ by

$$\mathbf{Ku} = \mathbf{p} \tag{a}$$

where

$$\mathbf{K} = \begin{bmatrix} 27.580 & 7.004 & -7.004 & 0.000 & 0.000 \\ 7.004 & 29.570 & -5.253 & 0.000 & -24.320 \\ -7.004 & -5.253 & 29.570 & 0.000 & 0.000 \\ 0.000 & 0.000 & 0.000 & 27.580 & -7.004 \\ 0.000 & -24.320 & 0.000 & -7.004 & 29.570 \end{bmatrix} \text{MN/m}$$

is called the *stiffness matrix* of the truss. If Eq. (a) is inverted by multiplying each side by $\mathbf{K}^{-1}$, we obtain $\mathbf{u} = \mathbf{K}^{-1}\mathbf{p}$, where $\mathbf{K}^{-1}$ is known as the *flexibility matrix*. The physical meaning of the elements of the flexibility matrix is: K_{ij}^{-1} = displacements u_i $(i = 1, 2, \ldots 5)$ produced by the unit load $p_j = 1$. Compute (a) the flexibility matrix of the truss; (b) the displacements of the joints due to the load $p_5 = -45$ kN (the load shown in Problem 14, Problem Set 2.2).

9. ■ Invert the matrices

$$A = \begin{bmatrix} 3 & -7 & 45 & 21 \\ 12 & 11 & 10 & 17 \\ 6 & 25 & -80 & -24 \\ 17 & 55 & -9 & 7 \end{bmatrix} \qquad B = \begin{bmatrix} 1 & 1 & 1 & 1 \\ 1 & 2 & 2 & 2 \\ 2 & 3 & 4 & 4 \\ 4 & 5 & 6 & 7 \end{bmatrix}$$

10. ■ Write a program for inverting a $n \times n$ lower triangular matrix. The inversion procedure should contain only forward substitution. Test the program by inverting the matrix

$$A = \begin{bmatrix} 36 & 0 & 0 & 0 \\ 18 & 36 & 0 & 0 \\ 9 & 12 & 36 & 0 \\ 5 & 4 & 9 & 36 \end{bmatrix}$$

Let the program also check the result by computing and printing $\mathbf{AA}^{-1}$.

11. Use the Gauss–Seidel method to solve

$$\begin{bmatrix} -2 & 5 & 9 \\ 7 & 1 & 1 \\ -3 & 7 & -1 \end{bmatrix} \begin{bmatrix} x_1 \\ x_2 \\ x_3 \end{bmatrix} = \begin{bmatrix} 1 \\ 6 \\ -26 \end{bmatrix}$$

12. Solve the following equations with the Gauss–Seidel method:

$$\begin{bmatrix} 12 & -2 & 3 & 1 \\ -2 & 15 & 6 & -3 \\ 1 & 6 & 20 & -4 \\ 0 & -3 & 2 & 9 \end{bmatrix} \begin{bmatrix} x_1 \\ x_2 \\ x_3 \\ x_4 \end{bmatrix} = \begin{bmatrix} 0 \\ 0 \\ 20 \\ 0 \end{bmatrix}$$

13. Use the Gauss–Seidel method with relaxation to solve $\mathbf{Ax} = \mathbf{b}$, where

$$\mathbf{A} = \begin{bmatrix} 4 & -1 & 0 & 0 \\ -1 & 4 & -1 & 0 \\ 0 & -1 & 4 & -1 \\ 0 & 0 & -1 & 3 \end{bmatrix} \qquad \mathbf{b} = \begin{bmatrix} 15 \\ 10 \\ 10 \\ 10 \end{bmatrix}$$

   Take $x_i = b_i/A_{ii}$ as the starting vector and use $\omega = 1.1$ for the relaxation factor.

14. Solve the equations

$$\begin{bmatrix} 2 & -1 & 0 \\ -1 & 2 & -1 \\ 0 & -1 & 1 \end{bmatrix} \begin{bmatrix} x_1 \\ x_2 \\ x_3 \end{bmatrix} = \begin{bmatrix} 1 \\ 1 \\ 1 \end{bmatrix}$$

   by the conjugate gradient method. Start with $\mathbf{x} = \mathbf{0}$.

15. Use the conjugate gradient method to solve

$$\begin{bmatrix} 3 & 0 & -1 \\ 0 & 4 & -2 \\ -1 & -2 & 5 \end{bmatrix} \begin{bmatrix} x_1 \\ x_2 \\ x_3 \end{bmatrix} = \begin{bmatrix} 4 \\ 10 \\ -10 \end{bmatrix}$$

   starting with $\mathbf{x} = \mathbf{0}$.

16. ■ Solve the simultaneous equations $\mathbf{Ax} = \mathbf{b}$ and $\mathbf{Bx} = \mathbf{b}$ by the Gauss–Seidel method with relaxation, where

$$\mathbf{b} = \begin{bmatrix} 10 & -8 & 10 & 10 & -8 & 10 \end{bmatrix}^T$$

$$
A = \begin{bmatrix}
3 & -2 & 1 & 0 & 0 & 0 \\
-2 & 4 & -2 & 1 & 0 & 0 \\
1 & -2 & 4 & -2 & 1 & 0 \\
0 & 1 & -2 & 4 & -2 & 1 \\
0 & 0 & 1 & -2 & 4 & -2 \\
0 & 0 & 0 & 1 & -2 & 3
\end{bmatrix}
$$

$$
B = \begin{bmatrix}
3 & -2 & 1 & 0 & 0 & 1 \\
-2 & 4 & -2 & 1 & 0 & 0 \\
1 & -2 & 4 & -2 & 1 & 0 \\
0 & 1 & -2 & 4 & -2 & 1 \\
0 & 0 & 1 & -2 & 4 & -2 \\
1 & 0 & 0 & 1 & -2 & 3
\end{bmatrix}
$$

Note that **A** is not diagonally dominant, but that does not necessarily preclude convergence.

17. ■ Modify the program in Example 2.17 (Gauss–Seidel method) so that it will solve the following equations:

$$
\begin{bmatrix}
4 & -1 & 0 & 0 & \cdots & 0 & 0 & 0 & 1 \\
-1 & 4 & -1 & 0 & \cdots & 0 & 0 & 0 & 0 \\
0 & -1 & 4 & -1 & \cdots & 0 & 0 & 0 & 0 \\
\vdots & \vdots & \vdots & \vdots & \cdots & \vdots & \vdots & \vdots & \vdots \\
0 & 0 & 0 & 0 & \cdots & -1 & 4 & -1 & 0 \\
0 & 0 & 0 & 0 & \cdots & 0 & -1 & 4 & -1 \\
1 & 0 & 0 & 0 & \cdots & 0 & 0 & -1 & 4
\end{bmatrix}
\begin{bmatrix}
x_1 \\ x_2 \\ x_3 \\ \vdots \\ x_{n-2} \\ x_{n-1} \\ x_n
\end{bmatrix}
=
\begin{bmatrix}
0 \\ 0 \\ 0 \\ \vdots \\ 0 \\ 0 \\ 100
\end{bmatrix}
$$

Run the program with $n = 20$ and compare the number of iterations with Example 2.17.

18. ■ Modify the program in Example 2.18 to solve the equations in Prob. 17 by the conjugate gradient method. Run the program with $n = 20$.

19. ■

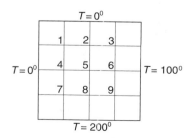

$T = 0^0$

$T = 0^0$ |1 2 3 / 4 5 6 / 7 8 9| $T = 100^0$

$T = 200^0$

The edges of the square plate are kept at the temperatures shown. Assuming steady-state heat conduction, the differential equation governing the temperature T in the interior is

$$\frac{\partial^2 T}{\partial x^2} + \frac{\partial^2 T}{\partial y^2} = 0$$

If this equation is approximated by finite differences using the mesh shown, we obtain the following algebraic equations for temperatures at the mesh points:

$$\begin{bmatrix} -4 & 1 & 0 & 1 & 0 & 0 & 0 & 0 & 0 \\ 1 & -4 & 1 & 0 & 1 & 0 & 0 & 0 & 0 \\ 0 & 1 & -4 & 0 & 0 & 1 & 0 & 0 & 0 \\ 1 & 0 & 0 & -4 & 1 & 0 & 1 & 0 & 0 \\ 0 & 1 & 0 & 1 & -4 & 1 & 0 & 1 & 0 \\ 0 & 0 & 1 & 0 & 1 & -4 & 0 & 0 & 1 \\ 0 & 0 & 0 & 1 & 0 & 0 & -4 & 1 & 0 \\ 0 & 0 & 0 & 0 & 1 & 0 & 1 & -4 & 1 \\ 0 & 0 & 0 & 0 & 0 & 1 & 0 & 1 & -4 \end{bmatrix} \begin{bmatrix} T_1 \\ T_2 \\ T_3 \\ T_4 \\ T_5 \\ T_6 \\ T_7 \\ T_8 \\ T_9 \end{bmatrix} = - \begin{bmatrix} 0 \\ 0 \\ 100 \\ 0 \\ 0 \\ 100 \\ 200 \\ 200 \\ 300 \end{bmatrix}$$

Solve these equations with the conjugate gradient method.

MATLAB Functions

`x = A\b` returns the solution x of $\mathbf{Ax} = \mathbf{b}$, obtained by Gauss elimination. If the equations are overdetermined (**A** has more rows than columns), the least-squares solution is computed.

`[L,U] = lu(A)` Doolittle's decomposition $\mathbf{A} = \mathbf{LU}$. On return, `U` is an upper triangular matrix and `L` contains a row-wise permutation of the lower triangular matrix.

`[M,U,P] = lu(A)` returns the same `U` as above, but now `M` is a lower triangular matrix and `P` is the permutation matrix so that `M = P*L`. Note that here `P*A = M*U`.

`L = chol(A)` Choleski's decomposition $\mathbf{A} = \mathbf{LL}^T$.

`B = inv(A)` returns `B` as the inverse of `A` (the method used is not specified).

`n = norm(A,1)` returns the norm $n = \max_j \sum_i |A_{ij}|$ (largest sum of elements in a column of **A**).

`c = cond(A)` returns the *condition number* of the matrix `A`.

MATLAB does not cater to banded matrices explicitly. However, banded matrices can be treated as a *sparse matrices* for which MATLAB provides extensive support. A banded matrix in sparse form can be created by the following command:

A = spdiags(B,d,n,n) creates a $n \times n$ sparse matrix from the columns of matrix B by placing the columns along the diagonals specified by d. The columns of B may be longer than the diagonals they represent. A diagonal in the upper part of A takes its elements from lower part of a column of B, while a lower diagonal uses the upper part of B.

Here is an example of creating the 5×5 tridiagonal matrix

$$\mathbf{A} = \begin{bmatrix} 2 & -1 & 0 & 0 & 0 \\ -1 & 2 & -1 & 0 & 0 \\ 0 & -1 & 2 & -1 & 0 \\ 0 & 0 & -1 & 2 & -1 \\ 0 & 0 & 0 & -1 & 2 \end{bmatrix}$$

```
>> c = ones(5,1);
>> A = spdiags([-c 2*c -c],[-1 0 1],5,5)
A =
   (1,1)        2
   (2,1)       -1
   (1,2)       -1
   (2,2)        2
   (3,2)       -1
   (2,3)       -1
   (3,3)        2
   (4,3)       -1
   (3,4)       -1
   (4,4)        2
   (5,4)       -1
   (4,5)       -1
   (5,5)        2
```

If the matrix is declared sparse, MATLAB stores only the nonzero elements of the matrix together with information locating the position of each element in the matrix. The printout of a sparse matrix displays the values of these elements and their indices (row and column numbers) in parentheses.

Almost all matrix functions, including the ones listed above, also work on sparse matrices. For example, [L,U] = lu(A) would return L and U in sparse matrix representation if A is a sparse matrix. There are many sparse matrix functions in MATLAB; here are just a few of them:

`A = full(S)` converts the sparse matrix S into a full matrix A.

`S = sparse(A)` converts the full matrix A into a sparse matrix S.

`x = lsqr(A,b)` conjugate gradient method for solving $\mathbf{Ax} = \mathbf{b}$.

`spy(S)` draws a map of the nonzero elements of S.

3 Interpolation and Curve Fitting

Given the n data points (x_i, y_i), $i = 1, 2, \ldots, n$, estimate $y(x)$.

3.1 Introduction

Discrete data sets, or tables of the form

x_1	x_2	x_3	$\cdots$	x_n
y_1	y_2	y_3	$\cdots$	y_n

are commonly involved in technical calculations. The source of the data may be experimental observations or numerical computations. There is a distinction between interpolation and curve fitting. In interpolation we construct a curve through the data points. In doing so, we make the implicit assumption that the data points are accurate and distinct. Curve fitting is applied to data that contain scatter (noise), usually due to measurement errors. Here we want to find a smooth curve that approximates the data in some sense. Thus the curve does not have to hit the data points. This difference between interpolation and curve fitting is illustrated in Fig. 3.1.

3.2 Polynomial Interpolation

Lagrange's Method

The simplest form of an interpolant is a polynomial. It is always possible to construct a unique polynomial $P_{n-1}(x)$ of degree $n - 1$ that passes through n distinct data points.

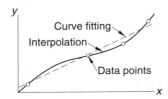

Figure 3.1. Interpolation and curve fitting of data.

One means of obtaining this polynomial is the *formula of Lagrange*

$$P_{n-1}(x) = \sum_{i=1}^{n} y_i \ell_i(x) \tag{3.1a}$$

where

$$\ell_i(x) = \frac{x - x_1}{x_i - x_1} \cdot \frac{x - x_2}{x_i - x_2} \cdots \frac{x - x_{i-1}}{x_i - x_{i-1}} \cdot \frac{x - x_{i+1}}{x_i - x_{i+1}} \cdots \frac{x - x_n}{x_i - x_n}$$

$$= \prod_{\substack{j=1 \\ j \neq i}}^{n} \frac{x - x_j}{x_i - x_j}, \quad i = 1, 2, \ldots, n \tag{3.1b}$$

are called the *cardinal functions.*

For example, if $n = 2$, the interpolant is the straight line $P_1(x) = y_1 \ell_1(x) + y_2 \ell_2(x)$, where

$$\ell_1(x) = \frac{x - x_2}{x_1 - x_2} \qquad \ell_2(x) = \frac{x - x_1}{x_2 - x_1}$$

With $n = 3$, interpolation is parabolic: $P_2(x) = y_1 \ell_1(x) + y_2 \ell_2(x) + y_3 \ell_3(x)$, where now

$$\ell_1(x) = \frac{(x - x_2)(x - x_3)}{(x_1 - x_2)(x_1 - x_3)}$$

$$\ell_2(x) = \frac{(x - x_1)(x - x_3)}{(x_2 - x_1)(x_2 - x_3)}$$

$$\ell_3(x) = \frac{(x - x_1)(x - x_2)}{(x_3 - x_1)(x_3 - x_2)}$$

The cardinal functions are polynomials of degree $n - 1$ and have the property

$$\ell_i(x_j) = \begin{cases} 0 & \text{if } i \neq j \\ 1 & \text{if } i = j \end{cases} = \delta_{ij} \tag{3.2}$$

where δ_{ij} is the Kronecker delta. This property is illustrated in Fig. 3.2 for three-point interpolation ($n = 3$) with $x_1 = 0$, $x_2 = 2$ and $x_3 = 3$.

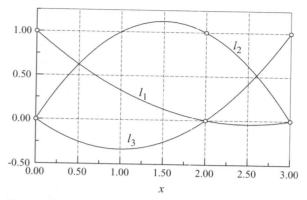

Figure 3.2. Example of quadratic cardinal functions.

To prove that the interpolating polynomial passes through the data points, we substitute $x = x_j$ into Eq. (3.1a) and then utilize Eq. (3.2). The result is

$$P_{n-1}(x_j) = \sum_{i=1}^{n} y_i \ell_i(x_j) = \sum_{i=1}^{n} y_i \delta_{ij} = y_j$$

It can be shown that the error in polynomial interpolation is

$$f(x) - P_{n-1}(x) = \frac{(x - x_1)(x - x_2) \ldots (x - x_n)}{n!} f^{(n)}(\xi) \tag{3.3}$$

where ξ lies somewhere in the interval (x_1, x_n); its value is otherwise unknown. It is instructive to note that the farther a data point is from x, the more it contributes to the error at x.

Newton's Method

Evaluation of polynomial

Although Lagrange's method is conceptually simple, it does not lend itself to an efficient algorithm. A better computational procedure is obtained with Newton's method, where the interpolating polynomial is written in the form

$$P_{n-1}(x) = a_1 + (x - x_1)a_2 + (x - x_1)(x - x_2)a_3 + \cdots + (x - x_1)(x - x_2) \cdots (x - x_{n-1})a_n$$

This polynomial lends itself to an efficient evaluation procedure. Consider, for example, four data points ($n = 4$). Here the interpolating polynomial is

$$P_3(x) = a_1 + (x - x_1)a_2 + (x - x_1)(x - x_2)a_3 + (x - x_1)(x - x_2)(x - x_3)a_4$$
$$= a_1 + (x - x_1) \{a_2 + (x - x_2) [a_3 + (x - x_3)a_4]\}$$

which can be evaluated backward with the following recurrence relations:

$$P_0(x) = a_4$$

$$P_1(x) = a_3 + (x - x_3)P_0(x)$$

$$P_2(x) = a_2 + (x - x_2)P_1(x)$$

$$P_3(x) = a_1 + (x - x_1)P_2(x)$$

For arbitrary n we have

$$P_0(x) = a_n \qquad P_k(x) = a_{n-k} + (x - x_{n-k})P_{k-1}(x), \quad k = 1, 2, \ldots, n-1 \tag{3.4}$$

■ newtonPoly

Denoting the x-coordinate array of the data points by xData, and the number of data points by n, we have the following algorithm for computing $P_{n-1}(x)$:

```
function p = newtonPoly(a,xData,x)
% Returns value of Newton's polynomial at x.
% USAGE: p = newtonPoly(a,xData,x)
% a      = coefficient array of the polynomial;
%          must be computed first by newtonCoeff.
% xData = x-coordinates of data points.

n = length(xData);
p = a(n);
for k = 1:n-1;
    p = a(n-k) + (x - xData(n-k))*p;
end
```

Computation of coefficients

The coefficients of $P_{n-1}(x)$ are determined by forcing the polynomial to pass through each data point: $y_i = P_{n-1}(x_i)$, $i = 1, 2, \ldots, n$. This yields the simultaneous equations

$$y_1 = a_1$$

$$y_2 = a_1 + (x_2 - x_1)a_2$$

$$y_3 = a_1 + (x_3 - x_1)a_2 + (x_3 - x_1)(x_3 - x_2)a_3 \tag{a}$$

$$\vdots$$

$$y_n = a_1 + (x_n - x_1)a_1 + \cdots + (x_n - x_1)(x_n - x_2)\cdots(x_n - x_{n-1})a_n$$

Introducing the *divided differences*

$$\nabla y_i = \frac{y_i - y_1}{x_i - x_1}, \quad i = 2, 3, \dots, n$$

$$\nabla^2 y_i = \frac{\nabla y_i - \nabla y_2}{x_i - x_2}, \quad i = 3, 4, \dots, n$$

$$\nabla^3 y_i = \frac{\nabla^2 y_i - \nabla^2 y_3}{x_i - x_3}, \quad i = 4, 5, \dots n \tag{3.5}$$

$$\vdots$$

$$\nabla^n y_n = \frac{\nabla^{n-1} y_n - \nabla^{n-1} y_{n-1}}{x_n - x_{n-1}}$$

the solution of Eqs. (a) is

$$a_1 = y_1 \qquad a_2 = \nabla y_2 \qquad a_3 = \nabla^2 y_3 \qquad \cdots \qquad a_n = \nabla^n y_n \tag{3.6}$$

If the coefficients are computed by hand, it is convenient to work with the format in Table 3.1 (shown for $n = 5$).

x_1	y_1				
x_2	y_2	∇y_2			
x_3	y_3	∇y_3	$\nabla^2 y_3$		
x_4	y_4	∇y_4	$\nabla^2 y_4$	$\nabla^3 y_4$	
x_5	y_5	∇y_5	$\nabla^2 y_5$	$\nabla^3 y_5$	$\nabla^4 y_5$

Table 3.1

The diagonal terms ($y_1, \nabla y_2, \nabla^2 y_3, \nabla^3 y_4$ and $\nabla^4 y_5$) in the table are the coefficients of the polynomial. If the data points are listed in a different order, the entries in the table will change, but the resultant polynomial will be the same—recall that a polynomial of degree $n - 1$ interpolating n distinct data points is unique.

■ newtonCoeff

Machine computations are best carried out within a one-dimensional array **a** employing the following algorithm:

```
function a = newtonCoeff(xData,yData)
% Returns coefficients of Newton's polynomial.
```

```
% USAGE: a = newtonCoeff(xData,yData)
% xData = x-coordinates of data points.
% yData = y-coordinates of data points.

n = length(xData);
a = yData;
for k = 2:n
    a(k:n) = (a(k:n) - a(k-1))./(xData(k:n) - xData(k-1));
end
```

Initially, **a** contains the y-values of the data, so that it is identical to the second column in Table 3.1. Each pass through the for-loop generates the entries in the next column, which overwrite the corresponding elements of **a**. Therefore, **a** ends up containing the diagonal terms of Table 3.1; i.e., the coefficients of the polynomial.

Neville's Method

Newton's method of interpolation involves two steps: computation of the coefficients, followed by evaluation of the polynomial. This works well if the interpolation is carried out repeatedly at different values of x using the same polynomial. If only one point is to be interpolated, a method that computes the interpolant in a single step, such as Neville's algorithm, is a better choice.

Let $P_k[x_i, x_{i+1}, \ldots, x_{i+k}]$ denote the polynomial of degree k that passes through the $k+1$ data points $(x_i, y_i), (x_{i+1}, y_{i+1}), \ldots, (x_{i+k}, y_{i+k})$. For a single data point, we have

$$P_0[x_i] = y_i \qquad (3.7)$$

The interpolant based on two data points is

$$P_1[x_i, x_{i+1}] = \frac{(x - x_{i+1}) P_0[x_i] + (x_i - x) P_0[x_{i+1}]}{x_i - x_{i+1}}$$

It is easily verified that $P_1[x_i, x_{i+1}]$ passes through the two data points; that is, $P_1[x_i, x_{i+1}] = y_i$ when $x = x_i$, and $P_1[x_i, x_{i+1}] = y_{i+1}$ when $x = x_{i+1}$.

The three-point interpolant is

$$P_2[x_i, x_{i+1}, x_{i+2}] = \frac{(x - x_{i+2}) P_1[x_i, x_{i+1}] + (x_i - x) P_1[x_{i+1}, x_{i+2}]}{x_i - x_{i+2}}$$

To show that this interpolant does intersect the data points, we first substitute $x = x_i$, obtaining

$$P_2[x_i, x_{i+1}, x_{i+2}] = P_1[x_i, x_{i+1}] = y_i$$

Similarly, $x = x_{i+2}$ yields

$$P_2[x_i, x_{i+1}, x_{i+2}] = P_1[x_{i+1}, x_{i+2}] = y_{i+2}$$

Finally, when $x = x_{i+1}$ we have

$$P_1[x_i, x_{i+1}] = P_1[x_{i+1}, x_{i+2}] = y_{i+1}$$

so that

$$P_2[x_i, x_{i+1}, x_{i+2}] = \frac{(x_{i+1} - x_{i+2})y_{i+1} + (x_i - x_{i+1})y_{i+1}}{x_i - x_{i+2}} = y_{i+1}$$

Having established the pattern, we can now deduce the general recursive formula:

$$P_k[x_i, x_{i+1}, \ldots, x_{i+k}] \tag{3.8}$$
$$= \frac{(x - x_{i+k})P_{k-1}[x_i, x_{i+1}, \ldots, x_{i+k-1}] + (x_i - x)P_{k-1}[x_{i+1}, x_{i+2}, \ldots, x_{i+k}]}{x_i - x_{i+k}}$$

Given the value of x, the computations can be carried out in the following tabular format (shown for four data points):

	$k = 0$	$k = 1$	$k = 2$	$k = 3$
x_1	$P_0[x_1] = y_1$	$P_1[x_1, x_2]$	$P_2[x_1, x_2, x_3]$	$P_3[x_1, x_2, x_3, x_4]$
x_2	$P_0[x_2] = y_2$	$P_1[x_2, x_3]$	$P_2[x_2, x_3, x_4]$	
x_3	$P_0[x_3] = y_3$	$P_1[x_3, x_4]$		
x_4	$P_0[x_4] = y_4$			

Table 3.2

■ neville

This algorithm works with the one-dimensional array **y**, which initially contains the y-values of the data (the second column in Table 3.2). Each pass through the for-loop computes the terms in next column of the table, which overwrite the previous elements of **y**. At the end of the procedure, **y** contains the diagonal terms of the table. The value of the interpolant (evaluated at x) that passes through all the data points is y_1, the first element of **y**.

```
function yInterp = neville(xData,yData,x)
% Neville's polynomial interpolation;
% returns the value of the interpolant at x.
```

```
% USAGE: yInterp = neville(xData,yData,x)
% xData = x-coordinates of data points.
% yData = y-coordinates of data points.

n = length(xData);
y = yData;
for k = 1:n-1
    y(1:n-k) = ((x - xData(k+1:n)).*y(1:n-k)...
               + (xData(1:n-k) - x).*y(2:n-k+1))...
               ./(xData(1:n-k) - xData(k+1:n));
end
yInterp = y(1);
```

Limitations of Polynomial Interpolation

Polynomial interpolation should be carried out with the fewest feasible number of data points. Linear interpolation, using the nearest two points, is often sufficient if the data points are closely spaced. Three to six nearest-neighbor points produce good results in most cases. An interpolant intersecting more than six points must be viewed with suspicion. The reason is that the data points that are far from the point of interest do not contribute to the accuracy of the interpolant. In fact, they can be detrimental.

The danger of using too many points is illustrated in Fig. 3.3. There are 11 equally spaced data points represented by the circles. The solid line is the interpolant, a polynomial of degree ten, that intersects all the points. As seen in the figure, a polynomial of such a high degree has a tendency to oscillate excessively between the data points. A much smoother result would be obtained by using a cubic interpolant spanning four nearest-neighbor points.

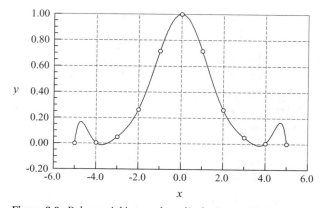

Figure 3.3. Polynomial interpolant displaying oscillations.

Polynomial extrapolation (interpolating outside the range of data points) is dangerous. As an example, consider Fig. 3.4. There are six data points, shown as circles. The fifth-degree interpolating polynomial is represented by the solid line. The interpolant looks fine within the range of data points, but drastically departs from the obvious trend when $x > 12$. Extrapolating y at $x = 14$, for example, would be absurd in this case.

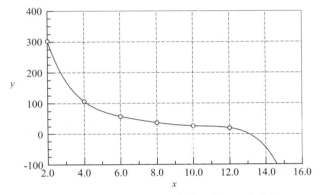

Figure 3.4. Extrapolation may not follow the trend of data.

If extrapolation cannot be avoided, the following two measures can be useful:

- Plot the data and visually verify that the extrapolated value makes sense.
- Use a low-order polynomial based on nearest-neighbor data points. A linear or quadratic interpolant, for example, would yield a reasonable estimate of $y(14)$ for the data in Fig. 3.4.
- Work with a plot of $\log x$ vs. $\log y$, which is usually much smoother than the x–y curve, and thus safer to extrapolate. Frequently this plot is almost a straight line. This is illustrated in Fig. 3.5, which represents the logarithmic plot of the data in Fig. 3.4.

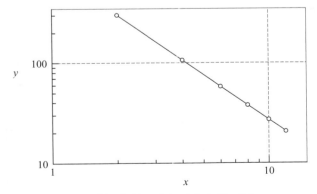

Figure 3.5. Logarithmic plot of the data in Fig. 3.4.

EXAMPLE 3.1

Given the data points

x	0	2	3
y	7	11	28

use Lagrange's method to determine y at $x = 1$.

Solution

$$\ell_1 = \frac{(x - x_2)(x - x_3)}{(x_1 - x_2)(x_1 - x_3)} = \frac{(1 - 2)(1 - 3)}{(0 - 2)(0 - 3)} = \frac{1}{3}$$

$$\ell_2 = \frac{(x - x_1)(x - x_3)}{(x_2 - x_1)(x_2 - x_3)} = \frac{(1 - 0)(1 - 3)}{(2 - 0)(2 - 3)} = 1$$

$$\ell_3 = \frac{(x - x_1)(x - x_2)}{(x_3 - x_1)(x_3 - x_2)} = \frac{(1 - 0)(1 - 2)}{(3 - 0)(3 - 2)} = -\frac{1}{3}$$

$$y = y_1\ell_1 + y_2\ell_2 + y_3\ell_3 = \frac{7}{3} + 11 - \frac{28}{3} = 4$$

EXAMPLE 3.2

The data points

x	-2	1	4	-1	3	-4
y	-1	2	59	4	24	-53

lie on a polynomial. Determine the degree of this polynomial by constructing the divided difference table, similar to Table 3.1.

Solution

i	x_i	y_i	∇y_i	$\nabla^2 y_i$	$\nabla^3 y_i$	$\nabla^4 y_i$	$\nabla^5 y_i$
1	-2	-1					
2	1	2	1				
3	4	59	10	3			
4	-1	4	5	-2	1		
5	3	24	5	2	1	0	
6	-4	-53	26	-5	1	0	0

Here are a few sample calculations used in arriving at the figures in the table:

$$\nabla y_3 = \frac{y_3 - y_1}{x_3 - x_1} = \frac{59 - (-1)}{4 - (-2)} = 10$$

$$\nabla^2 y_3 = \frac{\nabla y_3 - \nabla y_2}{x_3 - x_2} = \frac{10 - 1}{4 - 1} = 3$$

$$\nabla^3 y_6 = \frac{\nabla^2 y_6 - \nabla^2 y_3}{x_6 - x_3} = \frac{-5 - 3}{-4 - 4} = 1$$

From the table we see that the last nonzero coefficient (last nonzero diagonal term) of Newton's polynomial is $\nabla^3 y_3$, which is the coefficient of the cubic term. Hence the polynomial is a cubic.

EXAMPLE 3.3

Given the data points

x	4.0	3.9	3.8	3.7
y	-0.06604	-0.02724	0.01282	0.05383

determine the root of $y(x) = 0$ by Neville's method.

Solution This is an example of *inverse interpolation*, where the roles of x and y are interchanged. Instead of computing y at a given x, we are finding x that corresponds to a given y (in this case, $y = 0$). Employing the format of Table 3.2 (with x and y interchanged, of course), we obtain

i	y_i	$P_0[\,] = x_i$	$P_1[\,,]$	$P_2[\,,\,]$	$P_3[\,,\,,]$
1	-0.06604	4.0	3.8298	3.8316	3.8317
2	-0.02724	3.9	3.8320	3.8318	
3	0.01282	3.8	3.8313		
4	0.05383	3.7			

The following are a couple of sample computations used in the table:

$$P_1[y_1, y_2] = \frac{(y - y_2) P_0[y_1] + (y_1 - y) P_0[y_2]}{y_1 - y_2}$$

$$= \frac{(0 + 0.02724)(4.0) + (-0.06604 - 0)(3.9)}{-0.06604 + 0.02724} = 3.8298$$

$$P_2[y_2, y_3, y_4] = \frac{(y - y_4) P_1[y_2, y_3] + (y_2 - y) P_1[y_3, y_4]}{y_2 - y_4}$$

$$= \frac{(0 - 0.05383)(3.8320) + (-0.02724 - 0)(3.8313)}{-0.02724 - 0.05383} = 3.8318$$

All the *P*'s in the table are estimates of the root resulting from different orders of interpolation involving different data points. For example, $P_1[y_1, y_2]$ is the root obtained from linear interpolation based on the first two points, and $P_2[y_2, y_3, y_4]$ is the result from quadratic interpolation using the last three points. The root obtained from cubic interpolation over all four data points is $x = P_3[y_1, y_2, y_3, y_4] = 3.8317$.

EXAMPLE 3.4
The data points in the table lie on the plot of $f(x) = 4.8 \cos \pi x / 20$. Interpolate this data by Newton's method at $x = 0, 0.5, 1.0, \ldots, 8.0$ and compare the results with the "exact" values given by $y = f(x)$.

x	0.15	2.30	3.15	4.85	6.25	7.95
y	4.79867	4.49013	4.2243	3.47313	2.66674	1.51909

Solution
```
% Example 3.4 (Newton's interpolation)
xData = [0.15; 2.3; 3.15; 4.85; 6.25; 7.95];
yData = [4.79867; 4.49013; 4.22430; 3.47313;...
         2.66674; 1.51909];
a = newtonCoeff(xData,yData);
'       x        yInterp     yExact'
for x = 0: 0.5: 8
    y = newtonPoly(a,xData,x);
    yExact = 4.8*cos(pi*x/20);
    fprintf('%10.5f',x,y,yExact)
    fprintf('\n')
end
```

The results are:

```
ans =
        x        yInterp     yExact
   0.00000     4.80003     4.80000
   0.50000     4.78518     4.78520
   1.00000     4.74088     4.74090
   1.50000     4.66736     4.66738
   2.00000     4.56507     4.56507
   2.50000     4.43462     4.43462
   3.00000     4.27683     4.27683
   3.50000     4.09267     4.09267
```

4.00000	3.88327	3.88328
4.50000	3.64994	3.64995
5.00000	3.39411	3.39411
5.50000	3.11735	3.11735
6.00000	2.82137	2.82137
6.50000	2.50799	2.50799
7.00000	2.17915	2.17915
7.50000	1.83687	1.83688
8.00000	1.48329	1.48328

3.3 Interpolation with Cubic Spline

If there are more than a few data points, a cubic spline is hard to beat as a global interpolant. It is considerably "stiffer" than a polynomial in the sense that it has less tendency to oscillate between data points.

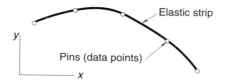

Figure 3.6. Mechanical model of natural cubic spline.

The mechanical model of a cubic spline is shown in Fig. 3.6. It is a thin, elastic strip that is attached with pins to the data points. Because the strip is unloaded between the pins, each segment of the spline curve is a cubic polynomial—recall from beam theory that the differential equation for the displacement of a beam is $d^4y/dx^4 = q/(EI)$, so that $y(x)$ is a cubic since the load q vanishes. At the pins, the slope and bending moment (and hence the second derivative) are continuous. There is no bending moment at the two end pins; hence the second derivative of the spline is zero at the end points. Since these end conditions occur naturally in the beam model, the resulting curve is known as the *natural cubic spline*. The pins, i.e., the data points, are called the *knots* of the spline.

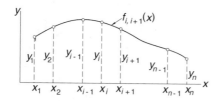

Figure 3.7. Cubic spline.

Figure 3.7 shows a cubic spline that spans n knots. We use the notation $f_{i,i+1}(x)$ for the cubic polynomial that spans the segment between knots i and $i+1$. Note that the spline is a *piecewise cubic* curve, put together from the $n-1$ cubics $f_{1,2}(x)$, $f_{2,3}(x)$, ..., $f_{n-1,n}(x)$, all of which have different coefficients.

If we denote the second derivative of the spline at knot i by k_i, continuity of second derivatives requires that

$$f''_{i-1,i}(x_i) = f''_{i,i+1}(x_i) = k_i \tag{a}$$

At this stage, each k is unknown, except for

$$k_1 = k_n = 0 \tag{3.9}$$

The starting point for computing the coefficients of $f_{i,i+1}(x)$ is the expression for $f''_{i,i+1}(x)$, which we know to be linear. Using Lagrange's two-point interpolation, we can write

$$f''_{i,i+1}(x) = k_i \ell_i(x) + k_{i+1}\ell_{i+1}(x)$$

where

$$\ell_i(x) = \frac{x - x_{i+1}}{x_i - x_{i+1}} \qquad \ell_{i+1}(x) = \frac{x - x_i}{x_{i+1} - x_i}$$

Therefore,

$$f''_{i,i+1}(x) = \frac{k_i(x - x_{i+1}) - k_{i+1}(x - x_i)}{x_i - x_{i+1}} \tag{b}$$

Integrating twice with respect to x, we obtain

$$f_{i,i+1}(x) = \frac{k_i(x - x_{i+1})^3 - k_{i+1}(x - x_i)^3}{6(x_i - x_{i+1})} + A(x - x_{i+1}) - B(x - x_i) \tag{c}$$

where A and B are constants of integration. The last two terms in Eq. (c) would usually be written as $Cx + D$. By letting $C = A - B$ and $D = -Ax_{i+1} + Bx_i$, we end up with the terms in Eq. (c), which are more convenient to use in the computations that follow.

Imposing the condition $f_{i,i+1}(x_i) = y_i$, we get from Eq. (c)

$$\frac{k_i(x_i - x_{i+1})^3}{6(x_i - x_{i+1})} + A(x_i - x_{i+1}) = y_i$$

Therefore,

$$A = \frac{y_i}{x_i - x_{i+1}} - \frac{k_i}{6}(x_i - x_{i+1}) \tag{d}$$

Similarly, $f_{i,i+1}(x_{i+1}) = y_{i+1}$ yields

$$B = \frac{y_{i+1}}{x_i - x_{i+1}} - \frac{k_{i+1}}{6}(x_i - x_{i+1}) \tag{e}$$

Substituting Eqs. (d) and (e) into Eq. (c) results in

$$
\begin{aligned}
f_{i,i+1}(x) = \frac{k_i}{6}\left[\frac{(x - x_{i+1})^3}{x_i - x_{i+1}} - (x - x_{i+1})(x_i - x_{i+1})\right] \\
- \frac{k_{i+1}}{6}\left[\frac{(x - x_i)^3}{x_i - x_{i+1}} - (x - x_i)(x_i - x_{i+1})\right] \\
+ \frac{y_i(x - x_{i+1}) - y_{i+1}(x - x_i)}{x_i - x_{i+1}}
\end{aligned}
\tag{3.10}
$$

The second derivatives k_i of the spline at the interior knots are obtained from the slope continuity conditions $f'_{i-1,i}(x_i) = f'_{i,i+1}(x_i)$, where $i = 2, 3, \ldots, n - 1$. After a little algebra, this results in the simultaneous equations

$$
\begin{aligned}
k_{i-1}(x_{i-1} - x_i) + 2k_i(x_{i-1} - x_{i+1}) + k_{i+1}(x_i - x_{i+1}) \\
= 6\left(\frac{y_{i-1} - y_i}{x_{i-1} - x_i} - \frac{y_i - y_{i+1}}{x_i - x_{i+1}}\right), \quad i = 2, 3, \ldots, n - 1
\end{aligned}
\tag{3.11}
$$

Because Eqs. (3.11) have a tridiagonal coefficient matrix, they can be solved economically with functions LUdec3 and LUsol3 described in Art. 2.4.

If the data points are evenly spaced at intervals h, then $x_{i-1} - x_i = x_i - x_{i+1} = -h$, and the Eqs. (3.11) simplify to

$$
k_{i-1} + 4k_i + k_{i+1} = \frac{6}{h^2}(y_{i-1} - 2y_i + y_{i+1}), \quad i = 2, 3, \ldots, n - 1
\tag{3.12}
$$

■ splineCurv

The first stage of cubic spline interpolation is to set up Eqs. (3.11) and solve them for the unknown k's (recall that $k_1 = k_n = 0$). This task is carried out by the function splineCurv:

```
function k = splineCurv(xData,yData)
% Returns curvatures of a cubic spline at the knots.
% USAGE: k = splineCurv(xData,yData)
% xData = x-coordinates of data points.
% yData = y-coordinates of data points.

n = length(xData);
c = zeros(n-1,1); d = ones(n,1);
e = zeros(n-1,1); k = zeros(n,1);
c(1:n-2) = xData(1:n-2) - xData(2:n-1);
d(2:n-1) = 2*(xData(1:n-2) - xData(3:n));
e(2:n-1) = xData(2:n-1) - xData(3:n);
k(2:n-1) = 6*(yData(1:n-2) - yData(2:n-1))...
           ./(xData(1:n-2) - xData(2:n-1))...
```

```
            - 6*(yData(2:n-1) - yData(3:n))...
              ./(xData(2:n-1) - xData(3:n));
[c,d,e] = LUdec3(c,d,e);
k = LUsol3(c,d,e,k);
```

■ splineEval

The function `splineEval` computes the interpolant at x from Eq. (3.10). The sub-function `findSeg` finds the segment of the spline that contains x by the method of bisection. It returns the segment number; that is, the value of the subscript i in Eq. (3.10).

```
function y = splineEval(xData,yData,k,x)
% Returns value of cubic spline interpolant at x.
% USAGE: y = splineEval(xData,yData,k,x)
% xData = x-coordinates of data points.
% yData = y-coordinates of data points.
% k     = curvatures of spline at the knots;
%         returned by function splineCurv.

i = findSeg(xData,x);
h = xData(i) - xData(i+1);
y = ((x - xData(i+1))^3/h - (x - xData(i+1))*h)*k(i)/6.0...
  - ((x - xData(i))^3/h - (x - xData(i))*h)*k(i+1)/6.0...
  + yData(i)*(x - xData(i+1))/h...
  - yData(i+1)*(x - xData(i))/h;

function i = findSeg(xData,x)
% Returns index of segment containing x.
iLeft = 1; iRight = length(xData);
while 1
    if(iRight - iLeft) <= 1
        i = iLeft; return
    end
    i = fix((iLeft + iRight)/2);
    if x < xData(i)
        iRight = i;
    else
        iLeft = i;
    end
end
```

EXAMPLE 3.5
Use natural cubic spline to determine y at $x = 1.5$. The data points are

x	1	2	3	4	5
y	0	1	0	1	0

Solution The five knots are equally spaced at $h = 1$. Recalling that the second derivative of a natural spline is zero at the first and last knot, we have $k_1 = k_5 = 0$. The second derivatives at the other knots are obtained from Eq. (3.12). Using $i = 2, 3, 4$ we get the simultaneous equations

$$0 + 4k_2 + k_3 = 6\,[0 - 2(1) + 0] = -12$$

$$k_2 + 4k_3 + k_4 = 6\,[1 - 2(0) + 1] = 12$$

$$k_3 + 4k_4 + 0 = 6\,[0 - 2(1) + 0] = -12$$

The solution is $k_2 = k_4 = -30/7$, $k_3 = 36/7$.

The point $x = 1.5$ lies in the segment between knots 1 and 2. The corresponding interpolant is obtained from Eq. (3.10) by setting $i = 1$. With $x_i - x_{i+1} = -h = -1$, we obtain

$$f_{1,2}(x) = -\frac{k_1}{6}\left[(x - x_2)^3 - (x - x_2)\right] + \frac{k_2}{6}\left[(x - x_1)^3 - (x - x_1)\right]$$
$$- \left[y_1(x - x_2) - y_2(x - x_1)\right]$$

Therefore,

$$y(1.5) = f_{1,2}(1.5) = 0 + \frac{1}{6}\left(-\frac{30}{7}\right)\left[(1.5 - 1)^3 - (1.5 - 1)\right] - [0 - 1(1.5 - 1)] = 0.7679$$

The plot of the interpolant, which in this case is made up of four cubic segments, is shown in the figure.

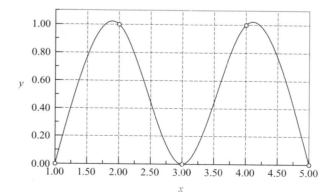

EXAMPLE 3.6

Sometimes it is preferable to replace one or both of the end conditions of the cubic spline with something other than the natural conditions. Use the end condition $f''_{1,2}(0) = 0$ (zero slope), rather than $f''_{1,2}(0) = 0$ (zero curvature), to determine the cubic spline interpolant at $x = 2.6$ based on the data points

x	0	1	2	3
y	1	1	0.5	0

Solution We must first modify Eqs. (3.12) to account for the new end condition. Setting $i = 1$ in Eq. (3.10) and differentiating, we get

$$f'_{1,2}(x) = \frac{k_1}{6}\left[3\frac{(x-x_2)^2}{x_1-x_2} - (x_1-x_2)\right] - \frac{k_2}{6}\left[3\frac{(x-x_1)^2}{x_1-x_2} - (x_1-x_2)\right] + \frac{y_1-y_2}{x_1-x_2}$$

Thus the end condition $f'_{1,2}(x_1) = 0$ yields

$$\frac{k_1}{3}(x_1-x_2) + \frac{k_2}{6}(x_1-x_2) + \frac{y_1-y_2}{x_1-x_2} = 0$$

or

$$2k_1 + k_2 = -6\frac{y_1-y_2}{(x_1-x_2)^2}$$

From the given data we see that $y_1 = y_2 = 1$, so that the last equation becomes

$$2k_1 + k_2 = 0 \tag{a}$$

The other equations in Eq. (3.12) are unchanged. Noting that $k_4 = 0$ and $h = 1$, we have

$$k_1 + 4k_2 + k_3 = 6[1 - 2(1) + 0.5] = -3 \tag{b}$$

$$k_2 + 4k_3 = 6[1 - 2(0.5) + 0] = 0 \tag{c}$$

The solution of Eqs. (a)–(c) is $k_1 = 0.4615$, $k_2 = -0.9231$, $k_3 = 0.2308$.

The interpolant can now be evaluated from Eq. (3.10). Substituting $i = 3$ and $x_i - x_{i+1} = -1$, we obtain

$$f_{3,4}(x) = \frac{k_3}{6}\left[-(x-x_4)^3 + (x-x_4)\right] - \frac{k_4}{6}\left[-(x-x_3)^3 + (x-x_3)\right]$$
$$- y_3(x-x_4) + y_4(x-x_3)$$

Therefore,

$$y(2.6) = f_{3,4}(2.6) = \frac{0.2308}{6}\left[-(-0.4)^3 + (-0.4)\right] + 0 - 0.5(-0.4) + 0 = 0.1871$$

EXAMPLE 3.7

Write a program that interpolates between given data points with the natural cubic spline. The program must be able to evaluate the interpolant for more than one value of x. As a test, use data points specified in Example 3.4 and compute the interpolant at $x = 1.5$ and $x = 4.5$ (due to symmetry, these values should be equal).

Solution The program below prompts for x; it is terminated by pressing the "return" key.

```
% Example 3.7 (Cubic spline)
xData = [1; 2; 3; 4; 5];
yData = [0; 1; 0; 1; 0];
k = splineCurv(xData,yData);
while 1
    x = input('x = ');
    if isempty(x)
        fprintf('Done'); break
    end
    y = splineEval(xData,yData,k,x)
    fprintf('\n')
end
```

Running the program produces the following results:

```
x = 1.5
y =
    0.7679


x = 4.5
y =
    0.7679


x =
Done
```

PROBLEM SET 3.1

1. Given the data points

x	-1.2	0.3	1.1
y	-5.76	-5.61	-3.69

   determine y at $x = 0$ using (a) Neville's method and (b) Lagrange's method.

2. Find the zero of $y(x)$ from the following data:

x	0	0.5	1	1.5	2	2.5	3
y	1.8421	2.4694	2.4921	1.9047	0.8509	−0.4112	−1.5727

Use Lagrange's interpolation over (a) three; and (b) four nearest-neighbor data points. *Hint*: after finishing part (a), part (b) can be computed with a relatively small effort.

3. The function $y(x)$ represented by the data in Prob. 2 has a maximum at $x = 0.7679$. Compute this maximum by Neville's interpolation over four nearest-neighbor data points.

4. Use Neville's method to compute y at $x = \pi/4$ from the data points

x	0	0.5	1	1.5	2
y	−1.00	1.75	4.00	5.75	7.00

5. Given the data

x	0	0.5	1	1.5	2
y	−0.7854	0.6529	1.7390	2.2071	1.9425

find y at $x = \pi/4$ and at $\pi/2$. Use the method that you consider to be most convenient.

6. The points

x	−2	1	4	−1	3	−4
y	−1	2	59	4	24	−53

lie on a polynomial. Use the divided difference table of Newton's method to determine the degree of the polynomial.

7. Use Newton's method to find the expression for the lowest-order polynomial that fits the following points:

x	−3	2	−1	3	1
y	0	5	−4	12	0

8. Use Neville's method to determine the equation of the quadratic that passes through the points

x	-1	1	3
y	17	-7	-15

9. The density of air ρ varies with elevation h in the following manner:

h (km)	0	3	6
ρ (kg/m^3)	1.225	0.905	0.652

Express $\rho(h)$ as a quadratic function using Lagrange's method.

10. Determine the natural cubic spline that passes through the data points

x	0	1	2
y	0	2	1

Note that the interpolant consists of two cubics, one valid in $0 \le x \le 1$, the other in $1 \le x \le 2$. Verify that these cubics have the same first and second derivatives at $x = 1$.

11. Given the data points

x	1	2	3	4	5
y	13	15	12	9	13

determine the natural cubic spline interpolant at $x = 3.4$.

12. Compute the zero of the function $y(x)$ from the following data:

x	0.2	0.4	0.6	0.8	1.0
y	1.150	0.855	0.377	-0.266	-1.049

Use inverse interpolation with the natural cubic spline. *Hint*: reorder the data so that the values of y are in ascending order.

13. Solve Example 3.6 with a cubic spline that has constant second derivatives within its first and last segments (the end segments are parabolic). The end conditions for this spline are $k_1 = k_2$ and $k_{n-1} = k_n$.

14. ■ Write a computer program for interpolation by Neville's method. The program must be able to compute the interpolant at several user-specified values of x. Test the program by determining y at $x = 1.1, 1.2$ and 1.3 from the following data:

x	−2.0	−0.1	−1.5	0.5
y	2.2796	1.0025	1.6467	1.0635

x	−0.6	2.2	1.0	1.8
y	1.0920	2.6291	1.2661	1.9896

(Answer: $y = 1.3262, 1.3938, 1.4693$)

15. ■ The specific heat c_p of aluminum depends on temperature T as follows:[4]

T (°C)	−250	−200	−100	0	100	300
c_p (kJ/kg·K)	0.0163	0.318	0.699	0.870	0.941	1.04

Determine c_p at $T = 200$°C and 400°C.

16. ■ Find y at $x = 0.46$ from the data

x	0	0.0204	0.1055	0.241	0.582	0.712	0.981
y	0.385	1.04	1.79	2.63	4.39	4.99	5.27

17. ■ The table shows the drag coefficient c_D of a sphere as a function of Reynolds number Re.[5] Use natural cubic spline to find c_D at $Re = 5, 50, 500$ and 5000. *Hint*: use log–log scale.

Re	0.2	2	20	200	2000	20 000
c_D	103	13.9	2.72	0.800	0.401	0.433

18. ■ Solve Prob. 17 using a polynomial interpolant intersecting four nearest-neighbor data points.

19. ■ The kinematic viscosity μ_k of water varies with temperature T in the following manner:

[4] Source: Black, Z.B., and Hartley, J.G., *Thermodynamics*, Harper & Row, 1985.
[5] Source: Kreith, F., *Principles of Heat Transfer*, Harper & Row, 1973.

T (°C)	0	21.1	37.8	54.4	71.1	87.8	100
μ_k (10^{-3} m^2/s)	1.79	1.13	0.696	0.519	0.338	0.321	0.296

Interpolate μ_k at $T = 10°, 30°, 60°$ and $90°$C.

20. ■ The table shows how the relative density ρ of air varies with altitude h. Determine the relative density of air at 10.5 km.

h (km)	0	1.525	3.050	4.575	6.10	7.625	9.150
ρ	1	0.8617	0.7385	0.6292	0.5328	0.4481	0.3741

3.4   Least-Squares Fit

Overview

If the data are obtained from experiments, they typically contain a significant amount of random noise due to measurement errors. The task of curve fitting is to find a smooth curve that fits the data points "on the average." This curve should have a simple form (e.g. a low-order polynomial), so as to not reproduce the noise.

Let

$$f(x) = f(x; a_1, a_2, \ldots, a_m)$$

be the function that is to be fitted to the n data points (x_i, y_i), $i = 1, 2, \ldots, n$. The notation implies that we have a function of x that contains the parameters a_j, $j = 1, 2, \ldots, m$, where $m < n$. The form of $f(x)$ is determined beforehand, usually from the theory associated with the experiment from which the data is obtained. The only means of adjusting the fit is the parameters. For example, if the data represent the displacements y_i of an overdamped mass–spring system at time t_i, the theory suggests the choice $f(t) = a_1 t e^{-a_2 t}$. Thus curve fitting consists of two steps: choosing the form of $f(x)$, followed by computation of the parameters that produce the best fit to the data.

This brings us to the question: what is meant by "best" fit? If the noise is confined to the y-coordinate, the most commonly used measure is the *least-squares fit*, which minimizes the function

$$S(a_1, a_2, \ldots, a_m) = \sum_{i=1}^{n} [y_i - f(x_i)]^2 \tag{3.13}$$

with respect to each a_j. Therefore, the optimal values of the parameters are given by the solution of the equations

$$\frac{\partial S}{\partial a_k} = 0, \quad k = 1, 2, \ldots, m \tag{3.14}$$

The terms $r_i = y_i - f(x_i)$ in Eq. (3.13) are called *residuals*; they represent the discrepancy between the data points and the fitting function at x_i. The function S to be minimized is thus the sum of the squares of the residuals. Equations (3.14) are generally nonlinear in a_j and may thus be difficult to solve. If the fitting function is chosen as a linear combination of specified functions $f_j(x)$:

$$f(x) = a_1 f_1(x) + a_2 f_2(x) + \cdots + a_m f_m(x)$$

then Eqs. (3.14) are linear. A typical example is a polynomial where $f_1(x) = 1$, $f_2(x) = x$, $f_3(x) = x^2$, etc.

The spread of the data about the fitting curve is quantified by the *standard deviation*, defined as

$$\sigma = \sqrt{\frac{S}{n-m}} \tag{3.15}$$

Note that if $n = m$, we have *interpolation*, not curve fitting. In that case, both the numerator and the denominator in Eq. (3.15) are zero, so that σ is meaningless, as it should be.

Fitting a Straight Line

Fitting a straight line

$$f(x) = a + bx \tag{3.16}$$

to data is also known as *linear regression*. In this case the function to be minimized is

$$S(a, b) = \sum_{i=1}^{n} (y_i - a - bx_i)^2$$

Equations (3.14) now become

$$\frac{\partial S}{\partial a} = \sum_{i=1}^{n} -2(y_i - a - bx_i) = 2\left(-\sum_{i=1}^{n} y_i + na + b\sum_{i=1}^{n} x_i\right) = 0$$

$$\frac{\partial S}{\partial b} = \sum_{i=1}^{n} -2(y_i - a - bx_i)x_i = 2\left(-\sum_{i=1}^{n} x_i y_i + a\sum_{i=1}^{n} x_i + b\sum_{i=1}^{n} x_i^2\right) = 0$$

Dividing both equations by $2n$ and rearranging terms, we get

$$a + \bar{x}b = \bar{y} \qquad a\bar{x} + \left(\frac{1}{n}\sum_{i=1}^{n} x_i^2\right)b = \frac{1}{n}\sum_{i=1}^{n} x_i y_i$$

where

$$\bar{x} = \frac{1}{n}\sum_{i=1}^{n} x_i \qquad \bar{y} = \frac{1}{n}\sum_{i=1}^{n} y_i \tag{3.17}$$

are the mean values of the x and y data. The solution for the parameters is

$$a = \frac{\bar{y}\sum x_i^2 - \bar{x}\sum x_i y_i}{\sum x_i^2 - n\bar{x}^2} \qquad b = \frac{\sum x_i y_i - n\bar{x}\bar{y}}{\sum x_i^2 - n\bar{x}^2} \tag{3.18}$$

These expressions are susceptible to roundoff errors (the two terms in each numerator as well as in each denominator can be roughly equal). It is better to compute the parameters from

$$b = \frac{\sum y_i(x_i - \bar{x})}{\sum x_i(x_i - \bar{x})} \qquad a = \bar{y} - \bar{x}b \tag{3.19}$$

which are equivalent to Eqs. (3.18), but much less affected by rounding off.

Fitting Linear Forms

Consider the least-squares fit of the *linear form*

$$f(x) = a_1 f_1(x) + a_2 f_2(x) + \cdots + a_m f_m(x) = \sum_{j=1}^{m} a_j f_j(x) \tag{3.20}$$

where each $f_j(x)$ is a predetermined function of x, called a *basis function*. Substitution into Eq. (3.13) yields

$$S = \sum_{i=1}^{n}\left[y_i - \sum_{j=1}^{m} a_j f_j(x_i) \right]^2 \tag{a}$$

Thus Eqs. (3.14) are

$$\frac{\partial S}{\partial a_k} = -2\left\{ \sum_{i=1}^{n}\left[y_i - \sum_{j=1}^{m} a_j f_j(x_i) \right] f_k(x_i) \right\} = 0, \quad k = 1, 2, \ldots, m$$

Dropping the constant (-2) and interchanging the order of summation, we get

$$\sum_{j=1}^{m}\left[\sum_{i=1}^{n} f_j(x_i) f_k(x_i) \right] a_j = \sum_{i=1}^{n} f_k(x_i) y_i, \quad k = 1, 2, \ldots, m$$

In matrix notation these equations are

$$\mathbf{Aa} = \mathbf{b} \tag{3.21a}$$

where

$$A_{kj} = \sum_{i=1}^{n} f_j(x_i) f_k(x_i) \qquad b_k = \sum_{i=1}^{n} f_k(x_i) y_i \tag{3.21b}$$

Equations (3.21a), known as the *normal equations* of the least-squares fit, can be solved with any of the methods discussed in Chapter 2. Note that the coefficient matrix is symmetric, i.e., $A_{kj} = A_{jk}$.

Polynomial Fit

A commonly used linear form is a polynomial. If the degree of the polynomial is $m - 1$, we have $f(x) = \sum_{j=1}^{m} a_j x^{j-1}$. Here the basis functions are

$$f_j(x) = x^{j-1}, \quad j = 1, 2, \ldots, m \tag{3.22}$$

so that Eqs. (3.21b) become

$$A_{kj} = \sum_{i=1}^{n} x_i^{j+k-2} \qquad b_k = \sum_{i=1}^{n} x_i^{k-1} y_i$$

or

$$\mathbf{A} = \begin{bmatrix} n & \sum x_i & \sum x_i^2 & \cdots & \sum x_i^m \\ \sum x_i & \sum x_i^2 & \sum x_i^3 & \cdots & \sum x_i^{m+1} \\ \vdots & \vdots & \vdots & \ddots & \vdots \\ \sum x_i^{m-1} & \sum x_i^m & \sum x_i^{m+1} & \cdots & \sum x_i^{2m-2} \end{bmatrix} \qquad \mathbf{b} = \begin{bmatrix} \sum y_i \\ \sum x_i y_i \\ \vdots \\ \sum x_i^{m-1} y_i \end{bmatrix} \tag{3.23}$$

where $\sum$ stands for $\sum_{i=1}^{n}$. The normal equations become progressively ill-conditioned with increasing m. Fortunately, this is of little practical consequence, because only low-order polynomials are useful in curve fitting. Polynomials of high order are not recommended, because they tend to reproduce the noise inherent in the data.

■ polynFit

The function `polynFit` computes the coefficients of a polynomial of degree $m - 1$ to fit n data points in the least-squares sense. To facilitate computations, the terms n, $\sum x_i, \sum x_i^2, \ldots, \sum x_i^{2m-2}$ that make up the coefficient matrix $\mathbf{A}$ in Eq. (3.23) are first stored in the vector s and then inserted into $\mathbf{A}$. The normal equations are solved for the coefficient vector `coeff` by Gauss elimination with pivoting. Since the elements of `coeff` emerging from the solution are not arranged in the usual order (the coefficient of the highest power of x first), the `coeff` array is "flipped" upside-down before returning to the calling program.

```
function coeff = polynFit(xData,yData,m)
% Returns the coefficients of the polynomial
% a(1)*x^(m-1) + a(2)*x^(m-2) + ... + a(m)
% that fits the data points in the least squares sense.
% USAGE: coeff = polynFit(xData,yData,m)
% xData = x-coordinates of data points.
% yData = y-coordinates of data points.
```

```
A = zeros(m); b = zeros(m,1); s = zeros(2*m-1,1);
for i = 1:length(xData)
    temp = yData(i);
    for j = 1:m
        b(j) = b(j) + temp;
        temp = temp*xData(i);
    end
    temp = 1;
    for j = 1:2*m-1
        s(j) = s(j) + temp;
        temp = temp*xData(i);
    end
end
for i = 1:m
    for j = 1:m
        A(i,j) = s(i+j-1);
    end
end
% Rearrange coefficients so that coefficient
%  of x^(m-1) is first
coeff = flipdim(gaussPiv(A,b),1);
```

■ stdDev

After the coefficients of the fitting polynomial have been obtained, the standard deviation σ can be computed with the function stdDev. The polynomial evaluation in stdDev is carried out by the subfunction polyEval which is described in Art. 4.7—see Eq. (4.10).

```
function sigma = stdDev(coeff,xData,yData)
% Returns the standard deviation between data
% points and the polynomial
% a(1)*x^(m-1) + a(2)*x^(m-2) + ... + a(m)
% USAGE: sigma = stdDev(coeff,xData,yData)
% coeff = coefficients of the polynomial.
% xData = x-coordinates of data points.
% yData = y-coordinates of data points.

m = length(coeff); n = length(xData);
```

```
sigma = 0;
for i =1:n
    y = polyEval(coeff,xData(i));
    sigma = sigma + (yData(i) - y)^2;
end
sigma =sqrt(sigma/(n - m));

function y = polyEval(coeff,x)
% Returns the value of the polynomial at x.
m = length(coeff);
y = coeff(1);
for j = 1:m-1
    y = y*x + coeff(j+1);
end
```

Weighting of Data

There are occasions when confidence in the accuracy of data varies from point to point. For example, the instrument taking the measurements may be more sensitive in a certain range of data. Sometimes the data represent the results of several experiments, each carried out under different circumstances. Under these conditions we may want to assign a confidence factor, or *weight*, to each data point and minimize the sum of the squares of the *weighted residuals* $r_i = W_i [y_i - f(x_i)]$, where W_i are the weights. Hence the function to be minimized is

$$S(a_1, a_2, \ldots, a_m) = \sum_{i=1}^{n} W_i^2 [y_i - f(x_i)]^2 \tag{3.24}$$

This procedure forces the fitting function $f(x)$ closer to the data points that have higher weights.

Weighted linear regression

If the fitting function is the straight line $f(x) = a + bx$, Eq. (3.24) becomes

$$S(a, b) = \sum_{i=1}^{n} W_i^2 (y_i - a - bx_i)^2 \tag{3.25}$$

The conditions for minimizing S are

$$\frac{\partial S}{\partial a} = -2 \sum_{i=1}^{n} W_i^2 (y_i - a - bx_i) = 0$$

$$\frac{\partial S}{\partial b} = -2 \sum_{i=1}^{n} W_i^2 (y_i - a - bx_i) x_i = 0$$

or

$$a \sum_{i=1}^{n} W_i^2 + b \sum_{i=1}^{n} W_i^2 x_i = \sum_{i=1}^{n} W_i^2 y_i \qquad (3.26a)$$

$$a \sum_{i=1}^{n} W_i^2 x_i + b \sum_{i=1}^{n} W_i^2 x_i^2 = \sum_{i=1}^{n} W_i^2 x_i y_i \qquad (3.26b)$$

Dividing Eq. (3.26a) by $\sum W_i^2$ and introducing the *weighted averages*

$$\hat{x} = \frac{\sum W_i^2 x_i}{\sum W_i^2} \qquad \hat{y} = \frac{\sum W_i^2 y_i}{\sum W_i^2} \qquad (3.27)$$

we obtain

$$a = \hat{y} - b\hat{x} \qquad (3.28a)$$

Substituting Eq. (3.28a) into Eq. (3.26b) and solving for b yields after some algebra

$$b = \frac{\sum_{i=1}^{n} W_i^2 y_i (x_i - \hat{x})}{\sum_{i=1}^{n} W_i^2 x_i (x_i - \hat{x})} \qquad (3.28b)$$

Note that Eqs. (3.28) are similar to Eqs. (3.19) for unweighted data.

Fitting exponential functions

A special application of weighted linear regression arises in fitting exponential functions to data. Consider as an example the fitting function

$$f(x) = ae^{bx}$$

Normally, the least-squares fit would lead to equations that are nonlinear in a and b. But if we fit $\ln y$ rather than y, the problem is transformed to linear regression: fit the function

$$F(x) = \ln f(x) = \ln a + bx$$

to the data points $(x_i, \ln y_i)$, $i = 1, 2, \ldots, n$. This simplification comes at a price: least-squares fit to the logarithm of the data is not the same as least-squares fit to the original data. The residuals of the logarithmic fit are

$$R_i = \ln y_i - F(x_i) = \ln y_i - \ln a - bx_i \qquad (3.29a)$$

whereas the residuals used in fitting the original data are

$$r_i = y_i - f(x_i) = y_i - ae^{bx_i} \qquad (3.29b)$$

This discrepancy can be largely eliminated by weighting the logarithmic fit. We note from Eq. (3.29b) that $\ln(r_i - y_i) = \ln(ae^{bx_i}) = \ln a + bx_i$, so that Eq. (3.29a) can be written as

$$R_i = \ln y_i - \ln(r_i - y_i) = \ln\left(1 - \frac{r_i}{y_i}\right)$$

If the residuals r_i are sufficiently small ($r_i << y_i$), we can use the approximation $\ln(1 - r_i/y_i) \approx r_i/y_i$, so that

$$R_i \approx r_i/y_i$$

We can now see that by minimizing $\sum R_i^2$, we inadvertently introduced the weights $1/y_i$. This effect can be negated if we apply the weights y_i when fitting $F(x)$ to $(\ln y_i, x_i)$; that is, by minimizing

$$S = \sum_{i=1}^{n} y_i^2 R_i^2 \qquad (3.30)$$

Other examples that also benefit from the weights $W_i = y_i$ are given in Table 3.3.

$f(x)$	$F(x)$	Data to be fitted by $F(x)$
axe^{bx}	$\ln[f(x)/x] = \ln a + bx$	$[x_i, \ln(y_i/x_i)]$
ax^b	$\ln f(x) = \ln a + b\ln(x)$	$(\ln x_i, \ln y_i)$

Table 3.3

EXAMPLE 3.8

Fit a straight line to the data shown and compute the standard deviation.

x	0.0	1.0	2.0	2.5	3.0
y	2.9	3.7	4.1	4.4	5.0

Solution The averages of the data are

$$\bar{x} = \frac{1}{5}\sum x_i = \frac{0.0 + 1.0 + 2.0 + 2.5 + 3.0}{5} = 1.7$$

$$\bar{y} = \frac{1}{5}\sum y_i = \frac{2.9 + 3.7 + 4.1 + 4.4 + 5.0}{5} = 4.02$$

The intercept a and slope b of the interpolant can now be determined from Eq. (3.19):

$$b = \frac{\sum y_i(x_i - \bar{x})}{\sum x_i(x_i - \bar{x})}$$

$$= \frac{2.9(-1.7) + 3.7(-0.7) + 4.1(0.3) + 4.4(0.8) + 5.0(1.3)}{0.0(-1.7) + 1.0(-0.7) + 2.0(0.3) + 2.5(0.8) + 3.0(1.3)}$$

$$= \frac{3.73}{5.8} = 0.6431$$

$$a = \bar{y} - \bar{x}b = 4.02 - 1.7(0.6431) = 2.927$$

Therefore, the regression line is $f(x) = 2.927 + 0.6431x$, which is shown in the figure together with the data points.

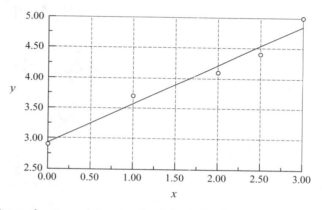

We start the evaluation of the standard deviation by computing the residuals:

y	2.900	3.700	4.100	4.400	5.000
$f(x)$	2.927	3.570	4.213	4.535	4.856
$y - f(x)$	−0.027	0.130	−0.113	−0.135	0.144

The sum of the squares of the residuals is

$$S = \sum [y_i - f(x_i)]^2$$

$$= (-0.027)^2 + (0.130)^2 + (-0.113)^2 + (-0.135)^2 + (0.144)^2 = 0.06936$$

so that the standard deviation in Eq. (3.15) becomes

$$\sigma = \sqrt{\frac{S}{n - m}} = \sqrt{\frac{0.06936}{5 - 2}} = 0.1520$$

EXAMPLE 3.9

Determine the parameters a and b so that $f(x) = ae^{bx}$ fits the following data in the least-squares sense.

x	1.2	2.8	4.3	5.4	6.8	7.9
y	7.5	16.1	38.9	67.0	146.6	266.2

Use two different methods: (1) fit $\ln y_i$; and (2) fit $\ln y_i$ with weights $W_i = y_i$. Compute the standard deviation in each case.

Solution of Part (1) The problem is to fit the function $\ln(ae^{bx}) = \ln a + bx$ to the data

x	1.2	2.8	4.3	5.4	6.8	7.9
$z = \ln y$	2.015	2.779	3.661	4.205	4.988	5.584

We are now dealing with linear regression, where the parameters to be found are $A = \ln a$ and b. Following the steps in Example 3.8, we get (skipping some of the arithmetic details)

$$\bar{x} = \frac{1}{6}\sum x_i = 4.733 \qquad \bar{z} = \frac{1}{6}\sum z_i = 3.872$$

$$b = \frac{\sum z_i(x_i - \bar{x})}{\sum x_i(x_i - \bar{x})} = \frac{16.716}{31.153} = 0.5366 \qquad A = \bar{z} - \bar{x}b = 1.3323$$

Therefore, $a = e^A = 3.790$ and the fitting function becomes $f(x) = 3.790e^{0.5366}$. The plots of $f(x)$ and the data points are shown in the figure.

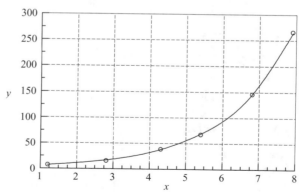

Here is the computation of standard deviation:

y	7.50	16.10	38.90	67.00	146.60	266.20
$f(x)$	7.21	17.02	38.07	68.69	145.60	262.72
$y - f(x)$	0.29	−0.92	0.83	−1.69	1.00	3.48

$$S = \sum [y_i - f(x_i)]^2 = 17.59$$

$$\sigma = \sqrt{\frac{S}{6-2}} = 2.10$$

As pointed out before, this is an approximate solution of the stated problem, since we did not fit y_i, but $\ln y_i$. Judging by the plot, the fit seems to be good.

Solution of Part (2) We again fit $\ln(ae^{bx}) = \ln a + bx$ to $z = \ln y$, but this time the weights $W_i = y_i$ are used. From Eqs. (3.27) the weighted averages of the data are (recall that we fit $z = \ln y$)

$$\hat{x} = \frac{\sum y_i^2 x_i}{\sum y_i^2} = \frac{737.5 \times 10^3}{98.67 \times 10^3} = 7.474$$

$$\hat{z} = \frac{\sum y_i^2 z_i}{\sum y_i^2} = \frac{528.2 \times 10^3}{98.67 \times 10^3} = 5.353$$

and Eqs. (3.28) yield for the parameters

$$b = \frac{\sum y_i^2 z_i (x_i - \hat{x})}{\sum y_i^2 x_i (x_i - \hat{x})} = \frac{35.39 \times 10^3}{65.05 \times 10^3} = 0.5440$$

$$\ln a = \hat{z} - b\hat{x} = 5.353 - 0.5440(7.474) = 1.287$$

Therefore,

$$a = e^{\ln a} = e^{1.287} = 3.622$$

so that the fitting function is $f(x) = 3.622e^{0.5440x}$. As expected, this result is somewhat different from that obtained in Part (1).

The computations of the residuals and standard deviation are as follows:

y	7.50	16.10	38.90	67.00	146.60	266.20
$f(x)$	6.96	16.61	37.56	68.33	146.33	266.20
$y - f(x)$	0.54	-0.51	1.34	-1.33	0.267	0.00

$$S = \sum [y_i - f(x_i)]^2 = 4.186$$

$$\sigma = \sqrt{\frac{S}{6-2}} = 1.023$$

Observe that the residuals and standard deviation are smaller than in Part (1), indicating a better fit, as expected.

It can be shown that fitting y_i directly (which involves the solution of a transcendental equation) results in $f(x) = 3.614e^{0.5442x}$. The corresponding standard deviation is $\sigma = 1.022$, which is very close to the result in Part (2).

EXAMPLE 3.10

Write a program that fits a polynomial of arbitrary degree k to the data points shown below. Use the program to determine k that best fits this data in the least-squares sense.

x	−0.04	0.93	1.95	2.90	3.83	5.00
y	−8.66	−6.44	−4.36	−3.27	−0.88	0.87
x	5.98	7.05	8.21	9.08	10.09	
y	3.31	4.63	6.19	7.40	8.85	

Solution The following program prompts for k. Execution is terminated by pressing "return."

```
% Example 3.10 (Polynomial curve fitting)
xData = [-0.04,0.93,1.95,2.90,3.83,5.0,...
                5.98,7.05,8.21,9.08,10.09]';
yData = [-8.66,-6.44,-4.36,-3.27,-0.88,0.87,...
                3.31,4.63,6.19,7.4,8.85]';
format short e
while 1
    k = input('degree of polynomial = ');
    if isempty(k)          % Loop is terminated
        fprintf('Done')    % by pressing ''return''
        break
    end
    coeff = polynFit(xData,yData,k+1)
    sigma = stdDev(coeff,xData,yData)
    fprintf('\n')
end
```

The results are:

```
Degree of polynomial = 1
coeff =
   1.7286e+000
  -7.9453e+000
```

```
sigma =
   5.1128e-001

degree of polynomial = 2
coeff =
  -4.1971e-002
   2.1512e+000
  -8.5701e+000
sigma =
   3.1099e-001

degree of polynomial = 3
coeff =
  -2.9852e-003
   2.8845e-003
   1.9810e+000
  -8.4660e+000
sigma =
   3.1948e-001

degree of polynomial =
Done
```

Because the quadratic $f(x) = -0.041971x^2 + 2.1512x - 8.5701$ produces the smallest standard deviation, it can be considered as the "best" fit to the data. But be warned—the standard deviation is not an infallible measure of the goodness-of-fit. It is always a good idea to plot the data points and $f(x)$ before final determination is made. The plot of our data indicates that the quadratic (solid line) is indeed a reasonable choice for the fitting function.

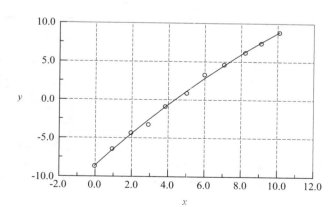

PROBLEM SET 3.2

Instructions Plot the data points and the fitting function whenever appropriate.

1. Show that the straight line obtained by least-squares fit of unweighted data always passes through the point $(\bar{x}, \bar{y})$.

2. Use linear regression to find the line that fits the data

x	−1.0	−0.5	0	0.5	1.0
y	−1.00	−0.55	0.00	0.45	1.00

and determine the standard deviation.

3. Three tensile tests were carried out on an aluminum bar. In each test the strain was measured at the same values of stress. The results were

Stress (MPa)	34.5	69.0	103.5	138.0
Strain (Test 1)	0.46	0.95	1.48	1.93
Strain (Test 2)	0.34	1.02	1.51	2.09
Strain (Test 3)	0.73	1.10	1.62	2.12

where the units of strain are mm/m. Use linear regression to estimate the modulus of elasticity of the bar (modulus of elasticity = stress/strain).

4. Solve Prob. 3 assuming that the third test was performed on an inferior machine, so that its results carry only half the weight of the other two tests.

5. ■ Fit a straight line to the following data and compute the standard deviation.

x	0	0.5	1	1.5	2	2.5
y	3.076	2.810	2.588	2.297	1.981	1.912
x	3	3.5	4	4.5	5	
y	1.653	1.478	1.399	1.018	0.794	

6. ■ The table displays the mass M and average fuel consumption ϕ of motor vehicles manufactured by Ford and Honda in 1999. Fit a straight line $\phi = a + bM$ to the data and compute the standard deviation.

Model	M (kg)	ϕ (km/liter)
Contour	1310	10.2
Crown Victoria	1810	8.1
Escort	1175	11.9
Expedition	2360	5.5
Explorer	1960	6.8
F-150	2020	6.8
Ranger	1755	7.7
Taurus	1595	8.9
Accord	1470	9.8
CR-V	1430	10.2
Civic	1110	13.2
Passport	1785	7.7

7. ■ The relative density ρ of air was measured at various altitudes h. The results were:

h (km)	0	1.525	3.050	4.575	6.10	7.625	9.150
ρ	1	0.8617	0.7385	0.6292	0.5328	0.4481	0.3741

Use a quadratic least-squares fit to determine the relative air density at $h = 10.5$ km. (This problem was solved by interpolation in Prob. 20, Problem Set 3.1.)

8. ■ Kinematic viscosity μ_k of water varies with temperature T as shown in the table. Determine the cubic that best fits the data, and use it to compute μ_k at $T = 10°, 30°, 60°$, and $90°$C. (This problem was solved in Prob. 19, Problem Set 3.1 by interpolation.)

T (°C)	0	21.1	37.8	54.4	71.1	87.8	100
μ_k (10^{-3} m^2/s)	1.79	1.13	0.696	0.519	0.338	0.321	0.296

9. ■ Fit a straight line and a quadratic to the data

x	1.0	2.5	3.5	4.0	1.1	1.8	2.2	3.7
y	6.008	15.722	27.130	33.772	5.257	9.549	11.098	28.828

Which is a better fit?

10. ■ The table displays thermal efficiencies of some early steam engines.[6] Determine the polynomial that provides the best fit to the data and use it to predict the thermal efficiency in the year 2000.

Year	Efficiency (%)	Type
1718	0.5	Newcomen
1767	0.8	Smeaton
1774	1.4	Smeaton
1775	2.7	Watt
1792	4.5	Watt
1816	7.5	Woolf compound
1828	12.0	Improved Cornish
1834	17.0	Improved Cornish
1878	17.2	Corliss compound
1906	23.0	Triple expansion

11. The table shows the variation of the relative thermal conductivity k of sodium with temperature T. Find the quadratic that fits the data in the least-squares sense.

T (°C)	79	190	357	524	690
k	1.00	0.932	0.839	0.759	0.693

12. Let $f(x) = ax^b$ be the least-squares fit of the data (x_i, y_i), $i = 1, 2, \ldots, n$, and let $F(x) = \ln a + b \ln x$ be the least-squares fit of $(\ln x_i, \ln y_i)$—see Table 3.3. Prove that $R_i \approx r_i/y_i$, where the residuals are $r_i = y_i - f(x_i)$ and $R_i = \ln y_i - F(x_i)$. Assume that $r_i \ll y_i$.

13. Determine a and b for which $f(x) = a \sin(\pi x/2) + b \cos(\pi x/2)$ fits the following data in the least-squares sense.

x	−0.5	−0.19	0.02	0.20	0.35	0.50
y	−3.558	−2.874	−1.995	−1.040	−0.068	0.677

---

[6] Source: Singer, C., Holmyard, E.J., Hall, A.R., and Williams, T.H., *A History of Technology*, Oxford University Press, 1958.

14. Determine a and b so that $f(x) = ax^b$ fits the following data in the least-squares sense.

x	0.5	1.0	1.5	2.0	2.5
y	0.49	1.60	3.36	6.44	10.16

15. Fit the function $f(x) = axe^{bx}$ to the data and compute the standard deviation.

x	0.5	1.0	1.5	2.0	2.5
y	0.541	0.398	0.232	0.106	0.052

16. ■ The intensity of radiation of a radioactive substance was measured at half-year intervals. The results were:

t (years)	0	0.5	1	1.5	2	2.5
γ	1.000	0.994	0.990	0.985	0.979	0.977
t (years)	3	3.5	4	4.5	5	5.5
γ	0.972	0.969	0.967	0.960	0.956	0.952

where γ is the relative intensity of radiation. Knowing that radioactivity decays exponentially with time: $\gamma(t) = ae^{-bt}$, estimate the radioactive half-life of the substance.

MATLAB Functions

y = interp1(xData,xData,x,method) returns the value of the interpolant y at point x according to the method specified: method = 'linear' uses linear interpolation between adjacent data points (this is the default); method = 'spline' carries out cubic spline interpolation. If x is an array, y is computed for all elements of x.

a = polyfit(xData,yData,m) returns the coefficients a of a polynomial of degree m that fits the data points in the least-squares sense.

y = polyval(a,x) evaluates a polynomial defined by its coefficients a at point x. If x is an array, y is computed for all elements of x.

s = std(x) returns the standard deviation of the elements of array x. If x is a matrix, s is computed for each column of x.

xbar = mean(x) computes the mean value of the elements of x. If x is a matrix, xbar is computed for each column of x.

Linear forms can be fitted to data by setting up the overdetermined equations in Eq. (3.22)

$$\mathbf{Fa} = \mathbf{y}$$

and solving them with the command a = F\y (recall that for overdetermined equations the backslash operator returns the least-squares solution). Here is an illustration how to fit

$$f(x) = a_1 + a_2 e^x + a_3 x e^{-x}$$

to the data in Example 3.9:

```
xData = [1.2; 2.8; 4.3; 5.4; 6.8; 7.0];
yData = [7.5; 16.1; 38.9; 67.0; 146.6; 266.2];
F = ones(length(xData),3);
F(:,2) = exp(xData(:));
F(:,3) = xData(:).*exp(-xData(:));
a = F\yData
```

4 Roots of Equations

Find the solutions of $f(x) = 0$, where the function f is given

4.1 Introduction

A common problem encountered in engineering analysis is this: given a function $f(x)$, determine the values of x for which $f(x) = 0$. The solutions (values of x) are known as the *roots* of the equation $f(x) = 0$, or the *zeroes* of the function $f(x)$.

Before proceeding further, it might be helpful to review the concept of a *function*. The equation

$$y = f(x)$$

contains three elements: an input value x, an output value y and the rule f for computing y. The function is said to be given if the rule f is specified. In numerical computing the rule is invariably a computer algorithm. It may be a function statement, such as

$$f(x) = \cosh(x)\cos(x) - 1$$

or a complex procedure containing hundreds or thousands of lines of code. As long as the algorithm produces an output y for each input x, it qualifies as a function.

The roots of equations may be real or complex. The complex roots are seldom computed, since they rarely have physical significance. An exception is the polynomial equation

$$a_1 x^n + a_2 x^{n-1} + \cdots + a_n x + a_{n+1} = 0$$

where the complex roots may be meaningful (as in the analysis of damped vibrations, for example). For the time being, we will concentrate on finding the real roots of equations. Complex zeroes of polynomials are treated near the end of this chapter.

In general, an equation may have any number of (real) roots, or no roots at all. For example,

$$\sin x - x = 0$$

has a single root, namely $x = 0$, whereas

$$\tan x - x = 0$$

has an infinite number of roots ($x = 0, \pm 4.493, \pm 7.725, \ldots$).

All methods of finding roots are iterative procedures that require a starting point, i.e., an estimate of the root. This estimate can be crucial; a bad starting value may fail to converge, or it may converge to the "wrong" root (a root different from the one sought). There is no universal recipe for estimating the value of a root. If the equation is associated with a physical problem, then the context of the problem (physical insight) might suggest the approximate location of the root. Otherwise, the function must be plotted, or a systematic numerical search for the roots can be carried out. One such search method is described in the next article.

It is highly advisable to go a step further and *bracket* the root (determine its lower and upper bounds) before passing the problem to a root-finding algorithm. Prior bracketing is, in fact, mandatory in the methods described in this chapter.

4.2   Incremental Search Method

The approximate locations of the roots are best determined by plotting the function. Often a very rough plot, based on a few points, is sufficient to give us reasonable starting values. Another useful tool for detecting and bracketing roots is the incremental search method. It can also be adapted for computing roots, but the effort would not be worthwhile, since other methods described in this chapter are more efficient for that.

The basic idea behind the incremental search method is simple: if $f(x_1)$ and $f(x_2)$ have opposite signs, then there is at least one root in the interval (x_1, x_2). If the interval is small enough, it is likely to contain a single root. Thus the zeroes of $f(x)$ can be detected by evaluating the function at intervals Δx and looking for change in sign.

There are several potential problems with the incremental search method:

- It is possible to miss two closely spaced roots if the search increment Δx is larger than the spacing of the roots.
- A double root (two roots that coincide) will not be detected.
- Certain singularities of $f(x)$ can be mistaken for roots. For example, $f(x) = \tan x$ changes sign at $x = \pm \frac{1}{2} n\pi$, $n = 1, 3, 5, \ldots$, as shown in Fig. 4.1. However, these locations are not true zeroes, since the function does not cross the x-axis.

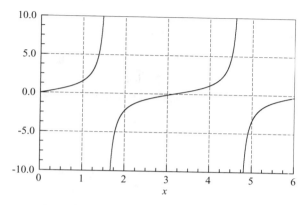

Figure 4.1. Plot of tan x.

■ rootsearch

The function `rootsearch` looks for a zero of the function $f(x)$ in the interval (a,b). The search starts at a and proceeds in steps `dx` toward b. Once a zero is detected, `rootsearch` returns its bounds (`x1`,`x2`) to the calling program. If a root was not detected, `x1` = `x2` = NaN is returned (in MATLAB NaN stands for "not a number"). After the first root (the root closest to a) has been bracketed, `rootsearch` can be called again with a replaced by `x2` in order to find the next root. This can be repeated as long as `rootsearch` detects a root.

```
function [x1,x2] = rootsearch(func,a,b,dx)
% Incremental search for a root of f(x).
% USAGE: [x1,x2] = rootsearch(func,a,d,dx)
% INPUT:
% func = handle of function that returns f(x).
% a,b  = limits of search.
% dx   = search increment.
% OUTPUT:
% x1,x2 = bounds on the smallest root in (a,b);
%          set to NaN if no root was detected

  x1 = a;      f1 = feval(func,x1);
  x2 = a + dx; f2 = feval(func,x2);
  while f1*f2  > 0.0
      if x1 >= b
          x1 = NaN; x2 = NaN; return
      end
      x1 = x2;      f1 = f2;
      x2 = x1 + dx; f2 = feval(func,x2);
  end
```

EXAMPLE 4.1

Use incremental search with $\Delta x = 0.2$ to bracket the smallest positive zero of $f(x) = x^3 - 10x^2 + 5$.

Solution We evaluate $f(x)$ at intervals $\Delta x = 0.2$, staring at $x = 0$, until the function changes its sign (value of the function is of no interest to us; only its sign is relevant). This procedure yields the following results:

x	$f(x)$
0.0	5.000
0.2	4.608
0.4	3.464
0.6	1.616
0.8	−0.888

From the sign change of the function we conclude that the smallest positive zero lies between $x = 0.6$ and $x = 0.8$.

4.3 Method of Bisection

After a root of $f(x) = 0$ has been bracketed in the interval (x_1, x_2), several methods can be used to close in on it. The method of bisection accomplishes this by successively halving the interval until it becomes sufficiently small. This technique is also known as the *interval halving method*. Bisection is not the fastest method available for computing roots, but it is the most reliable. Once a root has been bracketed, bisection will always close in on it.

The method of bisection uses the same principle as incremental search: if there is a root in the interval (x_1, x_2), then $f(x_1) \cdot f(x_2) < 0$. In order to halve the interval, we compute $f(x_3)$, where $x_3 = \frac{1}{2}(x_1 + x_2)$ is the midpoint of the interval. If $f(x_2) \cdot f(x_3) < 0$, then the root must be in (x_2, x_3) and we record this by replacing the original bound x_1 by x_3. Otherwise, the root lies in (x_1, x_3), in which case x_2 is replaced by x_3. In either case, the new interval (x_1, x_2) is half the size of the original interval. The bisection is repeated until the interval has been reduced to a small value ε, so that

$$|x_2 - x_1| \le \varepsilon$$

It is easy to compute the number of bisections required to reach a prescribed ε. The original interval Δx is reduced to $\Delta x/2$ after one bisection, $\Delta x/2^2$ after two bisections and after n bisections it is $\Delta x/2^n$. Setting $\Delta x/2^n = \varepsilon$ and solving for *n*, we get

$$n = \frac{\ln\left(|\Delta x| / \varepsilon\right)}{\ln 2}$$
(4.1)

■ bisect

This function uses the method of bisection to compute the root of $f(x) = 0$ that is known to lie in the interval (x1, x2). The number of bisections n required to reduce the interval to tol is computed from Eq. (4.1). The input argument filter controls the filtering of suspected singularities. By setting filter = 1, we force the routine to check whether the magnitude of $f(x)$ decreases with each interval halving. If it does not, the "root" may not be a root at all, but a singularity, in which case root = NaN is returned. Since this feature is not always desirable, the default value is filter = 0.

```
function root = bisect(func,x1,x2,filter,tol)
% Finds a bracketed zero of f(x) by bisection.
% USAGE: root = bisect(func,x1,x2,filter,tol)
% INPUT:
% func    = handle of function that returns f(x).
% x1,x2   = limits on interval containing the root.
% filter  = singularity filter: 0 = off (default), 1 = on.
% tol     = error tolerance (default is 1.0e4*eps).
% OUTPUT:
% root    = zero of f(x), or NaN if singularity suspected.

if nargin < 5; tol = 1.0e4*eps; end
if nargin < 4; filter = 0; end
f1 = feval(func,x1);
if f1 == 0.0; root = x1; return; end
f2 = feval(func,x2);
if f2 == 0.0; root = x2; return; end
if f1*f2 > 0;
    error('Root is not bracketed in (x1,x2)')
end
n = ceil(log(abs(x2 - x1)/tol)/log(2.0));
for i = 1:n
    x3 = 0.5*(x1 + x2);
    f3 = feval(func,x3);
    if(filter == 1) & (abs(f3) > abs(f1))...
                   & (abs(f3) > abs(f2))
```

```
      root = NaN; return
   end
   if f3 == 0.0
      root = x3; return
   end
   if f2*f3 < 0.0
      x1 = x3; f1 = f3;
   else
      x2 = x3; f2 = f3;
   end
end
root=(x1 + x2)/2;
```

EXAMPLE 4.2

Use bisection to find the root of $f(x) = x^3 - 10x^2 + 5 = 0$ that lies in the interval $(0.6, 0.8)$.

Solution The best way to implement the method is to use the table shown below. Note that the interval to be bisected is determined by the sign of $f(x)$, not its magnitude.

x	$f(x)$	Interval
0.6	1.616	—
0.8	−0.888	$(0.6, 0.8)$
$(0.6 + 0.8)/2 = 0.7$	0.443	$(0.7, 0.8)$
$(0.8 + 0.7)/2 = 0.75$	−0.203	$(0.7, 0.75)$
$(0.7 + 0.75)/2 = 0.725$	0.125	$(0.725, 0.75)$
$(0.75 + 0.725)/2 = 0.7375$	−0.038	$(0.725, 0.7375)$
$(0.725 + 0.7375)/2 = 0.73125$	0.044	$(0.7375, 0.73125)$
$(0.7375 + 0.73125)/2 = 0.73438$	0.003	$(0.7375, 0.73438)$
$(0.7375 + 0.73438)/2 = 0.73594$	−0.017	$(0.73438, 0.73594)$
$(0.73438 + 0.73594)/2 = 0.73516$	−0.007	$(0.73438, 0.73516)$
$(0.73438 + 0.73516)/2 = 0.73477$	−0.002	$(0.73438, 0.73477)$
$(0.73438 + 0.73477)/2 = 0.73458$	0.000	—

The final result $x = 0.7346$ is correct within four decimal places.

EXAMPLE 4.3

Find *all* the zeroes of $f(x) = x - \tan x$ in the interval (0, 20) by the method of bisection. Utilize the functions rootsearch and bisect.

Solution Note that $\tan x$ is singular and changes sign at $x = \pi/2, 3\pi/2, \ldots$. To prevent bisect from mistaking these point for roots, we set filter $= 1$. The closeness of roots to the singularities is another potential problem that can be alleviated by using small Δx in rootsearch. Choosing $\Delta x = 0.01$, we arrive at the following program:

```
% Example 4.3 (root finding with bisection)
a = 0.0; b = 20.0; dx = 0.01;
nroots = 0;
while 1
    [x1,x2] = rootsearch(@fex4_3,a,b,dx);
    if isnan(x1)
        break
    else
        a = x2;
        x = bisect(@fex4_3,x1,x2,1);
        if ~isnan(x)
            nroots = nroots + 1;
            root(nroots) = x;
        end
    end
end
root
```

Recall that in MATLAB the symbol @ before a function name creates a *handle* for the function. Thus the input argument @fex4_3 in rootsearch is a handle for the function fex4_3 listed below.

```
function y = fex4_3(x)
% Function used in Example4.3
y = x - tan(x);
```

Running the program resulted in the output

```
>> root =
        0    4.4934    7.7253    10.9041    14.0662    17.2208
```

4.4   Brent's Method

Brent's method[7] combines bisection and quadratic interpolation into an efficient root-finding algorithm. In most problems the method is much faster than bisection alone, but it can become sluggish if the function is not smooth. It is the recommended method of root finding if the derivative of the function is difficult or impossible to compute.

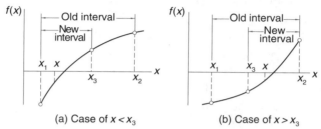

(a) Case of $x < x_3$ (b) Case of $x > x_3$

Figure 4.2. Inverse quadratic iteration.

Brent's method assumes that a root of $f(x) = 0$ has been initially bracketed in the interval (x_1, x_2). The root-finding process starts with a bisection step that halves the interval to either (x_1, x_3) or (x_3, x_2), where $x_3 = (x_1 + x_2)/2$, as shown in Figs. 4.2(a) and (b). In the course of bisection we had to compute $f_1 = f(x_1)$, $f_2 = f(x_2)$ and $f_3 = f(x_3)$, so that we now know three points on the $f(x)$ curve (the open circles in the figure). These points allow us to carry out the next iteration of the root by *inverse quadratic interpolation* (viewing x as a quadratic function of f). If the result x of the interpolation falls inside the latest bracket (as is the case in Figs. 4.2), we accept the result. Otherwise, another round of bisection is applied.

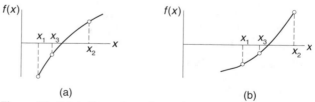

(a)   (b)

Figure 4.3. Relabeling points after an iteration.

The next step is to relabel x as x_3 and rename the limits of the new interval x_1 and x_2 ($x_1 < x_3 < x_2$), as indicated in Figs. 4.3. We have now recovered the original sequencing of points in Figs. 4.2, but the interval (x_1, x_2) containing the root

[7] Brent, R. P., *Algorithms for Minimization without Derivatives*, Prentice-Hall, 1973.

has been reduced. This completes the first iteration cycle. In the next cycle another inverse quadratic interpolation is attempted and the process is repeated until the convergence criterion $|x - x_3| < \varepsilon$ is satisfied, where ε is a prescribed error tolerance.

The inverse quadratic interpolation is carried out with Lagrange's three-point interpolant described in Art. 3.2. Interchanging the roles of x and f, we have

$$x(f) = \frac{(f - f_2)(f - f_3)}{(f_1 - f_2)(f_1 - f_3)}x_1 + \frac{(f - f_1)(f - f_3)}{(f_2 - f_1)(f_2 - f_3)}x_2 + \frac{(f - f_1)(f - f_2)}{(f_3 - f_1)(f_3 - f_2)}x_3$$

Setting $f = 0$ and simplifying, we obtain for the estimate of the root

$$x = x(0) = -\frac{f_2 f_3 x_1(f_2 - f_3) + f_3 f_1 x_2(f_3 - f_1) + f_1 f_2 x_3(f_1 - f_2)}{(f_1 - f_2)(f_2 - f_3)(f_3 - f_1)}$$

The change in the root is

$$\Delta x = x - x_3 = f_3 \frac{x_3(f_1 - f_2)(f_2 - f_3 + f_1) + f_2 x_1(f_2 - f_3) + f_1 x_2(f_3 - f_1)}{(f_2 - f_1)(f_3 - f_1)(f_2 - f_3)} \tag{4.2}$$

■ brent

The function `brent` listed below is a simplified version of the algorithm proposed by Brent. It omits some of Brent's safeguards against slow convergence; it also uses a less sophisticated convergence criterion.

```
function root = brent(func,a,b,tol)
% Finds a root of f(x) = 0 by combining quadratic
% interpolation with bisection (Brent's method).
% USAGE: root = brent(func,a,b,tol)
% INPUT:
% func = handle of function that returns f(x).
% a,b  = limits of the interval containing the root.
% tol  = error tolerance (default is 1.0e6*eps).
% OUTPUT:
% root = zero of f(x) (root = NaN if failed to converge).

if nargin < 4; tol = 1.0e6*eps; end
% First step is bisection
x1 = a; f1 = feval(func,x1);
if f1 == 0; root = x1; return; end
x2 = b; f2 = feval(func,x2);
if f2 == 0; root = x2; return; end
```

```
if f1*f2 > 0.0
    error('Root is not bracketed in (a,b)')
end
x3 = 0.5*(a + b);
% Beginning of iterative loop.
for i = 1:30
    f3 = feval(func,x3);
    if abs(f3) < tol
        root = x3; return
    end
    % Tighten brackets (a,b) on the root.
    if f1*f3 < 0.0; b = x3;
    else; a = x3;
    end
    if (b - a) < tol*max(abs(b),1.0)
        root = 0.5*(a + b); return
    end
    % Try quadratic interpolation.
    denom = (f2 - f1)*(f3 - f1)*(f2 - f3);
    numer = x3*(f1 - f2)*(f2 - f3 + f1)...
        + f2*x1*(f2 - f3) + f1*x2*(f3 - f1);
    % If division by zero, push x out of bracket
    % to force bisection.
    if denom == 0; dx = b - a;
    else; dx = f3*numer/denom;
    end
    x = x3 + dx;
    % If interpolation goes out of bracket, use bisection.
    if (b - x)*(x - a) < 0.0
        dx = 0.5*(b - a); x = a + dx;
    end
    % Let x3 <-- x & choose new x1, x2 so that x1 < x3 < x2.
    if x < x3
        x2 = x3; f2 = f3;
    else
        x1 = x3; f1 = f3;
    end
    x3 = x;
end
root = NaN;
```

EXAMPLE 4.4

Determine the root of $f(x) = x^3 - 10x^2 + 5 = 0$ that lies in $(0.6, 0.8)$ with Brent's method.

Solution

Bisection The starting points are

$$x_1 = 0.6 \qquad f_1 = 0.6^3 - 10(0.6)^2 + 5 = 1.616$$
$$x_2 = 0.8 \qquad f_2 = 0.8^3 - 10(0.8)^2 + 5 = -0.888$$

Bisection yields the point

$$x_3 = 0.7 \qquad f_3 = 0.7^3 - 10(0.7)^2 + 5 = 0.443$$

By inspecting the signs of f we conclude that the new brackets on the root are $(x_3, x_2) = (0.7, 0.8)$.

First interpolation cycle Substituting the above values of x and f into the numerator of the quotient in Eq. (4.2), we get

$$
\begin{aligned}
\text{num} &= x_3(f_1 - f_2)(f_2 - f_3 + f_1) + f_2 x_1 (f_2 - f_3) + f_1 x_2 (f_3 - f_1) \\
&= 0.7(1.616 + 0.888)(-0.888 - 0.443 + 1.616) \\
&\quad -0.888(0.6)(-0.888 - 0.443) + 1.616(0.8)(0.443 - 1.616) \\
&= -0.307\,75
\end{aligned}
$$

and the denominator becomes

$$
\begin{aligned}
\text{den} &= (f_2 - f_1)(f_3 - f_1)(f_2 - f_3) \\
&= (-0.888 - 1.616)(0.443 - 1.616)(-0.888 - 0.443) = -3.9094
\end{aligned}
$$

Therefore,

$$\Delta x = f_3 \frac{\text{num}}{\text{den}} = 0.443 \frac{(-0.307\,75)}{(-3.9094)} = 0.034\,87$$

and

$$x = x_3 + \Delta x = 0.7 + 0.034\,87 = 0.734\,87$$

Since the result is within the established brackets, we accept it.

Relabel points As $x > x_3$, the points are relabeled as illustrated in Figs. 4.2(b) and 4.3(b):

$$x_1 \leftarrow x_3 = 0.7$$
$$f_1 \leftarrow f_3 = 0.443$$

$$x_3 \leftarrow x = 0.734\,87$$

$$f_3 = 0.734\,87^3 - 10(0.734\,87)^2 + 5 = -0.00348$$

The new brackets on the root are $(x_1, x_3) = (0.7, 0.734\,87)$.

Second interpolation cycle Applying the interpolation in Eq. (4.2) again, we obtain (skipping the arithmetical details)

$$\Delta x = -0.000\,27$$

$$x = x_3 + \Delta x = 0.734\,87 - 0.000\,27 = 0.734\,60$$

Again x falls within the latest brackets, so the result is acceptable. At this stage, x is correct to five decimal places.

EXAMPLE 4.5

Compute the zero of

$$f(x) = x\,|\cos x| - 1$$

that lies in the interval $(0, 4)$ with Brent's method.

Solution

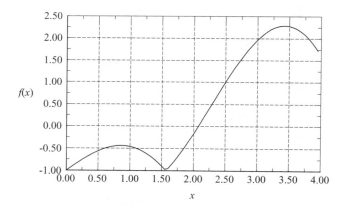

The plot of $f(x)$ shows that this is a rather nasty function within the specified interval, containing a slope discontinuity and two local maxima. The sensible approach is to avoid the potentially troublesome regions of the function by bracketing the root as tightly as possible from a visual inspection of the plot. In this case, the interval $(a, b) = (2.0, 2.2)$ would be a good starting point for Brent's algorithm.

Is Brent's method robust enough to handle the problem with the original brackets $(0, 4)$? Well, here is the MATLAB command and its output:

```
>> brent(@fex4_5,0.0,4.0)
ans =
    2.0739
```

The result was obtained after six iterations. The function defining $f(x)$ is

```
function y = fex4_5(x)
% Function used in Example 4.5
y = x*abs(cos(x)) - 1.0;
```

4.5 Newton–Raphson Method

The Newton–Raphson algorithm is the best-known method of finding roots for a good reason: it is simple and fast. The only drawback of the method is that it uses the derivative $f'(x)$ of the function as well as the function $f(x)$ itself. Therefore, the Newton–Raphson method is usable only in problems where $f'(x)$ can be readily computed.

The Newton–Raphson formula can be derived from the Taylor series expansion of $f(x)$ about x:

$$f(x_{i+1}) = f(x_i) + f'(x_i)(x_{i+1} - x_i) + O(x_{i+1} - x_i)^2 \tag{a}$$

If x_{i+1} is a root of $f(x) = 0$, Eq. (a) becomes

$$0 = f(x_i) + f'(x_i)(x_{i+1} - x_i) + O(x_{i+1} - x_i)^2 \tag{b}$$

Assuming that x_i is a close to x_{i+1}, we can drop the last term in Eq. (b) and solve for x_{i+1}. The result is the Newton–Raphson formula

$$x_{i+1} = x_i - \frac{f(x_i)}{f'(x_i)} \tag{4.3}$$

If x denotes the true value of the root, the error in x_i is $E_i = x - x_i$. It can be shown that if x_{i+1} is computed from Eq. (4.3), the corresponding error is

$$E_{i+1} = -\frac{f''(x_i)}{2 f'(x_i)} E_i^2$$

indicating that the Newton–Raphson method converges *quadratically* (the error is the square of the error in the previous step). As a consequence, the number of significant figures is roughly doubled in every iteration, provided that x_i is close to the root.

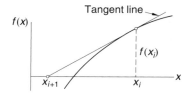

Figure 4.4. Graphical interpretation of the Newton–Raphson formula.

A graphical depiction of the Newton–Raphson formula is shown in Fig. 4.4. The formula approximates $f(x)$ by the straight line that is tangent to the curve at x_i. Thus x_{i+1} is at the intersection of the x-axis and the tangent line.

The algorithm for the Newton–Raphson method is simple: it repeatedly applies Eq. (4.3), starting with an initial value x_0, until the convergence criterion

$$|x_{i+1} - x_1| < \varepsilon$$

is reached, ε being the error tolerance. Only the latest value of x has to be stored. Here is the algorithm:

1. Let x be a guess for the root of $f(x) = 0$.
2. Compute $\Delta x = -f(x)/f'(x)$.
3. Let $x \leftarrow x + \Delta x$ and repeat steps 2-3 until $|\Delta x| < \varepsilon$.

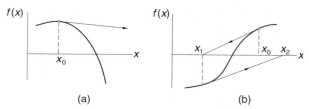

(a)   (b)

Figure 4.5. Examples where the Newton–Raphson method diverges.

Although the Newton–Raphson method converges fast near the root, its global convergence characteristics are poor. The reason is that the tangent line is not always an acceptable approximation of the function, as illustrated in the two examples in Fig. 4.5. But the method can be made nearly fail-safe by combining it with bisection, as in Brent's method.

■ newtonRaphson

The following *safe version* of the Newton–Raphson method assumes that the root to be computed is initially bracketed in (a,b). The midpoint of the bracket is used as the initial guess of the root. The brackets are updated after each iteration. If a

Newton–Raphson iteration does not stay within the brackets, it is disregarded and replaced with bisection. Since newtonRaphson uses the function f(x) as well as its derivative, function routines for both (denoted by func and dfunc in the listing) must be provided by the user.

```
function root = newtonRaphson(func,dfunc,a,b,tol)
% Newton-Raphson method combined with bisection for
% finding a root of f(x) = 0.
% USAGE: root = newtonRaphson(func,dfunc,a,b,tol)
% INPUT:
% func  = handle of function that returns f(x).
% dfunc = handle of function that returns f'(x).
% a,b   = brackets (limits) of the root.
% tol   = error tolerance (default is 1.0e6*eps).
% OUTPUT:
% root = zero of f(x) (root = NaN if no convergence).

if nargin < 5; tol = 1.0e6*eps; end
fa = feval(func,a); fb = feval(func,b);
if fa == 0; root = a; return; end
if fb == 0; root = b; return; end
if fa*fb > 0.0
    error('Root is not bracketed in (a,b)')
end
x = (a + b)/2.0;
for i = 1:30
    fx = feval(func,x);
    if abs(fx) < tol; root = x; return; end
    % Tighten brackets on the root
    if fa*fx < 0.0; b = x;
    else; a = x;
    end
    % Try Newton--Raphson step
    dfx = feval(dfunc,x);
    if abs(dfx) == 0; dx = b - a;
    else; dx = -fx/dfx;
    end
    x = x + dx;
    % If x not in bracket, use bisection
    if (b - x)*(x - a) < 0.0
```

```
        dx = (b - a)/2.0;
        x = a + dx;
    end
    % Check for convergence
    if abs(dx) < tol*max(b,1.0)
        root = x; return
    end
end
root = NaN
```

EXAMPLE 4.6

A root of $f(x) = x^3 - 10x^2 + 5 = 0$ lies close to $x = 0.7$. Compute this root with the Newton–Raphson method.

Solution The derivative of the function is $f'(x) = 3x^2 - 20x$, so that the Newton–Raphson formula in Eq. (4.3) is

$$x \leftarrow x - \frac{f(x)}{f'(x)} = x - \frac{x^3 - 10x^2 + 5}{3x^2 - 20x} = \frac{2x^3 - 10x^2 - 5}{x(3x - 20)}$$

It takes only two iterations to reach five decimal place accuracy:

$$x \leftarrow \frac{2(0.7)^3 - 10(0.7)^2 - 5}{0.7\,[3(0.7) - 20]} = 0.735\,36$$

$$x \leftarrow \frac{2(0.735\,36)^3 - 10(0.735\,36)^2 - 5}{0.735\,36\,[3(0.735\,36) - 20]} = 0.734\,60$$

EXAMPLE 4.7

Find the smallest positive zero of

$$f(x) = x^4 - 6.4x^3 + 6.45x^2 + 20.538x - 31.752$$

Solution

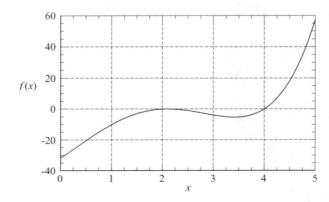

Inspecting the plot of the function, we suspect that the smallest positive zero is a double root near $x = 2$. Bisection and Brent's method would not work here, since they depend on the function changing its sign at the root. The same argument applies to the function newtonRaphson. But there no reason why the unrefined version of the Newton–Raphson method should not succeed. We used the following program, which prints the number of iterations in addition to the root:

```
function [root,numIter] = newton_simple(func,dfunc,x,tol)
% Simple version of Newton-Raphson method used in Example 4.7.

if nargin < 5; tol = 1.0e6*eps; end
for i = 1:30
    dx = -feval(func,x)/feval(dfunc,x);
    x = x + dx;
    if abs(dx) < tol
        root = x; numIter = i; return
    end
end
root = NaN
```

The two functions called by the program are

```
function y = fex4_7(x)
% Function used in Example 4.7.
y = x^4 - 6.4*x^3 + 6.45*x^2 + 20.538*x - 31.752;

function y = dfex4_7(x)
% Function used in Example 4.7.
y = 4.0*x^3 - 19.2*x^2 + 12.9*x + 20.538;
```

Here are the results:

```
>> [root,numIter] = newton_simple(@fex4_7,@dfex4_7,2.0)
root =
    2.1000
numIter =
    27
```

It can be shown that near a multiple root the convergence of the Newton–Raphson method is linear, rather than quadratic, which explains the large number of iterations. Convergence to a multiple root can be speeded up by replacing the Newton–Raphson difference approximation

$$\frac{\partial f_i}{\partial x_j} \approx \frac{f_i(\mathbf{x} + \mathbf{e}_j h) - f_i(\mathbf{x})}{h} \tag{4.8}$$

where h is a small increment and $\mathbf{e}_j$ represents a unit vector in the direction of x_j. This formula can be obtained from Eq. (4.5a) after dropping the terms of order Δx^2 and setting $\Delta \mathbf{x} = \mathbf{e}_j h$. We get away with the approximation in Eq. (4.8) because the Newton–Raphson method is rather insensitive to errors in $\mathbf{J}(\mathbf{x})$. By using this approximation, we also avoid the tedium of typing the expressions for $\partial f_i / \partial x_j$ into the computer code.

■ newtonRaphson2

This function is an implementation of the Newton–Raphson method. The nested function `jacobian` computes the Jacobian matrix from the finite difference approximation in Eq. (4.8). The simultaneous equations in Eq. (4.7) are solved by Gauss elimination with row pivoting using the function `gaussPivot.` listed in Section 2.5. The function subroutine `f` that returns the array $\mathbf{f}(\mathbf{x})$ must be supplied by the user.

```
## module newtonRaphson2
''' soln = newtonRaphson2(f,x,tol=1.0e-9).
    Solves the simultaneous equations f(x) = 0 by
    the Newton-Raphson method using {x} as the initial
    guess. Note that {f} and {x} are vectors.
'''
from numarray import zeros,Float64,dot
from gaussPivot import *
from math import sqrt

def newtonRaphson2(f,x,tol=1.0e-9):

    def jacobian(f,x):
        h = 1.0e-4
        n = len(x)
        jac = zeros((n,n),type=Float64)
        f0 = f(x)
        for i in range(n):
            temp = x[i]
            x[i] = temp + h
            f1 = f(x)
            x[i] = temp
```

```
        jac[:,i] = (f1 - f0)/h
    return jac,f0

for i in range(30):
    jac,f0 = jacobian(f,x)
    if sqrt(dot(f0,f0)/len(x)) < tol: return x
    dx = gaussPivot(jac,-f0)
    x = x + dx
    if sqrt(dot(dx,dx)) < tol*max(max(abs(x)),1.0): return x
print 'Too many iterations'
```

Note that the Jacobian matrix $J(x)$ is recomputed in each iterative loop. Since each calculation of $J(x)$ involves $n + 1$ evaluations of $f(x)$ (n is the number of equations), the expense of computation can be high depending on n and the complexity of $f(x)$. It is often possible to save computer time by neglecting the changes in the Jacobian matrix between iterations, thus computing $J(x)$ only once. This will work provided that the initial x is sufficiently close to the solution.

EXAMPLE 4.8
Determine the points of intersection between the circle $x^2 + y^2 = 3$ and the hyperbola $xy = 1$.

Solution The equations to be solved are

$$f_1(x, y) = x^2 + y^2 - 3 = 0 \tag{a}$$

$$f_2(x, y) = xy - 1 = 0 \tag{b}$$

The Jacobian matrix is

$$J(x, y) = \begin{bmatrix} \partial f_1/\partial x & \partial f_1/\partial y \\ \partial f_2/\partial x & \partial f_2/\partial y \end{bmatrix} = \begin{bmatrix} 2x & 2y \\ y & x \end{bmatrix}$$

Thus the linear equations $J(x)\Delta x = -f(x)$ associated with the Newton–Raphson method are

$$\begin{bmatrix} 2x & 2y \\ y & x \end{bmatrix} \begin{bmatrix} \Delta x \\ \Delta y \end{bmatrix} = \begin{bmatrix} -x^2 - y^2 + 3 \\ -xy + 1 \end{bmatrix} \tag{c}$$

By plotting the circle and the hyperbola, we see that there are four points of intersection. It is sufficient, however, to find only one of these points, as the others can be deduced from symmetry. From the plot we also get a rough estimate of the coordinates of an intersection point: $x = 0.5$, $y = 1.5$, which we use as the starting values.

in Eq. (4.8). The simultaneous equations in Eq. (4.7) are solved by using the left division operator of MATLAB. The function subroutine func that returns the array **f(x)** must be supplied by the user.

```
function root = newtonRaphson2(func,x,tol)
% Newton-Raphson method of finding a root of simultaneous
% equations fi(x1,x2,...,xn) = 0, i = 1,2,...,n.
% USAGE: root = newtonRaphson2(func,x,tol)
% INPUT:
% func = handle of function that returns[f1,f2,...,fn].
% x    = starting solution vector [x1,x2,...,xn].
% tol  = error tolerance (default is 1.0e4*eps).
% OUTPUT:
% root = solution vector.

if nargin == 2; tol = 1.0e4*eps; end
if size(x,1) == 1; x = x'; end    % x must be column vector
for i = 1:30
    [jac,f0] = jacobian(func,x);
    if sqrt(dot(f0,f0)/length(x)) < tol
        root = x; return
    end
    dx = jac\(-f0);
    x = x + dx;
    if sqrt(dot(dx,dx)/length(x)) < tol*max(abs(x),1.0)
        root = x; return
    end
end
error('Too many iterations')

function [jac,f0] = jacobian(func,x)
% Returns the Jacobian matrix and f(x).
h = 1.0e-4;
n = length(x);
jac = zeros(n);
f0 = feval(func,x);
for i =1:n
    temp = x(i);
    x(i) = temp + h;
    f1 = feval(func,x);
```

```
    x(i) = temp;
    jac(:,i) = (f1 - f0)/h;
end
```

Note that the Jacobian matrix $J(x)$ is recomputed in each iterative loop. Since each calculation of $J(x)$ involves $n + 1$ evaluations of $f(x)$ (n is the number of equations), the expense of computation can be high depending on n and the complexity of $f(x)$. It is often possible to save computer time by neglecting the changes in the Jacobian matrix between iterations, thus computing $J(x)$ only once. This will work provided that the initial x is sufficiently close to the solution.

EXAMPLE 4.8
Determine the points of intersection between the circle $x^2 + y^2 = 3$ and the hyperbola $xy = 1$.

Solution The equations to be solved are

$$f_1(x, y) = x^2 + y^2 - 3 = 0 \tag{a}$$

$$f_2(x, y) = xy - 1 = 0 \tag{b}$$

The Jacobian matrix is

$$J(x, y) = \begin{bmatrix} \partial f_1/\partial x & \partial f_1/\partial y \\ \partial f_2/\partial x & \partial f_2/\partial y \end{bmatrix} = \begin{bmatrix} 2x & 2y \\ y & x \end{bmatrix}$$

Thus the linear equations $J(x)\Delta x = -f(x)$ associated with the Newton–Raphson method are

$$\begin{bmatrix} 2x & 2y \\ y & x \end{bmatrix} \begin{bmatrix} \Delta x \\ \Delta y \end{bmatrix} = \begin{bmatrix} -x^2 - y^2 + 3 \\ -xy + 1 \end{bmatrix} \tag{c}$$

By plotting the circle and the hyperbola, we see that there are four points of intersection. It is sufficient, however, to find only one of these points, as the others can be deduced from symmetry. From the plot we also get a rough estimate of the coordinates of an intersection point: $x = 0.5$, $y = 1.5$, which we use as the starting values.

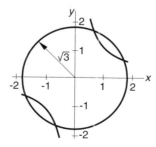

The computations then proceed as follows.

First iteration   Substituting $x = 0.5$, $y = 1.5$ in Eq. (c), we get

$$\begin{bmatrix} 1.0 & 3.0 \\ 1.5 & 0.5 \end{bmatrix} \begin{bmatrix} \Delta x \\ \Delta y \end{bmatrix} = \begin{bmatrix} 0.50 \\ 0.25 \end{bmatrix}$$

the solution of which is $\Delta x = \Delta y = 0.125$. Therefore, the improved coordinates of the intersection point are

$$x = 0.5 + 0.125 = 0.625 \qquad y = 1.5 + 0.125 = 1.625$$

Second iteration   Repeating the procedure using the latest values of x and y, we obtain

$$\begin{bmatrix} 1.250 & 3.250 \\ 1.625 & 0.625 \end{bmatrix} \begin{bmatrix} \Delta x \\ \Delta y \end{bmatrix} = \begin{bmatrix} -0.031250 \\ -0.015625 \end{bmatrix}$$

which yields $\Delta x = \Delta y = -0.00694$. Thus

$$x = 0.625 - 0.006\,94 = 0.618\,06 \qquad y = 1.625 - 0.006\,94 = 1.618\,06$$

Third iteration   Substitution of the latest x and y into Eq. (c) yields

$$\begin{bmatrix} 1.236\,12 & 3.23612 \\ 1.618\,06 & 0.61806 \end{bmatrix} \begin{bmatrix} \Delta x \\ \Delta y \end{bmatrix} = \begin{bmatrix} -0.000\,116 \\ -0.000\,058 \end{bmatrix}$$

The solution is $\Delta x = \Delta y = -0.00003$, so that

$$x = 0.618\,06 - 0.000\,03 = 0.618\,03$$
$$y = 1.618\,06 - 0.000\,03 = 1.618\,03$$

Subsequent iterations would not change the results within five significant figures. Therefore, the coordinates of the four intersection points are

$$\pm(0.618\,03, 1.618\,03) \text{ and } \pm(1.618\,03, 0.618\,03)$$

Alternate solution   If there are only a few equations, it may be possible to eliminate all but one of the unknowns. Then we would be left with a single equation which can be solved by the methods described in Arts. 4.2–4.5. In this problem, we obtain from Eq. (b)

$$y = \frac{1}{x}$$

which upon substitution into Eq. (a) yields $x^2 + 1/x^2 - 3 = 0$, or

$$x^4 - 3x^2 + 1 = 0$$

The solutions of this biquadratic equation: $x = \pm 0.618\,03$ and $\pm 1.618\,03$ agree with the results obtained by the Newton–Raphson method.

EXAMPLE 4.9

Find a solution of

$$\sin x + y^2 + \ln z - 7 = 0$$
$$3x + 2^y - z^3 + 1 = 0$$
$$x + y + z - 5 = 0$$

using `newtonRaphson2`. Start with the point $(1, 1, 1)$.

Solution Letting $x = x_1$, $y = x_2$ and $z = x_3$, the code defining the function array f(x) is

```
function y = fex4_9(x)
% Function used in Example 4.9
y = [sin(x(1)) + x(2)^2 + log(x(3)) - 7; ...
        3*x(1) + 2^x(2) - x(3)^3 + 1;    ...
        x(1) + x(2) + x(3) - 5];
```

The solution can now be obtained with the single command

```
>> newtonRaphson2(@fex4_9,[1;1;1])
```

which results in

```
ans =
    0.5991
    2.3959
    2.0050
```

Hence the solution is $x = 0.5991$, $y = 2.3959$ and $z = 2.0050$.

PROBLEM SET 4.1

1. Use the Newton–Raphson method and a four-function calculator $(+ - \times \div$ operations only) to compute $\sqrt[3]{75}$ with four significant figure accuracy.

2. Find the smallest positive (real) root of $x^3 - 3.23x^2 - 5.54x + 9.84 = 0$ by the method of bisection.

3. The smallest positive, nonzero root of $\cosh x \cos x - 1 = 0$ lies in the interval $(4, 5)$. Compute this root by Brent's method.

4. Solve Prob. 3 by the Newton–Raphson method.

5. A root of the equation $\tan x - \tanh x = 0$ lies in $(7.0, 7.4)$. Find this root with three decimal place accuracy by the method of bisection.

6. Determine the two roots of $\sin x + 3 \cos x - 2 = 0$ that lie in the interval $(-2, 2)$. Use the Newton–Raphson method.

7. A popular method of hand computation is the *secant formula* where the improved estimate of the root (x_{i+1}) is obtained by linear interpolation based two previous estimates $(x_i$ and $x_{i-1})$:

$$x_{i+1} = x_i - \frac{x_i - x_{i-1}}{f(x_i) - f(x_{i-1})} f(x_i)$$

Solve Prob. 6 using the secant formula.

8. Draw a plot of $f(x) = \cosh x \cos x - 1$ in the range $4 \le x \le 8$. (a) Verify from the plot that the smallest positive, nonzero root of $f(x) = 0$ lies in the interval $(4, 5)$. (b) Show graphically that the Newton–Raphson formula would not converge to this root if it is started with $x = 4$.

9. The equation $x^3 - 1.2x^2 - 8.19x + 13.23 = 0$ has a double root close to $x = 2$. Determine this root with the Newton–Raphson method within four decimal places.

10. ■ Write a program that computes all the roots of $f(x) = 0$ in a given interval with Brent's method. Utilize the functions `rootsearch` and `brent`. You may use the program in Example 4.3 as a model. Test the program by finding the roots of $x \sin x + 3 \cos x - x = 0$ in $(-6, 6)$.

11. ■ Solve Prob. 10 with the Newton–Raphson method.

12. ■ Determine all real roots of $x^4 + 0.9x^3 - 2.3x^2 + 3.6x - 25.2 = 0$.

13. ■ Compute all positive real roots of $x^4 + 2x^3 - 7x^2 + 3 = 0$.

14. ■ Find all positive, nonzero roots of $\sin x - 0.1x = 0$.

15. ■ The natural frequencies of a uniform cantilever beam are related to the roots β_i of the frequency equation $f(\beta) = \cosh \beta \cos \beta + 1 = 0$, where

$$\beta_i^4 = (2\pi f_i)^2 \frac{mL^3}{EI}$$

$f_i = i$th natural frequency (cps)

$m =$ mass of the beam

$L =$ length of the beam

$E =$ modulus of elasticity

$I =$ moment of inertia of the cross section

Determine the lowest two frequencies of a steel beam 0.9 m long, with a rectangular cross section 25 mm wide and 2.5 mm in. high. The mass density of steel is 7850 kg/m^3 and $E = 200$ GPa.

16. ■

A steel cable of length s is suspended as shown in the figure. The maximum tensile stress in the cable, which occurs at the supports, is

$$\sigma_{max} = \sigma_0 \cosh \beta$$

where

$$\beta = \frac{\gamma L}{2\sigma_0}$$

σ_0 = tensile stress in the cable at O
γ = weight of the cable per unit volume
L = horizontal span of the cable

The length to span ratio of the cable is related to β by

$$\frac{s}{L} = \frac{1}{\beta} \sinh \beta$$

Find σ_{max} if $\gamma = 77 \times 10^3 \text{ N/m}^3$ (steel), $L = 1000$ m and $s = 1100$ m.

17. ■

The aluminum W310 × 202 (wide flange) column is subjected to an eccentric axial load P as shown. The maximum compressive stress in the column is given by the so-called *secant formula*:

$$\sigma_{max} = \bar{\sigma} \left[1 + \frac{ec}{r^2} \sec \left(\frac{L}{2r} \sqrt{\frac{\bar{\sigma}}{E}} \right) \right]$$

where

$\bar{\sigma} = P/A$ = average stress
$A = 25\,800 \text{ mm}^2$ = cross-sectional area of the column
$e = 85$ mm = eccentricity of the load
$c = 170$ mm = half-depth of the column
$r = 142$ mm = radius of gyration of the cross section

$$L = 7100 \text{ mm} = \text{length of the column}$$

$$E = 71 \times 10^9 \text{ Pa} = \text{modulus of elasticity}$$

Determine the maximum load P that the column can carry if the maximum stress is not to exceed 120×10^6 Pa.

18. ■

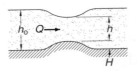

Bernoulli's equation for fluid flow in an open channel with a small bump is

$$\frac{Q^2}{2gb^2h_0^2} + h_0 = \frac{Q^2}{2gb^2h^2} + h + H$$

where

$$Q = 1.2 \text{ m}^3/\text{s} = \text{volume rate of flow}$$

$$g = 9.81 \text{ m/s}^2 = \text{gravitational acceleration}$$

$$b = 1.8 \text{ m} = \text{width of channel}$$

$$h_0 = 0.6 \text{ m} = \text{upstream water level}$$

$$H = 0.075 \text{ m} = \text{height of bump}$$

$$h = \text{water level above the bump}$$

Determine h.

19. ■ The speed v of a Saturn V rocket in vertical flight near the surface of earth can be approximated by

$$v = u \ln \frac{M_0}{M_0 - \dot{m}t} - gt$$

where

$$u = 2510 \text{ m/s} = \text{velocity of exhaust relative to the rocket}$$

$$M_0 = 2.8 \times 10^6 \text{ kg} = \text{mass of rocket at liftoff}$$

$$\dot{m} = 13.3 \times 10^3 \text{ kg/s} = \text{rate of fuel consumption}$$

$$g = 9.81 \text{ m/s}^2 = \text{gravitational acceleration}$$

$$t = \text{time measured from liftoff}$$

Determine the time when the rocket reaches the speed of sound (335 m/s).

20. ■

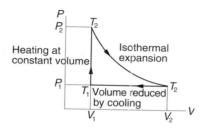

The figure shows the thermodynamic cycle of an engine. The efficiency of this engine for monatomic gas is

$$\eta = \frac{\ln(T_2/T_1) - (1 - T_1/T_2)}{\ln(T_2/T_1) + (1 - T_1/T_2)/(\gamma - 1)}$$

where T is the absolute temperature and $\gamma = 5/3$. Find T_2/T_1 that results in 30% efficiency ($\eta = 0.3$).

21. ■ Gibb's free energy of one mole of hydrogen at temperature T is

$$G = -RT \ln\left[(T/T_0)^{5/2}\right] \text{ J}$$

where $R = 8.314\,41$ J/K is the gas constant and $T_0 = 4.444\,18$ K. Determine the temperature at which $G = -10^5$ J.

22. ■ The chemical equilibrium equation in the production of methanol from CO and H_2 is[8]

$$\frac{\xi(3 - 2\xi)^2}{(1 - \xi)^3} = 249.2$$

where ξ is the *equilibrium extent of the reaction*. Determine ξ.

23. ■ Determine the coordinates of the two points where the circles $(x - 2)^2 + y^2 = 4$ and $x^2 + (y - 3)^2 = 4$ intersect. Start by estimating the locations of the points from a sketch of the circles, and then use the Newton–Raphson method to compute the coordinates.

24. ■ The equations

$$\sin x + 3\cos x - 2 = 0$$
$$\cos x - \sin y + 0.2 = 0$$

have a solution in the vicinity of the point $(1, 1)$. Use the Newton–Raphson method to refine the solution.

---

[8] From Alberty, R.A., *Physical Chemistry*, 7th ed., Wiley, 1987.

25. ■ Use any method to find *all* real solutions in $0 < x < 1.5$ of the simultaneous equations

$$\tan x - y = 1$$
$$\cos x - 3\sin y = 0$$

26. ■ The equation of a circle is

$$(x - a)^2 + (y - b)^2 = R^2$$

where R is the radius and (a, b) are the coordinates of the center. If the coordinates of three points on the circle are

x	8.21	0.34	5.96
y	0.00	6.62	−1.12

determine R, a and b.

27. ■

The trajectory of a satellite orbiting the earth is

$$R = \frac{C}{1 + e\sin(\theta + \alpha)}$$

where (R, θ) are the polar coordinates of the satellite, and C, e and α are constants (e is known as the eccentricity of the orbit). If the satellite was observed at the following three positions

θ	−30°	0°	30°
R (km)	6870	6728	6615

determine the smallest R of the trajectory and the corresponding value of θ.

28. ■

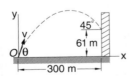

A projectile is launched at O with the velocity v at the angle θ to the horizontal. The parametric equation of the trajectory is

$$x = (v\cos\theta)t$$

$$y = -\frac{1}{2}gt^2 + (v\sin\theta)t$$

where t is the time measured from the instant of launch, and $g = 9.81$ m/s^2 represents the gravitational acceleration. If the projectile is to hit the target at the 45° angle shown in the figure, determine v, θ and the time of flight.

29. ■

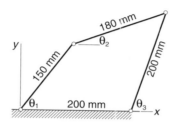

The three angles shown in the figure of the four-bar linkage are related by

$$150\cos\theta_1 + 180\cos\theta_2 - 200\cos\theta_3 = 200$$
$$150\sin\theta_1 + 180\sin\theta_2 - 200\sin\theta_3 = 0$$

Determine θ_1 and θ_2 when $\theta_3 = 75°$. Note that there are two solutions.

*## *4.7 Zeroes of Polynomials

Introduction

A polynomial of degree n has the form

$$P_n(x) = a_1 x^n + a_2 x^{n-1} + \cdots + a_n x + a_{n+1} \qquad (4.9)$$

where the coefficients a_i may be real or complex. We will concentrate on polynomials with real coefficients, but the algorithms presented in this article also work with complex coefficients.

The polynomial equation $P_n(x) = 0$ has exactly n roots, which may be real or complex. If the coefficients are real, the complex roots always occur in conjugate pairs $(x_r + ix_i,\ x_r - ix_i)$, where x_r and x_i are the real and imaginary parts, respectively. For real coefficients, the number of real roots can be estimated from the *rule of Descartes*:

- The number of positive, real roots equals the number of sign changes in the expression for $P_n(x)$, or less by an even number.
- The number of negative, real roots is equal to the number of sign changes in $P_n(-x)$, or less by an even number.

As an example, consider $P_3(x) = x^3 - 2x^2 - 8x + 27$. Since the sign changes twice, $P_3(x) = 0$ has either two or zero positive real roots. On the other hand, $P_3(-x) = -x^3 - 2x^2 + 8x + 27$ contains a single sign change; hence $P_3(x)$ possesses one negative real zero.

The real zeroes of polynomials with real coefficients can always be computed by one of the methods already described. But if complex roots are to be computed, it is best to use a method that specializes in polynomials. Here we present a method due to Laguerre, which is reliable and simple to implement. Before proceeding to Laguerre's method, we must first develop two numerical tools that are needed in any method capable of determining the zeroes of a polynomial. The first of these is an efficient algorithm for evaluating a polynomial and its derivatives. The second algorithm we need is for the *deflation* of a polynomial, i.e., for dividing the $P_n(x)$ by $x - r$, where r is a root of $P_n(x) = 0$.

Evaluation of Polynomials

It is tempting to evaluate the polynomial in Eq. (4.9) from left to right by the following algorithm (we assume that the coefficients are stored in the array **a**):

```
p = 0.0
for i = 1:n+1
    p = p + a(i)*x^(n-i+1)
end
```

Since x^k is evaluated as $x \times x \times \cdots \times x$ ($k - 1$ multiplications), we deduce that the number of multiplications in this algorithm is

$$1 + 2 + 3 + \cdots + n - 1 = \frac{1}{2}n(n - 1)$$

If n is large, the number of multiplications can be reduced considerably if we evaluate the polynomial from right to left. For an example, take

$$P_4(x) = a_1 x^4 + a_2 x^3 + a_3 x^2 + a_4 x + a_5$$

which can be rewritten as

$$P_4(x) = a_5 + x\{a_4 + x[a_3 + x(a_2 + xa_1)]\}$$

We now see that an efficient computational sequence for evaluating the polynomial is

$$\begin{aligned}
P_0(x) &= a_1 \\
P_1(x) &= a_2 + xP_0(x) \\
P_2(x) &= a_3 + xP_1(x) \\
P_3(x) &= a_4 + xP_2(x) \\
P_4(x) &= a_5 + xP_3(x)
\end{aligned}$$

For a polynomial of degree n, the procedure can be summarized as

$$P_0(x) = a_1$$
$$P_i(x) = a_{n+i} + xP_{i-1}(x), \quad i = 1, 2, \ldots, n \tag{4.10}$$

leading to the algorithm

```
p = a(1);
for i = 1:n
    p = p*x + a(i+1)
end
```

The last algorithm involves only n multiplications, making it more efficient for $n > 3$. But computational economy is not the prime reason why this algorithm should be used. Because the result of each multiplication is rounded off, the procedure with the least number of multiplications invariably accumulates the smallest roundoff error.

Some root-finding algorithms, including Laguerre's method, also require evaluation of the first and second derivatives of $P_n(x)$. From Eq. (4.10) we obtain by differentiation

$$P_0'(x) = 0 \quad P_i'(x) = P_{i-1}(x) + xP_{i-1}'(x), \quad i = 1, 2, \ldots, n \tag{4.11a}$$

$$P_0''(x) = 0 \quad P_i''(x) = 2P_{i-1}'(x) + xP_{i-1}''(x), \quad i = 1, 2, \ldots, n \tag{4.11b}$$

■ evalPoly

Here is the function that evaluates a polynomial and its derivatives:

```
function [p,dp,ddp] = evalpoly(a,x)
% Evaluates the polynomial
% p = a(1)*x^n + a(2)*x^(n-1) + ... + a(n+1)
% and its first two derivatives dp and ddp.
% USAGE: [p,dp,ddp] = evalpoly(a,x)

n = length(a) - 1;
p = a(1); dp = 0.0; ddp = 0.0;
for i = 1:n
    ddp = ddp*x + 2.0*dp;
    dp = dp*x + p;
    p = p*x + a(i+1);
end
```

Deflation of Polynomials

After a root r of $P_n(x) = 0$ has been computed, it is desirable to factor the polynomial as follows:

$$P_n(x) = (x - r)P_{n-1}(x) \tag{4.12}$$

This procedure, known as deflation or *synthetic division*, involves nothing more than computing the coefficients of $P_{n-1}(x)$. Since the remaining zeros of $P_n(x)$ are also the zeros of $P_{n-1}(x)$, the root-finding procedure can now be applied to $P_{n-1}(x)$ rather than $P_n(x)$. Deflation thus makes it progressively easier to find successive roots, because the degree of the polynomial is reduced every time a root is found. Moreover, by eliminating the roots that have already been found, the chances of computing the same root more than once are eliminated.

If we let

$$P_{n-1}(x) = b_1 x^{n-1} + b_2 x^{n-2} + \cdots + b_{n-1} x + b_n$$

then Eq. (4.12) becomes

$$a_1 x^n + a_2 x^{n-1} + \cdots + a_n x + a_{n+1}$$
$$= (x - r)(b_1 x^{n-1} + b_2 x^{n-2} + \cdots + b_{n-1} x + b_n)$$

Equating the coefficients of like powers of x, we obtain

$$b_1 = a_1 \qquad b_2 = a_2 + rb_1 \quad \cdots \quad b_n = a_n + rb_{n-1} \tag{4.13}$$

which leads to *Horner's deflation algorithm*:

```
b(1) = a(1);
for i = 2:n
    b(i) = a(i) + r*b(i-1);
end
```

Laguerre's Method

Laguerre's formulas are not easily derived for a general polynomial $P_n(x)$. However, the derivation is greatly simplified if we consider the special case where the polynomial has a zero at $x = r$ and $(n - 1)$ zeros at $x = q$. If the zeros were known, this polynomial can be written as

$$P_n(x) = (x - r)(x - q)^{n-1} \tag{a}$$

Our problem is now this: given the polynomial in Eq. (a) in the form

$$P_n(x) = a_1 x^n + a_2 x^{n-1} + \cdots + a_n x + a_{n+1}$$

determine r (note that q is also unknown). It turns out that the result, which is exact for the special case considered here, works well as an iterative formula with any polynomial.

Differentiating Eq. (a) with respect to x, we get

$$P_n'(x) = (x - q)^{n-1} + (n - 1)(x - r)(x - q)^{n-2}$$

$$= P_n(x) \left(\frac{1}{x - r} + \frac{n - 1}{x - q} \right)$$

Thus

$$\frac{P_n'(x)}{P_n(x)} = \frac{1}{x - r} + \frac{n - 1}{x - q} \tag{b}$$

which upon differentiation yields

$$\frac{P_n''(x)}{P_n(x)} - \left[\frac{P_n'(x)}{P_n(x)} \right]^2 = -\frac{1}{(x - r)^2} - \frac{n - 1}{(x - q)^2} \tag{c}$$

It is convenient to introduce the notation

$$G(x) = \frac{P_n'(x)}{P_n(x)} \qquad H(x) = G^2(x) - \frac{P_n''(x)}{P_n(x)} \tag{4.14}$$

so that Eqs. (b) and (c) become

$$G(x) = \frac{1}{x - r} + \frac{n - 1}{x - q} \tag{4.15a}$$

$$H(x) = \frac{1}{(x - r)^2} + \frac{n - 1}{(x - q)^2} \tag{4.15b}$$

If we solve Eq. (4.15a) for $x - q$ and substitute the result into Eq. (4.15b), we obtain a quadratic equation for $x - r$. The solution of this equation is the *Laguerre's formula*

$$x - r = \frac{n}{G(x) \pm \sqrt{(n - 1)\left[nH(x) - G^2(x) \right]}} \tag{4.16}$$

The procedure for finding a zero of a general polynomial by Laguerre's formula is:

1. Let x be a guess for the root of $P_n(x) = 0$ (any value will do).
2. Evaluate $P_n(x)$, $P_n'(x)$ and $P_n''(x)$ using the procedure outlined in Eqs. (4.10) and (4.11).
3. Compute $G(x)$ and $H(x)$ from Eqs. (4.14).
4. Determine the improved root r from Eq. (4.16) choosing the sign that results in the *larger magnitude of the denominator* (this can be shown to improve convergence).
5. Let $x \leftarrow r$ and repeat steps 2–5 until $|P_n(x)| < \varepsilon$ or $|x - r| < \varepsilon$, where ε is the error tolerance.

One nice property of Laguerre's method is that converges to a root, with very few exceptions, from any starting value of x.

■ polyRoots

The function `polyRoots` in this module computes all the roots of $P_n(x) = 0$, where the polynomial $P_n(x)$ defined by its coefficient array $\mathbf{a} = [a_1, a_2, a_3, \ldots, a_{n+1}]$. After the first root is computed by the subfunction `laguerre`, the polynomial is deflated using `deflPoly` and the next zero computed by applying `laguerre` to the deflated polynomial. This process is repeated until all n roots have been found. If a computed root has a very small imaginary part, it is very likely that it represents roundoff error. Therefore, `polyRoots` replaces a tiny imaginary part by zero.

```
function root = polyroots(a,tol)
% Returns all the roots of the polynomial
% a(1)*x^n + a(2)*x^(n-1) + ... + a(n+1).
% USAGE: root = polyroots(a,tol).
% tol = error tolerance (default is 1.0e4*eps).

if nargin == 1; tol = 1.0e-6; end
n = length(a) - 1;
root = zeros(n,1);
for i = 1:n
    x = laguerre(a,tol);
    if abs(imag(x)) < tol; x = real(x); end
    root(i) = x;
    a = deflpoly(a,x);
end

function x = laguerre(a,tol)
% Returns a root of the polynomial
% a(1)*x^n + a(2)*x^(n-1) + ... + a(n+1).
x = randn;                 % Start with random number
n = length(a) - 1;
for i = 1:30
    [p,dp,ddp] = evalpoly(a,x);
    if abs(p) < tol; return; end
    g = dp/p; h = g*g - ddp/p;
    f = sqrt((n - 1)*(n*h - g*g));
```

```
      if abs(g + f) >= abs(g - f); dx = n/(g + f);
      else; dx = n/(g - f); end
      x = x - dx;
      if abs(dx) < tol; return; end
end
error('Too many iterations in laguerre')


function b = deflpoly(a,r)
% Horner's deflation:
% a(1)*x^n + a(2)*x^(n-1) + ... + a(n+1)
% = (x - r)[b(1)*x^(n-1) + b(2)*x^(n-2) + ...+ b(n)].
n = length(a) - 1;
b = zeros(n,1);
b(1) = a(1);
for i = 2:n; b(i) = a(i) + r*b(i-1); end
```

Since the roots are computed with finite accuracy, each deflation introduces small errors in the coefficients of the deflated polynomial. The accumulated roundoff error increases with the degree of the polynomial and can become severe if the polynomial is ill-conditioned (small changes in the coefficients produce large changes in the roots). Hence the results should be viewed with caution when dealing with polynomials of high degree.

The errors caused by deflation can be reduced by recomputing each root using the original, undeflated polynomial. The roots obtained previously in conjunction with deflation are employed as the starting values.

EXAMPLE 4.10

A zero of the polynomial $P_4(x) = 3x^4 - 10x^3 - 48x^2 - 2x + 12$ is $x = 6$. Deflate the polynomial with Horner's algorithm, i.e., find $P_3(x)$ so that $(x - 6)P_3(x) = P_4(x)$.

Solution With $r = 6$ and $n = 4$, Eqs. (4.13) become

$$b_1 = a_1 = 3$$
$$b_2 = a_2 + 6b_1 = -10 + 6(3) = 8$$
$$b_3 = a_3 + 6b_2 = -48 + 6(8) = 0$$
$$b_4 = a_4 + 6b_3 = -2 + 6(0) = -2$$

Therefore,

$$P_3(x) = 3x^3 + 8x^2 - 2$$

EXAMPLE 4.11

A root of the equation $P_3(x) = x^3 - 4.0x^2 - 4.48x + 26.1$ is approximately $x = 3 - i$. Find a more accurate value of this root by one application of Laguerre's iterative formula.

Solution Use the given estimate of the root as the starting value. Thus

$$x = 3 - i \qquad x^2 = 8 - 6i \qquad x^3 = 18 - 26i$$

Substituting these values in $P_3(x)$ and its derivatives, we get

$$P_3(x) = x^3 - 4.0x^2 - 4.48x + 26.1$$
$$= (18 - 26i) - 4.0(8 - 6i) - 4.48(3 - i) + 26.1 = -1.34 + 2.48i$$
$$P_3'(x) = 3.0x^2 - 8.0x - 4.48$$
$$= 3.0(8 - 6i) - 8.0(3 - i) - 4.48 = -4.48 - 10.0i$$
$$P_3''(x) = 6.0x - 8.0 = 6.0(3 - i) - 8.0 = 10.0 - 6.0i$$

Equations (4.14) then yield

$$G(x) = \frac{P_3'(x)}{P_3(x)} = \frac{-4.48 - 10.0i}{-1.34 + 2.48i} = -2.36557 + 3.08462i$$

$$H(x) = G^2(x) - \frac{P_3''(x)}{P_3(x)} = (-2.36557 + 3.08462i)^2 - \frac{10.0 - 6.0i}{-1.34 + 2.48i}$$
$$= 0.35995 - 12.48452i$$

The term under the square root sign of the denominator in Eq. (4.16) becomes

$$F(x) = \sqrt{(n-1)\left[n\,H(x) - G^2(x)\right]}$$
$$= \sqrt{2\left[3(0.35995 - 12.48452i) - (-2.36557 + 3.08462i)^2\right]}$$
$$= \sqrt{5.67822 - 45.71946i} = 5.08670 - 4.49402i$$

Now we must find which sign in Eq. (4.16) produces the larger magnitude of the denominator:

$$|G(x) + F(x)| = |(-2.36557 + 3.08462i) + (5.08670 - 4.49402i)|$$
$$= |2.72113 - 1.40940i| = 3.06448$$

$$|G(x) - F(x)| = |(-2.36557 + 3.08462i) - (5.08670 - 4.49402i)|$$
$$= |-7.45227 + 7.57864i| = 10.62884$$

Using the minus sign, we obtain from Eq. (4.16) the following improved approximation for the root

$$r = x - \frac{n}{G(x) - F(x)} = (3 - i) - \frac{3}{-7.45227 + 7.57864i}$$
$$= 3.19790 - 0.79875i$$

Thanks to the good starting value, this approximation is already quite close to the exact value $r = 3.20 - 0.80i$.

EXAMPLE 4.12

Use `polyRoots` to compute *all* the roots of $x^4 - 5x^3 - 9x^2 + 155x - 250 = 0$.

Solution The command

```
>> polyroots([1 -5 -9 155 -250])
```

results in

```
ans =
   2.0000
   4.0000 - 3.0000i
   4.0000 + 3.0000i
  -5.0000
```

There are two real roots ($x = 2$ and -5) and a pair of complex conjugate roots ($x = 4 \pm 3i$).

PROBLEM SET 4.2

Problems 1–5 A zero $x = r$ of $P_n(x)$ is given. Verify that r is indeed a zero, and then deflate the polynomial, i.e., find $P_{n-1}(x)$ so that $P_n(x) = (x - r) P_{n-1}(x)$.

1. $P_3(x) = 3x^3 + 7x^2 - 36x + 20, r = -5$.
2. $P_4(x) = x^4 - 3x^2 + 3x - 1, r = 1$.
3. $P_5(x) = x^5 - 30x^4 + 361x^3 - 2178x^2 + 6588x - 7992, r = 6$.
4. $P_4(x) = x^4 - 5x^3 - 2x^2 - 20x - 24, r = 2i$.
5. $P_3(x) = 3x^3 - 19x^2 + 45x - 13, r = 3 - 2i$.

Problems 6–9 A zero $x = r$ of $P_n(x)$ is given. Determine all the other zeroes of $P_n(x)$ by using a calculator. You should need no tools other than deflation and the quadratic formula.

6. $P_3(x) = x^3 + 1.8x^2 - 9.01x - 13.398, r = -3.3$.
7. $P_3(x) = x^3 - 6.64x^2 + 16.84x - 8.32, r = 0.64$.
8. $P_3(x) = 2x^3 - 13x^2 + 32x - 13, r = 3 - 2i$.
9. $P_4(x) = x^4 - 3x^3 + 10x^2 - 6x - 20, r = 1 + 3i$.

Problems 10–16 Find all the zeroes of the given $P_n(x)$.

10. ■$P_4(x) = x^4 + 2.1x^3 - 2.52x^2 + 2.1x - 3.52$.
11. ■$P_5(x) = x^5 - 156x^4 - 5x^3 + 780x^2 + 4x - 624$.

12. ■ $P_6(x) = x^6 + 4x^5 - 8x^4 - 34x^3 + 57x^2 + 130x - 150$.

13. ■ $P_7(x) = 8x^7 + 28x^6 + 34x^5 - 13x^4 - 124x^3 + 19x^2 + 220x - 100$.

14. ■ $P_8(x) = x^8 - 7x^7 + 7x^6 + 25x^5 + 24x^4 - 98x^3 - 472x^2 + 440x + 800$.

15. ■ $P_4(x) = x^4 + (5 + i)x^3 - (8 - 5i)x^2 + (30 - 14i)x - 84$.

16. ■

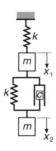

The two blocks of mass m each are connected by springs and a dashpot. The stiffness of each spring is k, and c is the coefficient of damping of the dashpot. When the system is displaced and released, the displacement of each block during the ensuing motion has the form

$$x_k(t) = A_k e^{\omega_r t} \cos(\omega_i t + \phi_k), \quad k = 1, 2$$

where A_k and ϕ_k are constants, and $\omega = \omega_r \pm i\omega_i$ are the roots of

$$\omega^4 + 2\frac{c}{m}\omega^3 + 3\frac{k}{m}\omega^2 + \frac{c}{m}\frac{k}{m}\omega + \left(\frac{k}{m}\right)^2 = 0$$

Determine the two possible combinations of ω_r and ω_i if $c/m = 12 \text{ s}^{-1}$ and $k/m = 1500 \text{ s}^{-2}$.

MATLAB Functions

x = fzero(@func,x0) returns the zero of the function func closest to x0.

x = fzero(@func,[a b]) can be used when the root has been bracketed in (a,b).

The algorithm used for fzero is Brent's method.

x = roots(a) returns the zeros of the polynomial $P_n(x) = a_1 x^n + \cdots + a_n x + a_{n+1}$.

The zeros are obtained by calculating the eigenvalues of the $n \times n$ "companion matrix"

$$\mathbf{A} = \begin{bmatrix} -a_2/a_1 & -a_3/a_1 & \cdots & -a_n/a_1 & -a_{n+1}/a_1 \\ 1 & 0 & \cdots & 0 & 0 \\ 0 & 1 & & 0 & 0 \\ \vdots & \vdots & \ddots & \vdots & \vdots \\ 0 & 0 & \cdots & 1 & 0 \end{bmatrix}$$

The characteristic equation (see Art. 9.1) of this matrix is

$$x^n + \frac{a_2}{a_1}x^{n-1} + \cdots \frac{a_n}{a_1}x + \frac{a_{n+1}}{a_1} = 0$$

which is equivalent to $P_n(x) = 0$. Thus the eigenvalues of $\mathbf{A}$ are the zeroes of $P_n(x)$. The eigenvalue method is robust, but considerably slower than Laguerre's method.

5 Numerical Differentiation

Given the function $f(x)$, compute $d^n f/dx^n$ at given x

5.1 Introduction

Numerical differentiation deals with the following problem: we are given the function $y = f(x)$ and wish to obtain one of its derivatives at the point $x = x_k$. The term "given" means that we either have an algorithm for computing the function, or possess a set of discrete data points (x_i, y_i), $i = 1, 2, \ldots, n$. In either case, we have access to a finite number of (x, y) data pairs from which to compute the derivative. If you suspect by now that numerical differentiation is related to interpolation, you are right—one means of finding the derivative is to approximate the function locally by a polynomial and then differentiate it. An equally effective tool is the Taylor series expansion of $f(x)$ about the point x_k. The latter has the advantage of providing us with information about the error involved in the approximation.

Numerical differentiation is not a particularly accurate process. It suffers from a conflict between roundoff errors (due to limited machine precision) and errors inherent in interpolation. For this reason, a derivative of a function can never be computed with the same precision as the function itself.

5.2 Finite Difference Approximations

The derivation of the finite difference approximations for the derivatives of $f(x)$ are based on forward and backward Taylor series expansions of $f(x)$ about x, such as

$$f(x + h) = f(x) + hf'(x) + \frac{h^2}{2!} f''(x) + \frac{h^3}{3!} f'''(x) + \frac{h^4}{4!} f^{(4)}(x) + \cdots \qquad \text{(a)}$$

$$f(x - h) = f(x) - hf'(x) + \frac{h^2}{2!} f''(x) - \frac{h^3}{3!} f'''(x) + \frac{h^4}{4!} f^{(4)}(x) - \cdots \qquad \text{(b)}$$

$$f(x + 2h) = f(x) + 2hf'(x) + \frac{(2h)^2}{2!} f''(x) + \frac{(2h)^3}{3!} f'''(x) + \frac{(2h)^4}{4!} f^{(4)}(x) + \cdots \qquad \text{(c)}$$

$$f(x - 2h) = f(x) - 2hf'(x) + \frac{(2h)^2}{2!} f''(x) - \frac{(2h)^3}{3!} f'''(x) + \frac{(2h)^4}{4!} f^{(4)}(x) - \cdots \qquad \text{(d)}$$

We also record the sums and differences of the series:

$$f(x + h) + f(x - h) = 2 f(x) + h^2 f''(x) + \frac{h^4}{12} f^{(4)}(x) + \cdots \qquad \text{(e)}$$

$$f(x + h) - f(x - h) = 2hf'(x) + \frac{h^3}{3} f'''(x) + \cdots \qquad \text{(f)}$$

$$f(x + 2h) + f(x - 2h) = 2 f(x) + 4h^2 f''(x) + \frac{4h^4}{3} f^{(4)}(x) + \cdots \qquad \text{(g)}$$

$$f(x + 2h) - f(x - 2h) = 4hf'(x) + \frac{8h^3}{3} f'''(x) + \cdots \qquad \text{(h)}$$

Note that the sums contain only even derivatives, while the differences retain just the odd derivatives. Equations (a)–(h) can be viewed as simultaneous equations that can be solved for various derivatives of $f(x)$. The number of equations involved and the number of terms kept in each equation depend on the order of the derivative and the desired degree of accuracy.

First Central Difference Approximations

The solution of Eq. (f) for $f'(x)$ is

$$f'(x) = \frac{f(x + h) - f(x - h)}{2h} - \frac{h^2}{6} f'''(x) - \cdots$$

Keeping only the first term on the right-hand side, we have

$$f'(x) = \frac{f(x + h) - f(x - h)}{2h} + \mathcal{O}(h^2) \qquad (5.1)$$

which is called the *first central difference approximation* for $f'(x)$. The term $\mathcal{O}(h^2)$ reminds us that the truncation error behaves as h^2.

From Eq. (e) we obtain

$$f''(x) = \frac{f(x + h) - 2 f(x) + f(x - h)}{h^2} + \frac{h^2}{12} f^{(4)}(x) + \cdots$$

or

$$f''(x) = \frac{f(x + h) - 2 f(x) + f(x - h)}{h^2} + \mathcal{O}(h^2) \qquad (5.2)$$

Central difference approximations for other derivatives can be obtained from Eqs. (a)–(h) in a similar manner. For example, eliminating $f'(x)$ from Eqs. (f) and (h) and solving for $f'''(x)$ yield

$$f'''(x) = \frac{f(x+2h) - 2f(x+h) + 2f(x-h) - f(x-2h)}{2h^3} + \mathcal{O}(h^2) \qquad (5.3)$$

The approximation

$$f^{(4)}(x) = \frac{f(x+2h) - 4f(x+h) + 6f(x) - 4f(x-h) + f(x-2h)}{h^4} + \mathcal{O}(h^2) \qquad (5.4)$$

is available from Eq. (e) and (g) after eliminating $f''(x)$. Table 5.1 summarizes the results.

	$f(x-2h)$	$f(x-h)$	$f(x)$	$f(x+h)$	$f(x+2h)$
$2hf'(x)$		-1	0	1	
$h^2 f''(x)$		1	-2	1	
$2h^3 f'''(x)$	-1	2	0	-2	1
$h^4 f^{(4)}(x)$	1	-4	6	-4	1

Table 5.1. Coefficients of central finite difference approximations of $\mathcal{O}(h^2)$

First Noncentral Finite Difference Approximations

Central finite difference approximations are not always usable. For example, consider the situation where the function is given at the n discrete points $x_1, x_2, \dots, x_n$. Since central differences use values of the function on each side of x, we would be unable to compute the derivatives at x_1 and x_n. Clearly, there is a need for finite difference expressions that require evaluations of the function only on one side of x. These expressions are called *forward* and *backward* finite difference approximations.

Noncentral finite differences can also be obtained from Eqs. (a)–(h). Solving Eq. (a) for $f'(x)$ we get

$$f'(x) = \frac{f(x+h) - f(x)}{h} - \frac{h}{2} f''(x) - \frac{h^2}{6} f'''(x) - \frac{h^3}{4!} f^{(4)}(x) - \cdots$$

Keeping only the first term on the right-hand side leads to the *first forward difference approximation*

$$f'(x) = \frac{f(x+h) - f(x)}{h} + \mathcal{O}(h) \qquad (5.5)$$

Similarly, Eq. (b) yields the *first backward difference approximation*

$$f'(x) = \frac{f(x) - f(x-h)}{h} + \mathcal{O}(h) \qquad (5.6)$$

Note that the truncation error is now $\mathcal{O}(h)$, which is not as good as the $\mathcal{O}(h^2)$ error in central difference approximations.

We can derive the approximations for higher derivatives in the same manner. For example, Eqs. (a) and (c) yield

$$f''(x) = \frac{f(x+2h) - 2f(x+h) + f(x)}{h^2} + \mathcal{O}(h) \tag{5.7}$$

The third and fourth derivatives can be derived in a similar fashion. The results are shown in Tables 5.2a and 5.2b.

	$f(x)$	$f(x+h)$	$f(x+2h)$	$f(x+3h)$	$f(x+4h)$
$hf'(x)$	-1	1			
$h^2 f''(x)$	1	-2	1		
$h^3 f'''(x)$	-1	3	-3	1	
$h^4 f^{(4)}(x)$	1	-4	6	-4	1

Table 5.2a. Coefficients of forward finite difference approximations of $\mathcal{O}(h)$

	$f(x-4h)$	$f(x-3h)$	$f(x-2h)$	$f(x-h)$	$f(x)$
$hf'(x)$				-1	1
$h^2 f''(x)$			1	-2	1
$h^3 f'''(x)$		-1	3	-3	1
$h^4 f^{(4)}(x)$	1	-4	6	-4	1

Table 5.2b. Coefficients of backward finite difference approximations of $\mathcal{O}(h)$

Second Noncentral Finite Difference Approximations

Finite difference approximations of $\mathcal{O}(h)$ are not popular due to reasons that will be explained shortly. The common practice is to use expressions of $\mathcal{O}(h^2)$. To obtain noncentral difference formulas of this order, we have to retain more terms in the Taylor series. As an illustration, we will derive the expression for $f'(x)$. We start with Eqs. (a) and (c), which are

$$f(x+h) = f(x) + hf'(x) + \frac{h^2}{2}f''(x) + \frac{h^3}{6}f'''(x) + \frac{h^4}{24}f^{(4)}(x) + \cdots$$

$$f(x+2h) = f(x) + 2hf'(x) + 2h^2 f''(x) + \frac{4h^3}{3}f'''(x) + \frac{2h^4}{3}f^{(4)}(x) + \cdots$$

We eliminate $f''(x)$ by multiplying the first equation by 4 and subtracting it from the second equation. The result is

$$f(x + 2h) - 4f(x + h) = -3f(x) - 2hf'(x) + \frac{2h^2}{3} f'''(x) + \cdots$$

Therefore,

$$f'(x) = \frac{-f(x + 2h) + 4f(x + h) - 3f(x)}{2h} + \frac{h^2}{3} f'''(x) + \cdots$$

or

$$f'(x) \frac{-f(x + 2h) + 4f(x + h) - 3f(x)}{2h} + \mathcal{O}(h^2) \qquad (5.8)$$

Equation (5.8) is called the *second forward finite difference approximation*.

Derivation of finite difference approximations for higher derivatives involve additional Taylor series. Thus the forward difference approximation for $f''(x)$ utilizes series for $f(x + h)$, $f(x + 2h)$ and $f(x + 3h)$; the approximation for $f'''(x)$ involves Taylor expansions for $f(x + h)$, $f(x + 2h)$, $f(x + 3h)$ and $f(x + 4h)$, etc. As you can see, the computations for high-order derivatives can become rather tedious. The results for both the forward and backward finite differences are summarized in Tables 5.3a and 5.3b.

	$f(x)$	$f(x + h)$	$f(x + 2h)$	$f(x + 3h)$	$f(x + 4h)$	$f(x + 5h)$
$2hf'(x)$	-3	4	-1			
$h^2 f''(x)$	2	-5	4	-1		
$2h^3 f'''(x)$	-5	18	-24	14	-3	
$h^4 f^{(4)}(x)$	3	-14	26	-24	11	-2

Table 5.3a. Coefficients of forward finite difference approximations of $\mathcal{O}(h^2)$

	$f(x - 5h)$	$f(x - 4h)$	$f(x - 3h)$	$f(x - 2h)$	$f(x - h)$	$f(x)$
$2hf'(x)$				1	-4	3
$h^2 f''(x)$			-1	4	-5	2
$2h^3 f'''(x)$		3	-14	24	-18	5
$h^4 f^{(4)}(x)$	-2	11	-24	26	-14	3

Table 5.3b. Coefficients of backward finite difference approximations of $\mathcal{O}(h^2)$

Errors in Finite Difference Approximations

Observe that in all finite difference expressions the sum of the coefficients is zero. The effect on the *roundoff error* can be profound. If h is very small, the values of $f(x)$, $f(x \pm h)$, $f(x \pm 2h)$, etc. will be approximately equal. When they are multiplied by the coefficients in the finite difference formulas and added, several significant figures can be lost. On the other hand, we cannot make h too large, because then the *truncation error* would become excessive. This unfortunate situation has no remedy, but we can obtain some relief by taking the following precautions:

- Use double-precision arithmetic.
- Employ finite difference formulas that are accurate to at least $\mathcal{O}(h^2)$.

To illustrate the errors, let us compute the second derivative of $f(x) = e^{-x}$ at $x = 1$ from the central difference formula, Eq. (5.2). We carry out the calculations with six- and eight-digit precision, using different values of h. The results, shown in Table 5.4, should be compared with $f''(1) = e^{-1} = 0.367\,879\,44$.

h	6-digit precision	8-digit precision
0.64	0.380 610	0.380 609 11
0.32	0.371 035	0.371 029 39
0.16	0.368 711	0.368 664 84
0.08	0.368 281	0.368 076 56
0.04	0.368 75	0.367 831 25
0.02	0.37	0.3679
0.01	0.38	0.3679
0.005	0.40	0.3676
0.0025	0.48	0.3680
0.00125	1.28	0.3712

Table 5.4. $(e^{-x})''$ at $x = 1$ from central finite difference approximation

In the six-digit computations, the optimal value of h is 0.08, yielding a result accurate to three significant figures. Hence three significant figures are lost due to a combination of truncation and roundoff errors. Above optimal h, the dominant error is due to truncation; below it, the roundoff error becomes pronounced. The best result obtained with the eight-digit computation is accurate to four significant figures. Because the extra precision decreases the roundoff error, the optimal h is smaller (about 0.02) than in the six-figure calculations.

5.3 Richardson Extrapolation

Richardson extrapolation is a simple method for boosting the accuracy of certain numerical procedures, including finite difference approximations (we will also use it later in numerical integration).

Suppose that we have an approximate means of computing some quantity G. Moreover, assume that the result depends on a parameter h. Denoting the approximation by $g(h)$, we have $G = g(h) + E(h)$, where $E(h)$ represents the error. Richardson extrapolation can remove the error, provided that it has the form $E(h) = ch^p$, c and p being constants. We start by computing $g(h)$ with some value of h, say $h = h_1$. In that case we have

$$G = g(h_1) + ch_1^p \tag{i}$$

Then we repeat the calculation with $h = h_2$, so that

$$G = g(h_2) + ch_2^p \tag{j}$$

Eliminating c and solving for G, we obtain from Eqs. (i) and (j)

$$G = \frac{(h_1/h_2)^p g(h_2) - g(h_1)}{(h_1/h_2)^p - 1} \tag{5.9a}$$

which is the *Richardson extrapolation formula*. It is common practice to use $h_2 = h_1/2$, in which case Eq. (5.9a) becomes

$$G = \frac{2^p g(h_1/2) - g(h_1)}{2^p - 1} \tag{5.9b}$$

Let us illustrate Richardson extrapolation by applying it to the finite difference approximation of $(e^{-x})''$ at $x = 1$. We work with six-digit precision and utilize the results in Table 5.4. Since the extrapolation works only on the truncation error, we must confine h to values that produce negligible roundoff. Choosing $h_1 = 0.64$ and letting $g(h)$ be the approximation of $f''(1)$ obtained with h, we get from Table 5.4

$$g(h_1) = 0.380\,610 \qquad g(h_1/2) = 0.371\,035$$

The truncation error in the central difference approximation is $E(h) = \mathcal{O}(h^2) = c_1 h^2 + c_2 h^4 + c_3 h^6 + \cdots$. Therefore, we can eliminate the first (dominant) error term if we substitute $p = 2$ and $h_1 = 0.64$ in Eq. (5.9b). The result is

$$G = \frac{2^2 g(0.32) - g(0.64)}{2^2 - 1} = \frac{4(0.371\,035) - 0.380\,610}{3} = 0.367\,843$$

which is an approximation of $(e^{-x})''$ with the error $\mathcal{O}(h^4)$. Note that it is as accurate as the best result obtained with eight-digit computations in Table 5.4.

EXAMPLE 5.1

Given the evenly spaced data points

x	0	0.1	0.2	0.3	0.4
$f(x)$	0.0000	0.0819	0.1341	0.1646	0.1797

compute $f'(x)$ and $f''(x)$ at $x = 0$ and 0.2 using finite difference approximations of $\mathcal{O}(h^2)$.

Solution From the forward difference formulas in Table 5.3a we get

$$f'(0) = \frac{-3f(0) + 4f(0.1) - f(0.2)}{2(0.1)} = \frac{-3(0) + 4(0.0819) - 0.1341}{0.2} = 0.967$$

$$f''(0) = \frac{2f(0) - 5f(0.1) + 4f(0.2) - f(0.3)}{(0.1)^2}$$

$$= \frac{2(0) - 5(0.0819) + 4(0.1341) - 0.1646}{(0.1)^2} = -3.77$$

The central difference approximations in Table 5.1 yield

$$f'(0.2) = \frac{-f(0.1) + f(0.3)}{2(0.1)} = \frac{-0.0819 + 0.1646}{0.2} = 0.4135$$

$$f''(0.2) = \frac{f(0.1) - 2f(0.2) + f(0.3)}{(0.1)^2} = \frac{0.0819 - 2(0.1341) + 0.1646}{(0.1)^2} = -2.17$$

EXAMPLE 5.2

Use the data in Example 5.1 to compute $f'(0)$ as accurately as you can.

Solution One solution is to apply Richardson extrapolation to finite difference approximations. We start with two forward difference approximations for $f'(0)$: one using $h = 0.2$ and the other one using $h = 0.1$. Referring to the formulas of $\mathcal{O}(h^2)$ in Table 5.3a, we get

$$g(0.2) = \frac{-3f(0) + 4f(0.2) - f(0.4)}{2(0.2)} = \frac{3(0) + 4(0.1341) - 0.1797}{0.4} = 0.8918$$

$$g(0.1) = \frac{-3f(0) + 4f(0.1) - f(0.2)}{2(0.1)} = \frac{-3(0) + 4(0.0819) - 0.1341}{0.2} = 0.9675$$

where g denotes the finite difference approximation of $f'(0)$. Recalling that the error in both approximations is of the form $E(h) = c_1 h^2 + c_2 h^4 + c_3 h^6 + \cdots$, we can use Richardson extrapolation to eliminate the dominant error term. With $p = 2$ we obtain from Eq. (5.9)

$$f'(0) \approx G = \frac{2^2 g(0.1) - g(0.2)}{2^2 - 1} = \frac{4(0.9675) - 0.8918}{3} = 0.9927$$

which is a finite difference approximation of $O(h^4)$.

EXAMPLE 5.3

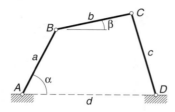

The linkage shown has the dimensions $a = 100$ mm, $b = 120$ mm, $c = 150$ mm and $d = 180$ mm. It can be shown by geometry that the relationship between the angles α and β is

$$(d - a\cos\alpha - b\cos\beta)^2 + (a\sin\alpha + b\sin\beta)^2 - c^2 = 0$$

For a given value of α, we can solve this transcendental equation for β by one of the root-finding methods in Chapter 4. This was done with $\alpha = 0°, 5°, 10°, \ldots, 30°$, the results being

α (deg)	0	5	10	15	20	25	30
β (rad)	1.6595	1.5434	1.4186	1.2925	1.1712	1.0585	0.9561

If link AB rotates with the constant angular velocity of 25 rad/s, use finite difference approximations of $O(h^2)$ to tabulate the angular velocity $d\beta/dt$ of link BC against α.

Solution The angular speed of BC is

$$\frac{d\beta}{dt} = \frac{d\beta}{d\alpha}\frac{d\alpha}{dt} = 25\frac{d\beta}{d\alpha} \text{ rad/s}$$

where $d\beta/d\alpha$ is computed from finite difference approximations using the data in the table. Forward and backward differences of $O(h^2)$ are used at the endpoints, central differences elsewhere. Note that the increment of α is

$$h = (5 \text{ deg})\left(\frac{\pi}{180}\text{rad/deg}\right) = 0.087266 \text{ rad}$$

The computations yield

$$\dot{\beta}(0°) = 25\frac{-3\beta(0°) + 4\beta(5°) - \beta(10°)}{2h} = 25\frac{-3(1.6595) + 4(1.5434) - 1.4186}{2(0.087266)}$$

$$= -32.01 \text{ rad/s}$$

$$\dot{\beta}(5°) = 25\frac{\beta(10°) - \beta(0°)}{2h} = 25\frac{1.4186 - 1.6595}{2(0.087266)} = -34.51 \text{ rad/s}$$

etc.

The complete set of results is

α (deg)	0	5	10	15	20	25	30
$\dot{\beta}$ (rad/s)	-32.01	-34.51	-35.94	-35.44	-33.52	-30.81	-27.86

5.4   Derivatives by Interpolation

If $f(x)$ is given as a set of discrete data points, interpolation can be a very effective means of computing its derivatives. The idea is to approximate the derivative of $f(x)$ by the derivative of the interpolant. This method is particularly useful if the data points are located at uneven intervals of x, when the finite difference approximations listed in the last article are not applicable.[9]

Polynomial Interpolant

The idea here is simple: fit the polynomial of degree $n - 1$

$$P_{n-1}(x) = a_1 x^{n-1} + a_2 x^{n-2} + \cdots + a_n \tag{a}$$

through n data points and then evaluate its derivatives at the given x. As pointed out in Art. 3.2, it is generally advisable to limit the degree of the polynomial to less than six in order to avoid spurious oscillations of the interpolant. Since these oscillations are magnified with each differentiation, their effect can be devastating. In view of the above limitation, the interpolation should usually be a local one, involving no more than a few nearest-neighbor data points.

For evenly spaced data points, polynomial interpolation and finite difference approximations produce identical results. In fact, the finite difference formulas are equivalent to polynomial interpolation.

Several methods of polynomial interpolation were introduced in Art. 3.2. Unfortunately, none of them is suited for the computation of derivatives. The method that we need is one that determines the coefficients $a_1, a_2, \ldots, a_n$ of the polynomial in Eq. (a). There is only one such method discussed in Chapter 3—the *least-squares fit*. Although this method is designed mainly for smoothing of data, it will carry out interpolation if we use $m = n$ in Eq. (3.22). If the data contains noise, then the least-squares fit should be used in the smoothing mode, that is, with $m < n$. After the coefficients of

[9] It is possible to derive finite difference approximations for unevenly spaced data, but they would not be as accurate as the formulas derived in Art. 5.2.

compute $f'(2)$ and $f''(2)$ using (1) polynomial interpolation over three nearest-neighbor points, and (2) natural cubic spline interpolant spanning all the data points.

Solution of Part (1) The interpolant is $P_2(x) = a_0 + a_1 x + a_2 x^2$ passing through the points at $x = 1.9, 2.1$ and 2.4. The normal equations, Eqs. (3.23), of the least-squares fit are

$$
\begin{bmatrix} n & \sum x_i & \sum x_i^2 \\ \sum x_i & \sum x_i^2 & \sum x_i^3 \\ \sum x_i^2 & \sum x_i^3 & \sum x_i^4 \end{bmatrix} \begin{bmatrix} a_0 \\ a_1 \\ a_2 \end{bmatrix} = \begin{bmatrix} \sum y_i \\ \sum y_i x_i \\ \sum y_i x_i^2 \end{bmatrix}
$$

After substituting the data, we get

$$
\begin{bmatrix} 3 & 6.4 & 13.78 \\ 6.4 & 13.78 & 29.944 \\ 13.78 & 29.944 & 65.6578 \end{bmatrix} \begin{bmatrix} a_0 \\ a_1 \\ a_2 \end{bmatrix} = \begin{bmatrix} 4.6742 \\ 10.0571 \\ 21.8385 \end{bmatrix}
$$

which yields $\mathbf{a} = \begin{bmatrix} -0.7714 & 1.5075 & -0.1930 \end{bmatrix}^T$.

The derivatives of the interpolant are $P_2'(x) = a_1 + 2a_2 x$ and $P_2''(x) = 2a_2$. Therefore,

$$
f'(2) \approx P_2'(2) = 1.5075 + 2(-0.1930)(2) = 0.7355
$$
$$
f''(2) \approx P_2''(2) = 2(-0.1930) = -0.3860
$$

Solution of Part (2) We must first determine the second derivatives k_i of the spline at its knots, after which the derivatives of $f(x)$ can be computed from Eqs. (5.10) and (5.11). The first part can be carried out by the following small program:

```
#!/usr/bin/python
## example5_4
from cubicSpline import curvatures
from LUdecomp3 import *
from numarray import array

xData = array([1.5, 1.9, 2.1, 2.4, 2.6, 3.1])
yData = array([1.0628, 1.3961, 1.5432, 1.7349, 1.8423, 2.0397])
print curvatures(xData,yData)
raw_input(''Press return to exit'')
```

The output of the program, consisting of k_0 to k_5, is

```
[ 0.   -0.4258431 -0.37744139 -0.38796663 -0.55400477   0.   ]
Press return to exit
```

which yields $\mathbf{a} = \begin{bmatrix} -0.7714 & 1.5075 & -0.1930 \end{bmatrix}^T$. Thus the interpolant and its derivatives are

$$P_2(x) = -0.1903x^2 + 1.5075x - 0.7714$$

$$P_2'(x) = -0.3860x + 1.5075$$

$$P_2''(x) = -0.3860$$

which gives us

$$f'(2) \approx P_2'(2) = -0.3860(2) + 1.5075 = 0.7355$$

$$f''(2) \approx P_2''(2) = -0.3860$$

Solution of Part (2) We must first determine the second derivatives k_i of the spline at its knots, after which the derivatives of $f(x)$ can be computed from Eqs. (5.10) and (5.11). The first part can be carried out by the following small program:

```
% Example 5.4 (Curvatures of cubic spline at the knots)
xData = [1.5; 1.9; 2.1; 2.4; 2.6; 3.1];
yData = [1.0628; 1.3961; 1.5432; 1.7349; 1.8423; 2.0397];
k = splineCurv(xData,yData)
```

The output of the program, consisting of k_1 to k_6, is

```
>> k =
         0
   -0.4258
   -0.3774
   -0.3880
   -0.5540
         0
```

Since $x = 2$ lies between knots 2 and 3, we must use Eqs. (5.10) and (5.11) with $i = 2$. This yields

$$f'(2) \approx f_{2,3}'(2) = \frac{k_2}{6} \left[\frac{3(x - x_3)^2}{x_2 - x_3} - (x_1 - x_3) \right]$$

$$- \frac{k_3}{6} \left[\frac{3(x - x_2)^2}{x_2 - x_3} - (x_2 - x_3) \right] + \frac{y_2 - y_3}{x_2 - x_3}$$

$$= \frac{(-0.4258)}{6} \left[\frac{3(2-2.1)^2}{(-0.2)} - (-0.2) \right]$$

$$- \frac{(-0.3774)}{6} \left[\frac{3(2-1.9)^2}{(-0.2)} - (-0.2) \right] + \frac{1.3961 - 1.5432}{(-0.2)}$$

$$= 0.7351$$

$$f''(2) \approx f''_{2,3}(2) = k_2 \frac{x - x_3}{x_2 - x_3} - k_3 \frac{x - x_2}{x_2 - x_3}$$

$$= (-0.4258) \frac{2 - 2.1}{(-0.2)} - (-0.3774) \frac{2 - 1.9}{(-0.2)} = -0.4016$$

Note that the solutions for $f'(2)$ in parts (1) and (2) differ only in the fourth significant figure, but the values of $f''(2)$ are much farther apart. This is not unexpected, considering the general rule: the higher the order of the derivative, the lower the precision with which it can be computed. It is impossible to tell which of the two results is better without knowing the expression for $f(x)$. In this particular problem, the data points fall on the curve $f(x) = x^2 e^{-x/2}$, so that the "correct" values of the derivatives are $f'(2) = 0.7358$ and $f''(2) = -0.3679$.

EXAMPLE 5.5
Determine $f'(0)$ and $f'(1)$ from the following noisy data

x	0	0.2	0.4	0.6
$f(x)$	1.9934	2.1465	2.2129	2.1790
x	0.8	1.0	1.2	1.4
$f(x)$	2.0683	1.9448	1.7655	1.5891

Solution We used the program listed in Example 3.10 to find the best polynomial fit (in the least-squares sense) to the data. The results were:

```
degree of polynomial = 2
coeff =
 -7.0240e-001
  6.4704e-001
  2.0262e+000
sigma =
  3.6097e-002

degree of polynomial = 3
coeff =
  4.0521e-001
```

```
      -1.5533e+000
       1.0928e+000
       1.9921e+000
sigma =
       8.2604e-003

degree of polynomial = 4
coeff =
      -1.5329e-002
       4.4813e-001
      -1.5906e+000
       1.1028e+000
       1.9919e+000
sigma =
       9.5193e-003

degree of polynomial =
Done
```

   Based on standard deviation, the cubic seems to be the best candidate for the interpolant. Before accepting the result, we compare the plots of the data points and the interpolant—see the figure. The fit does appear to be satisfactory.

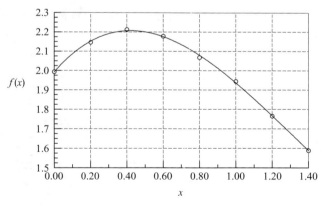

Approximating $f(x)$ by the interpolant, we have

$$f(x) \approx a_1 x^3 + a_2 x^2 + a_3 x + a^4$$

so that

$$f'(x) \approx 3a_1 x^2 + 2a_2 x + a_3$$

Therefore,

$$f'(0) \approx a_3 = 1.093$$

$$f'(1) = 3a_1 + 2a_2 + a_3 = 3(0.405) + 2(-1.553) + 1.093 = -0.798$$

In general, derivatives obtained from noisy data are at best rough approximations. In this problem, the data represent $f(x) = (x + 2)/\cosh x$ with added random noise. Thus $f'(x) = [1 - (x + 2)\tanh x]/\cosh x$, so that the "correct" derivatives are $f'(0) = 1.000$ and $f'(1) = -0.833$.

PROBLEM SET 5.1

1. Given the values of $f(x)$ at the points x, $x - h_1$ and $x + h_2$, determine the finite difference approximation for $f''(x)$. What is the order of the truncation error?

2. Given the first backward finite difference approximations for $f'(x)$ and $f''(x)$, derive the first backward finite difference approximation for $f'''(x)$ using the operation $f'''(x) = [f''(x)]'$.

3. Derive the central difference approximation for $f''(x)$ accurate to $\mathcal{O}(h^4)$ by applying Richardson extrapolation to the central difference approximation of $\mathcal{O}(h^2)$.

4. Derive the second forward finite difference approximation for $f'''(x)$ from the Taylor series.

5. Derive the first central difference approximation for $f^{(4)}(x)$ from the Taylor series.

6. Use finite difference approximations of $\mathcal{O}(h^2)$ to compute $f'(2.36)$ and $f''(2.36)$ from the data

x	2.36	2.37	2.38	2.39
$f(x)$	0.85866	0.86289	0.86710	0.87129

7. Estimate $f'(1)$ and $f''(1)$ from the following data:

x	0.97	1.00	1.05
$f(x)$	0.85040	0.84147	0.82612

8. Given the data

x	0.84	0.92	1.00	1.08	1.16
$f(x)$	0.431711	0.398519	0.367879	0.339596	0.313486

calculate $f''(1)$ as accurately as you can.

9. Use the data in the table to compute $f'(0.2)$ as accurately as possible.

x	0	0.1	0.2	0.3	0.4
$f(x)$	0.000 000	0.078 348	0.138 910	0.192 916	0.244 981

10. Using five significant figures in the computations, determine $d(\sin x)/dx$ at $x = 0.8$ from (a) the first forward difference approximation, and (b) the first central difference approximation. In each case, use h that gives the most accurate result (this requires experimentation).

11. ■ Use polynomial interpolation to compute f' and f'' at $x = 0$, using the data

x	−2.2	−0.3	0.8	1.9
$f(x)$	15.180	10.962	1.920	−2.040

12. ■

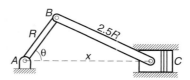

The crank AB of length $R = 90$ mm is rotating at the constant angular speed of $d\theta/dt = 5000$ rev/min. The position of the piston C can be shown to vary with the angle θ as

$$x = R\left(\cos\theta + \sqrt{2.5^2 - \sin^2\theta}\right)$$

Write a program that computes the acceleration of the piston at $\theta = 0°, 5°, 10°, \ldots, 180°$ by numerical differentiation.

13. ■

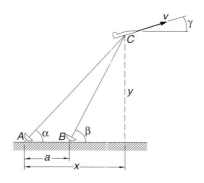

The radar stations A and B, separated by the distance $a = 500$ m, track the plane C by recording the angles α and β at one-second intervals. If three successive readings are

t (s)	9	10	11
α	54.80°	54.06°	53.34°
β	65.59°	64.59°	63.62°

calculate the speed v of the plane and the climb angle γ at $t = 10$ s. The coordinates of the plane can be shown to be

$$x = a\frac{\tan \beta}{\tan \beta - \tan \alpha} \qquad y = a\frac{\tan \alpha \tan \beta}{\tan \beta - \tan \alpha}$$

14. ■

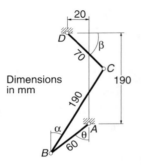

Dimensions in mm

Geometric analysis of the linkage shown resulted in the following table relating the angles θ and β:

θ (deg)	0	30	60	90	120	150
β (deg)	59.96	56.42	44.10	25.72	−0.27	−34.29

Assuming that member AB of the linkage rotates with the constant angular velocity $d\theta/dt = 1$ rad/s, compute $d\beta/dt$ in rad/s at the tabulated values of θ. Use cubic spline interpolation.

MATLAB Functions

`d = diff(y)` returns the differences `d(i) = y(i+1) - y(i)`. Note that `length(d) = length(y) - 1`.

$dn = diff(y,n)$ returns the nth differences; e.g., $d2(i) = d(i+1) - d(i)$, $d3(i) = d2(i+1) - d2(i)$, etc. Here $length(dn) = length(y) - n$.

$d = gradient(y,h)$ returns the finite difference approximation of dy/dx at each point, where h is the spacing between the points.

$d2 = del2(y,h)$ returns the finite difference approximation of $(d^2y/dx^2)/4$ at each point, where h is the spacing between the points.

6 Numerical Integration

Compute $\int_a^b f(x)\,dx$, where $f(x)$ is a given function

6.1 Introduction

Numerical integration, also known as *quadrature*, is intrinsically a much more accurate procedure than numerical differentiation. Quadrature approximates the definite integral

$$\int_a^b f(x)\,dx$$

by the sum

$$I = \sum_{i=1}^{n} A_i f(x_i)$$

where the *nodal abscissas* x_i and *weights* A_i depend on the particular rule used for the quadrature. All rules of quadrature are derived from polynomial interpolation of the integrand. Therefore, they work best if $f(x)$ can be approximated by a polynomial.

Methods of numerical integration can be divided into two groups: Newton–Cotes formulas and Gaussian quadrature. Newton–Cotes formulas are characterized by equally spaced abscissas, and include well-known methods such as the trapezoidal rule and Simpson's rule. They are most useful if $f(x)$ has already been computed at equal intervals, or can be computed at low cost. Since Newton–Cotes formulas are based on local interpolation, they require only a piecewise fit to a polynomial.

In Gaussian quadrature the locations of the abscissas are chosen to yield the best possible accuracy. Because Gaussian quadrature requires fewer evaluations of the integrand for a given level of precision, it is popular in cases where $f(x)$ is expensive to evaluate. Another advantage of Gaussian quadrature is its ability to handle integrable singularities, enabling us to evaluate expressions such as

$$\int_0^1 \frac{g(x)}{\sqrt{1-x^2}} dx$$

provided that $g(x)$ is a well-behaved function.

6.2 Newton–Cotes Formulas

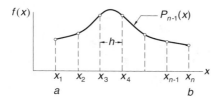

Figure 6.1. Polynomial approximation of $f(x)$.

Consider the definite integral

$$\int_a^b f(x)\, dx \tag{6.1}$$

We divide the range of integration (a, b) into $n-1$ equal intervals of length $h = (b-a)/(n-1)$ each, as shown in Fig. 6.1, and denote the abscissas of the resulting nodes by $x_1, x_2, \ldots, x_n$. Next we approximate $f(x)$ by a polynomial of degree $n-1$ that intersects all the nodes. Lagrange's form of this polynomial, Eq. (3.1a), is

$$P_{n-1}(x) = \sum_{i=1}^{n} f(x_i)\ell_i(x)$$

where $\ell_i(x)$ are the cardinal functions defined in Eq. (3.1b). Therefore, an approximation to the integral in Eq. (6.1) is

$$I = \int_a^b P_{n-1}(x)dx = \sum_{i=1}^{n}\left[f(x_i)\int_a^b \ell_i(x)dx \right] = \sum_{i=1}^{n} A_i f(x_i) \tag{6.2a}$$

where

$$A_i = \int_a^b \ell_i(x)dx, \quad i = 1, 2, \ldots, n \tag{6.2b}$$

Equations (6.2) are the *Newton–Cotes formulas*. Classical examples of these formulas are the *trapezoidal rule* ($n = 2$), *Simpson's rule* ($n = 3$) and *Simpson's 3/8 rule* ($n = 4$). The most important of these is the trapezoidal rule. It can be combined with Richardson extrapolation into an efficient algorithm known as *Romberg integration*, which makes the other classical rules somewhat redundant.

Trapezoidal Rule

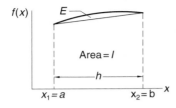

Figure 6.2. Trapezoidal rule.

If $n = 2$, we have $\ell_1 = (x - x_2)/(x_1 - x_2) = -(x - b)/h$. Therefore,

$$A_1 = -\frac{1}{h}\int_a^b (x - b)\,dx = \frac{1}{2h}(b - a)^2 = \frac{h}{2}$$

Also $\ell_2 = (x - x_1)/(x_2 - x_1) = (x - a)/h$, so that

$$A_2 = \frac{1}{h}\int_a^b (x - a)\,dx = \frac{1}{2h}(b - a)^2 = \frac{h}{2}$$

Substitution in Eq. (6.2a) yields

$$I = [f(a) + f(b)]\frac{h}{2} \tag{6.3}$$

which is known as the *trapezoidal rule*. It represents the area of the trapezoid in Fig. 6.2.

The error in the trapezoidal rule

$$E = \int_a^b f(x)dx - I$$

is the area of the region between $f(x)$ and the straight-line interpolant, as indicated in Fig. 6.2. It can be obtained by integrating the interpolation error in Eq. (4.3):

$$E = \frac{1}{2!}\int_a^b (x - x_1)(x - x_2)\,f''(\xi)dx = \frac{1}{2}f''(\xi)\int_a^b (x - a)(x - b)dx$$

$$= -\frac{1}{12}(b - a)^3\,f''(\xi) = -\frac{h^3}{12}f''(\xi) \tag{6.4}$$

Composite Trapezoidal Rule

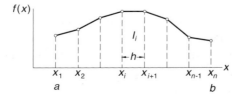

Figure 6.3. Composite trapezoidal rule.

In practice the trapezoidal rule is applied in a piecewise fashion. Figure 6.3 shows the region (a, b) divided into $n - 1$ panels, each of width h. The function $f(x)$ to be integrated is approximated by a straight line in each panel. From the trapezoidal rule we obtain for the approximate area of a typical (ith) panel

$$I_i = [f(x_i) + f(x_{i+1})] \frac{h}{2}$$

Hence total area, representing $\int_a^b f(x)\, dx$, is

$$I = \sum_{i=1}^{n-1} I_i = [f(x_1) + 2f(x_2) + 2f(x_3) + \cdots + 2f(x_{n-1}) + f(x_n)] \frac{h}{2} \qquad (6.5)$$

which is the *composite trapezoidal rule*.

The truncation error in the area of a panel is from Eq. (6.4),

$$E_i = -\frac{h^3}{12} f''(\xi_i)$$

where ξ_i lies in (x_i, x_{i+1}). Hence the truncation error in Eq. (6.5) is

$$E = \sum_{i=1}^{n-1} E_i = -\frac{h^3}{12} \sum_{i=1}^{n-1} f''(\xi_i) \qquad (a)$$

But

$$\sum_{i=1}^{n-1} f''(\xi_i) = (n - 1)\, \bar{f}''$$

where $\bar{f}''$ is the arithmetic mean of the second derivatives. If $f''(x)$ is continuous, there must be a point ξ in (a, b) at which $f''(\xi) = \bar{f}''$, enabling us to write

$$\sum_{i=1}^{n-1} f''(\xi_i) = (n - 1)\, f''(\xi) = \frac{b - a}{h} f''(\xi)$$

Therefore, Eq. (a) becomes

$$E = -\frac{(b - a)h^2}{12} f''(\xi) \qquad (6.6)$$

It would be incorrect to conclude from Eq. (6.6) that $E = ch^2$ (c being a constant), because $f''(\xi)$ is not entirely independent of h. A deeper analysis of the error[10] shows that if $f(x)$ and its derivatives are finite in (a, b), then

$$E = c_1 h^2 + c_2 h^4 + c_3 h^6 + \cdots \qquad (6.7)$$

---

[10] The analysis requires familiarity with the *Euler–Maclaurin summation formula*, which is covered in advanced texts.

Recursive Trapezoidal Rule

Let I_k be the integral evaluated with the composite trapezoidal rule using 2^{k-1} panels. Note that if k is increased by one, the number of panels is doubled. Using the notation

$$H = b - a$$

we obtain from Eq. (6.5) the following results for $k = 1, 2$ and 3.

$k = 1$ (1 panel):

$$I_1 = [f(a) + f(b)] \frac{H}{2} \tag{6.8}$$

$k = 2$ (2 panels):

$$I_2 = \left[f(a) + 2f \left(a + \frac{H}{2} \right) + f(b) \right] \frac{H}{4} = \frac{1}{2} I_1 + f \left(a + \frac{H}{2} \right) \frac{H}{2}$$

$k = 3$ (4 panels):

$$I_3 = \left[f(a) + 2f \left(a + \frac{H}{4} \right) + 2f \left(a + \frac{H}{2} \right) + 2f \left(a + \frac{3H}{4} \right) + f(b) \right] \frac{H}{8}$$

$$= \frac{1}{2} I_2 + \left[f \left(a + \frac{H}{4} \right) + f \left(a + \frac{3H}{4} \right) \right] \frac{H}{4}$$

We can now see that for arbitrary $k > 1$ we have

$$I_k = \frac{1}{2} I_{k-1} + \frac{H}{2^{k-1}} \sum_{i=1}^{2^{k-2}} f \left[a + \frac{(2i-1)H}{2^{k-1}} \right], \quad k = 2, 3, \ldots \tag{6.9a}$$

which is the *recursive trapezoidal rule*. Observe that the summation contains only the new nodes that were created when the number of panels was doubled. Therefore, the computation of the sequence $I_1, I_2, I_3, \ldots, I_k$ from Eqs. (6.8) and (6.9) involves the same amount of algebra as the calculation of I_k directly from Eq. (6.5). The advantage of using the recursive trapezoidal rule is that it allows us to monitor convergence and terminate the process when the difference between I_{k-1} and I_k becomes sufficiently small. A form of Eq. (6.9a) that is easier to remember is

$$I(h) = \frac{1}{2} I(2h) + h \sum f(x_{\text{new}}) \tag{6.9b}$$

where $h = H/(n-1)$ is the width of each panel.

■ trapezoid

The function trapezoid computes $I(h)$, given $I(2h)$ from Eqs. (6.8) and (6.9). We can compute $\int_a^b f(x) \, dx$ by calling trapezoid repeatedly with $k = 1, 2, \ldots$ until the desired precision is attained.

```
function Ih = trapezoid(func,a,b,I2h,k)
% Recursive trapezoidal rule.
% USAGE: Ih = trapezoid(func,a,b,I2h,k)
% func = handle of function being integrated.
% a,b  = limits of integration.
% I2h  = integral with 2^(k-1) panels.
% Ih = integral with 2^k panels.

if k == 1
    fa = feval(func,a);  fb = feval(func,b);
    Ih = (fa + fb)*(b - a)/2.0;
else
    n = 2^(k -2 );        % Number of new points
    h = (b - a)/n ;       % Spacing of new points
    x = a + h/2.0;        % Coord. of 1st new point
    sum = 0.0;
    for i = 1:n
        fx = feval(func,x);
        sum = sum + fx;
        x = x + h;
    end
    Ih = (I2h + h*sum)/2.0;
end
```

Simpson's Rules

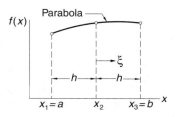

Figure 6.4. Simpson's 1/3 rule.

Simpson's 1/3 rule can be obtained from Newton–Cotes formulas with $n = 3$; that is, by passing a parabolic interpolant through three adjacent nodes, as shown in Fig. 6.4. The area under the parabola, which represents an approximation of $\int_a^b f(x)\,dx$, is (see derivation in Example 6.1)

$$I = \left[f(a) + 4f\left(\frac{a+b}{2} \right) + f(b) \right] \frac{h}{3} \tag{a}$$

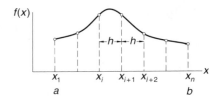

Figure 6.5. Composite Simpson's 1/3 rule.

To obtain the *composite Simpson's 1/3 rule*, the integration range (a, b) is divided into $n - 1$ panels (n odd) of width $h = (b - a)/(n - 1)$ each, as indicated in Fig. 6.5. Applying Eq. (a) to two adjacent panels, we have

$$\int_{x_i}^{x_{i+2}} f(x)\, dx \approx [f(x_i) + 4 f(x_{i+1}) + f(x_{i+2})] \frac{h}{3} \tag{b}$$

Substituting Eq. (b) into

$$\int_a^b f(x)dx = \int_{x_1}^{x_n} f(x)\, dx = \sum_{i=1,3,\dots}^{n-2} \left[\int_{x_i}^{x_{i+2}} f(x)dx\right]$$

yields

$$\int_a^b f(x)\, dx \approx I = [f(x_1) + 4 f(x_2) + 2 f(x_3) + 4 f(x_4) + \cdots \tag{6.10}$$

$$\cdots + 2 f(x_{n-2}) + 4 f(x_{n-1}) + f(x_n)] \frac{h}{3}$$

The composite Simpson's 1/3 rule in Eq. (6.10) is perhaps the best-known method of numerical integration. Its reputation is somewhat undeserved, since the trapezoidal rule is more robust, and Romberg integration is more efficient.

The error in the composite Simpson's rule is

$$E = \frac{(b - a)h^4}{180} f^{(4)}(\xi) \tag{6.11}$$

from which we conclude that Eq. (6.10) is exact if $f(x)$ is a polynomial of degree three or less.

Simpson's 1/3 rule requires the number of panels to be even. If this condition is not satisfied, we can integrate over the first (or last) three panels with *Simpson's 3/8 rule*:

$$I = [f(x_1) + 3 f(x_2) + 3 f(x_3) + f(x_4)] \frac{3h}{8} \tag{6.12}$$

and use Simpson's 1/3 rule for the remaining panels. The error in Eq. (6.12) is of the same order as in Eq. (6.10).

EXAMPLE 6.1
Derive Simpson's 1/3 rule from Newton–Cotes formulas.

Solution Referring to Fig. 6.4, we see that Simpson's 1/3 rule uses three nodes located at $x_1 = a$, $x_2 = (a + b)/2$ and $x_3 = b$. The spacing of the nodes is $h = (b - a)/2$. The cardinal functions of Lagrange's three-point interpolation are (see Art. 3.2)

$$\ell_1(x) = \frac{(x - x_2)(x - x_3)}{(x_1 - x_2)(x_1 - x_3)} \qquad \ell_2(x) = \frac{(x - x_1)(x - x_3)}{(x_2 - x_1)(x_2 - x_3)}$$

$$\ell_3(x) = \frac{(x - x_1)(x - x_2)}{(x_3 - x_1)(x_3 - x_2)}$$

The integration of these functions is easier if we introduce the variable ξ with origin at x_2. Then the coordinates of the nodes are $\xi_1 = -h$, $\xi_2 = 0$, $\xi_3 = h$ and Eq. (6.2b) becomes $A_i = \int_a^b \ell_i(x)dx = \int_{-h}^h \ell_i(\xi)d\xi$. Therefore,

$$A_1 = \int_{-h}^h \frac{(\xi - 0)(\xi - h)}{(-h)(-2h)} d\xi = \frac{1}{2h^2} \int_{-h}^h (\xi^2 - h\xi)d\xi = \frac{h}{3}$$

$$A_2 = \int_{-h}^h \frac{(\xi + h)(\xi - h)}{(h)(-h)} d\xi = -\frac{1}{h^2} \int_{-h}^h (\xi^2 - h^2)d\xi = \frac{4h}{3}$$

$$A_3 = \int_{-h}^h \frac{(\xi + h)(\xi - 0)}{(2h)(h)} d\xi = \frac{1}{2h^2} \int_{-h}^h (\xi^2 + h\xi)d\xi = \frac{h}{3}$$

Equation (6.2a) then yields

$$I = \sum_{i=1}^3 A_i f(x_i) = \left[f(a) + 4f\left(\frac{a + b}{2}\right) + f(b) \right] \frac{h}{3}$$

which is Simpson's 1/3 rule.

EXAMPLE 6.2
Evaluate the bounds on $\int_0^\pi \sin(x)\, dx$ with the composite trapezoidal rule using (1) eight panels and (2) sixteen panels.

Solution of Part (1) With 8 panels there are 9 nodes spaced at $h = \pi/8$. The abscissas of the nodes are $x_i = (i - 1)\pi/8$, $i = 1, 2, \ldots, 9$. From Eq. (6.5) we get

$$I = \left[\sin 0 + 2 \sum_{i=2}^8 \sin \frac{i\pi}{8} + \sin \pi \right] \frac{\pi}{16} = 1.97423$$

The error is given by Eq. (6.6):

$$E = -\frac{(b - a)h^2}{12} f''(\xi) = -\frac{(\pi - 0)(\pi/8)^2}{12}(-\sin \xi) = \frac{\pi^3}{768} \sin \xi$$

where $0 < \xi < \pi$. Since we do not know the value of ξ, we cannot evaluate E, but we can determine its bounds:

$$E_{\min} = \frac{\pi^3}{768} \sin(0) = 0 \qquad E_{\max} = \frac{\pi^3}{768} \sin \frac{\pi}{2} = 0.040\,37$$

Therefore, $I + E_{\min} < \int_0^\pi \sin(x)\,dx < I + E_{\max}$, or

$$1.974\,23 < \int_0^\pi \sin(x)\,dx < 2.014\,60$$

The exact integral is, of course, $I = 2$.

Solution of Part (2) The new nodes created by the doubling of panels are located at midpoints of the old panels. Their abscissas are

$$x_j = \frac{\pi}{16} + (j-1)\frac{\pi}{8} = (2j-1)\frac{\pi}{16}, \quad j = 1, 2, \ldots, 8$$

Using the recursive trapezoidal rule in Eq. (6.9b), we get

$$I = \frac{1.974\,23}{2} + \frac{\pi}{16} \sum_{j=1}^{8} \sin\frac{(2j-1)\pi}{16} = 1.993\,58$$

and the bounds on the error become (note that E is quartered when h is halved) $E_{\min} = 0$, $E_{\max} = 0.040\,37/4 = 0.010\,09$. Hence

$$1.993\,58 < \int_0^\pi \sin(x)\,dx < 2.003\,67$$

EXAMPLE 6.3
Estimate $\int_0^{2.5} f(x)\,dx$ from the data

x	0	0.5	1.0	1.5	2.0	2.5
$f(x)$	1.5000	2.0000	2.0000	1.6364	1.2500	0.9565

Solution We will use Simpson's rules, since they are more accurate than the trapezoidal rule. Because the number of panels is odd, we compute the integral over the first three panels by Simpson's 3/8 rule, and use the 1/3 rule for the last two panels:

$$I = [f(0) + 3f(0.5) + 3f(1.0) + f(1.5)]\frac{3(0.5)}{8}$$

$$+ [f(1.5) + 4f(2.0) + f(2.5)]\frac{0.5}{3}$$

$$= 2.8381 + 1.2655 = 4.1036$$

EXAMPLE 6.4
Use the recursive trapezoidal rule to evaluate $\int_0^\pi \sqrt{x}\cos x\,dx$ to six decimal places. How many function evaluations are required to achieve this result?

Solution The program listed below utilizes the function `trapezoid`. Apart from the value of the integral, it displays the number of function evaluations used in the computation.

```
% Example 6.4 (Recursive trapezoidal rule)
format long    % Display extra precision
I2h = 0;
for k = 1:20
    Ih = trapezoid(@fex6_4,0,pi,I2h,k);
    if (k > 1 & abs(Ih - I2h) < 1.0e-6)
        Integral = Ih
        No_of_func_evaluations = 2^(k-1) + 1
        return
    end
    I2h = Ih;
end
error('Too many iterations')
```

The M-file containing the function to be integrated is

```
function y = fex6_4(x)
% Function used in Example 6.4
y = sqrt(x)*cos(x);
```

Here is the output:

```
>> Integral =
  -0.89483166485329
No_of_func_evaluations =
       32769
```

Rounding to six decimal places, we have $\int_0^\pi \sqrt{x}\cos x\,dx = -0.894\,832$

The number of function evaluations is unusually large in this problem. The slow convergence is the result of the derivatives of $f(x)$ being singular at $x = 0$. Consequently, the error does not behave as shown in Eq. (6.7): $E = c_1 h^2 + c_2 h^4 + \cdots$, but is unpredictable. Difficulties of this nature can often be remedied by a change in variable. In this case, we introduce $t = \sqrt{x}$, so that $dt = dx/(2\sqrt{x}) = dx/(2t)$, or $dx = 2t\,dt$. Thus

$$\int_0^\pi \sqrt{x}\cos x\,dx = \int_0^{\sqrt{\pi}} 2t^2 \cos t^2\,dt$$

Evaluation of the integral on the right-hand side would require 4097 function evaluations.

6.3   Romberg Integration

Romberg integration combines the composite trapezoidal rule with Richardson extrapolation (see Art. 5.3). Let us first introduce the notation

$$R_{i,1} = I_i$$

where, as before, I_i represents the approximate value of $\int_a^b f(x)dx$ computed by the recursive trapezoidal rule using 2^{i-1} panels. Recall that the error in this approximation is $E = c_1 h^2 + c_2 h^4 + \cdots$, where

$$h = \frac{b-a}{2^{i-1}}$$

is the width of a panel.

Romberg integration starts with the computation of $R_{1,1} = I_1$ (one panel) and $R_{2,1} = I_2$ (two panels) from the trapezoidal rule. The leading error term $c_1 h^2$ is then eliminated by Richardson extrapolation. Using $p = 2$ (the exponent in the error term) in Eq. (5.9) and denoting the result by $R_{2,2}$, we obtain

$$R_{2,2} = \frac{2^2 R_{2,1} - R_{1,1}}{2^2 - 1} = \frac{4}{3} R_{2,1} - \frac{1}{3} R_{1,1} \qquad \text{(a)}$$

It is convenient to store the results in an array of the form

$$\begin{bmatrix} R_{1,1} \\ R_{2,1} & R_{2,2} \end{bmatrix}$$

The next step is to calculate $R_{3,1} = I_3$ (four panels) and repeat Richardson extrapolation with $R_{2,1}$ and $R_{3,1}$, storing the result as $R_{3,2}$:

$$R_{3,2} = \frac{4}{3} R_{3,1} - \frac{1}{3} R_{2,1} \qquad \text{(b)}$$

The elements of array **R** calculated so far are

$$\begin{bmatrix} R_{1,1} \\ R_{2,1} & R_{2,2} \\ R_{3,1} & R_{3,2} \end{bmatrix}$$

Both elements of the second column have an error of the form $c_2 h^4$, which can also be eliminated with Richardson extrapolation. Using $p = 4$ in Eq. (5.9), we get

$$R_{3,3} = \frac{2^4 R_{3,2} - R_{2,2}}{2^4 - 1} = \frac{16}{15} R_{3,2} - \frac{1}{15} R_{2,2} \qquad \text{(c)}$$

This result has an error of $\mathcal{O}(h^6)$. The array has now expanded to

$$\begin{bmatrix} R_{1,1} \\ R_{2,1} & R_{2,2} \\ R_{3,1} & R_{3,2} & R_{3,3} \end{bmatrix}$$

After another round of calculations we get

$$\begin{bmatrix} R_{1,1} \\ R_{2,1} & R_{2,2} \\ R_{3,1} & R_{3,2} & R_{3,3} \\ R_{4,1} & R_{4,2} & R_{4,3} & R_{4,4} \end{bmatrix}$$

where the error in $R_{4,4}$ is $\mathcal{O}(h^8)$. Note that the most accurate estimate of the integral is always the last diagonal term of the array. This process is continued until the difference between two successive diagonal terms becomes sufficiently small. The general extrapolation formula used in this scheme is

$$R_{i,j} = \frac{4^{j-1} R_{i,j-1} - R_{i-1,j-1}}{4^{j-1} - 1}, \quad i > 1, \quad j = 2, 3, \ldots, i \qquad (6.13a)$$

A pictorial representation of Eq. (6.13a) is

$$\boxed{R_{i-1,j-1}}$$

$$\searrow$$
$$\alpha \qquad\qquad (6.13b)$$
$$\searrow$$

$$\boxed{R_{i,j-1}} \ \to \beta \to \ \boxed{R_{i,j}}$$

where the multipliers α and β depend on j in the following manner:

j	2	3	4	5	6
α	$-1/3$	$-1/15$	$-1/63$	$-1/255$	$-1/1023$
β	$4/3$	$16/15$	$64/63$	$256/255$	$1024/1023$

(6.13c)

The triangular array is convenient for hand computations, but computer implementation of the Romberg algorithm can be carried out within a one-dimensional array **r**. After the first extrapolation—see Eq. (a)—$R_{1,1}$ is never used again, so that it can be replaced with $R_{2,2}$. As a result, we have the array

$$\begin{bmatrix} r_1 = R_{2,2} \\ r_2 = R_{2,1} \end{bmatrix}$$

In the second extrapolation round, defined by Eqs. (b) and (c), $R_{3,2}$ overwrites $R_{2,1}$, and $R_{3,3}$ replaces $R_{2,2}$, so that the array now contains

$$\begin{bmatrix} r_1 = R_{3,3} \\ r_2 = R_{3,2} \\ r_3 = R_{3,1} \end{bmatrix}$$

and so on. In this manner, r_1 always contains the best current result. The extrapolation formula for the kth round is

$$r_j = \frac{4^{k-j} r_{j+1} - r_j}{4^{k-j} - 1}, \quad j = k-1, k-2, \ldots, 1 \qquad (6.14)$$

■ romberg

The algorithm for Romberg integration is implemented in the function `romberg`. It returns the value of the integral and the required number of function evaluations. Richardson's extrapolation is performed by the subfunction `richardson`.

```
function [I,numEval] = romberg(func,a,b,tol,kMax)
% Romberg integration.
% USAGE: [I,numEval] = romberg(func,a,b,tol,kMax)
% INPUT:
% func    = handle of function being integrated.
% a,b     = limits of integration.
% tol     = error tolerance (default is 1.0e-8).
% kMax    = limit on the number of panel doublings
%           (default is 20).
% OUTPUT:
% I       = value of the integral.
% numEval = number of function evaluations.

if nargin < 5; kMax = 20; end
if nargin < 4; tol = 1.0e-8; end
r = zeros(kMax);
r(1) = trapezoid(func,a,b,0,1);
rOld = r(1);
for k = 2:kMax
    r(k) = trapezoid(func,a,b,r(k-1),k);
    r = richardson(r,k);
    if abs(r(1) - rOld) < tol
        numEval = 2^(k-1) + 1; I = r(1);
        return
    end
    rOld = r(1);
end
error('Failed to converge')

function r = richardson(r,k)
% Richardson's extrapolation in Eq. (6.14).
for j = k-1:-1:1
    c = 4^(k-j); r(j) = (c*r(j+1) - r(j))/(c-1);
end
```

EXAMPLE 6.5
Show that $R_{k,2}$ in Romberg integration is identical to the composite Simpson's 1/3 rule in Eq. (6.10) with 2^{k-1} panels.

Solution Recall that in Romberg integration $R_{k,1} = I_k$ denoted the approximate integral obtained by the composite trapezoidal rule with 2^{k-1} panels. Denoting the abscissas of the nodes by $x_1, x_2, \ldots, x_n$, we have from the composite trapezoidal rule in Eq. (6.5)

$$R_{k,1} = I_k = \left[f(x_1) + 2 \sum_{i=2}^{n-1} f(x_i) + \frac{1}{2} f(x_n) \right] \frac{h}{2}$$

When we halve the number of panels (panel width $2h$), only the odd-numbered abscissas enter the composite trapezoidal rule, yielding

$$R_{k-1,1} = I_{k-1} = \left[f(x_1) + 2 \sum_{i=3,5,\ldots}^{n-2} f(x_i) + f(x_n) \right] h$$

Applying Richardson extrapolation yields

$$R_{k,2} = \frac{4}{3} R_{k,1} - \frac{1}{3} R_{k-1,1}$$

$$= \left[\frac{1}{3} f(x_1) + \frac{4}{3} \sum_{i=2,4,\ldots}^{n-1} f(x_i) + \frac{2}{3} \sum_{i=3,5,\ldots}^{n-2} f(x_i) + \frac{1}{3} f(x_n) \right] h$$

which agrees with Simpson's rule in Eq. (6.10).

EXAMPLE 6.6
Use Romberg integration to evaluate $\int_0^\pi f(x)\,dx$, where $f(x) = \sin x$. Work with four decimal places.

Solution From the recursive trapezoidal rule in Eq. (6.9b) we get

$$R_{1,1} = I(\pi) = \frac{\pi}{2} [f(0) + f(\pi)] = 0$$

$$R_{2,1} = I(\pi/2) = \frac{1}{2} I(\pi) + \frac{\pi}{2} f(\pi/2) = 1.5708$$

$$R_{3,1} = I(\pi/4) = \frac{1}{2} I(\pi/2) + \frac{\pi}{4} [f(\pi/4) + f(3\pi/4)] = 1.8961$$

$$R_{4,1} = I(\pi/8) = \frac{1}{2} I(\pi/4) + \frac{\pi}{8} [f(\pi/8) + f(3\pi/8) + f(5\pi/8) + f(7\pi/8)]$$
$$= 1.9742$$

Using the extrapolation formulas in Eqs. (6.13), we can now construct the following table:

$$
\begin{bmatrix}
R_{1,1} & & & \\
R_{2,1} & R_{2,2} & & \\
R_{3,1} & R_{3,2} & R_{3,3} & \\
R_{4,1} & R_{4,2} & R_{4,3} & R_{4,4}
\end{bmatrix}
=
\begin{bmatrix}
0 & & & \\
1.5708 & 2.0944 & & \\
1.8961 & 2.0046 & 1.9986 & \\
1.9742 & 2.0003 & 2.0000 & 2.0000
\end{bmatrix}
$$

It appears that the procedure has converged. Therefore, $\int_0^\pi \sin x\, dx = R_{4,4} = 2.0000$, which is, of course, the correct result.

EXAMPLE 6.7

Use Romberg integration to evaluate $\int_0^{\sqrt{\pi}} 2x^2 \cos x^2\, dx$ and compare the results with Example 6.4.

Solution

```
>> format long
>> [Integral,numEval] = romberg(@fex6_7,0,sqrt(pi))
Integral =
  -0.89483146948416
numEval =
   257
>>
```

Here the M-file defining the function to be integrated is

```
function y = fex6_7(x)
% Function used in Example 6.7
y = 2*(x^2)*cos(x^2);
```

It is clear that Romberg integration is considerably more efficient than the trapezoidal rule. It required 257 function evaluations as compared to 4097 evaluations with the composite trapezoidal rule in Example 6.4.

PROBLEM SET 6.1

1. Use the recursive trapezoidal rule to evaluate $\int_0^{\pi/4} \ln(1 + \tan x)\,dx$. Explain the results.

2. The table shows the power P supplied to the driving wheels of a car as a function of the speed v. If the mass of the car is $m = 2000$ kg, determine the time Δt it takes for the car to accelerate from 1 m/s to 6 m/s. Use the trapezoidal rule for

integration. *Hint*:

$$\Delta t = m \int_{1s}^{6s} (v/P)\, dv$$

which can be derived from Newton's law $F = m(dv/dt)$ and the definition of power $P = Fv$.

v (m/s)	0	1.0	1.8	2.4	3.5	4.4	5.1	6.0
P (kW)	0	4.7	12.2	19.0	31.8	40.1	43.8	43.2

3. Evaluate $\int_{-1}^{1} \cos(2\cos^{-1} x)\,dx$ with Simpson's 1/3 rule using 2, 4 and 6 panels. Explain the results.

4. Determine $\int_{1}^{\infty} (1 + x^4)^{-1}\,dx$ with the trapezoidal rule using five panels and compare the result with the "exact" integral 0.243 75. *Hint*: use the transformation $x^3 = 1/t$.

5.

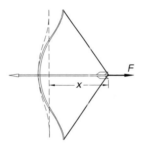

The table below gives the pull F of the bow as a function of the draw x. If the bow is drawn 0.5 m, determine the speed of the 0.075-kg arrow when it leaves the bow. *Hint:* the kinetic energy of arrow equals the work done in drawing the bow; that is, $mv^2/2 = \int_0^{0.5\,m} F\,dx$.

x (m)	0.00	0.05	0.10	0.15	0.20	0.25
F (N)	0	37	71	104	134	161
x (m)	0.30	0.35	0.40	0.45	0.50	
F (N)	185	207	225	239	250	

6. Evaluate $\int_0^2 (x^5 + 3x^3 - 2)\,dx$ by Romberg integration.

7. Estimate $\int_0^\pi f(x)\,dx$ as accurately as possible, where $f(x)$ is defined by the data

x	0	$\pi/4$	$\pi/2$	$3\pi/4$	π
$f(x)$	1.0000	0.3431	0.2500	0.3431	1.0000

8. Evaluate

$$\int_0^1 \frac{\sin x}{\sqrt{x}}\, dx$$

with Romberg integration. *Hint*: use transformation of variable to eliminate the indeterminacy at $x = 0$.

9. Show that if $y = f(x)$ is approximated by a natural cubic spline with evenly spaced knots at $x_1, x_2, \ldots, x_n$, the quadrature formula becomes

$$I = \frac{h}{2}\,(y_1 + 2y_2 + 2y_3 + \cdots + 2y_{n-1} + y_n)$$
$$-\frac{h^3}{24}\,(k_1 + 2k_2 + k_3 + \cdots + 2k_{n-1} + k_n)$$

where h is the spacing of the knots and $k = y''$. Note that the first part is the composite trapezoidal rule; the second part may be viewed as a "correction" for curvature.

10. ■ Use a computer program to evaluate

$$\int_0^{\pi/4} \frac{dx}{\sqrt{\sin x}}$$

with Romberg integration. *Hint*: use the transformation $\sin x = t^2$.

11. ■ The period of a simple pendulum of length L is $\tau = 4\sqrt{L/g}\; h(\theta_0)$, where g is the gravitational acceleration, θ_0 represents the angular amplitude and

$$h(\theta_0) = \int_0^{\pi/2} \frac{d\theta}{\sqrt{1 - \sin^2(\theta_0/2)\sin^2\theta}}$$

Compute $h(15°)$, $h(30°)$ and $h(45°)$, and compare these values with $h(0) = \pi/2$ (the approximation used for small amplitudes).

12. ■

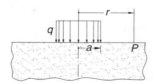

The figure shows an elastic half-space that carries uniform loading of intensity q over a circular area of radius a. The vertical displacement of the surface at point P can be shown to be

$$w(r) = w_0 \int_0^{\pi/2} \frac{\cos^2\theta}{\sqrt{(r/a)^2 - \sin^2\theta}}\, d\theta \qquad r \geq a$$

where w_0 is the displacement at $r = a$. Use numerical integration to determine w/w_0 at $r = 2a$.

13. ■

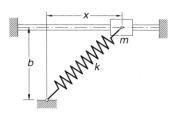

The mass m is attached to a spring of free length b and stiffness k. The coefficient of friction between the mass and the horizontal rod is μ. The acceleration of the mass can be shown to be (you may wish to prove this) $\ddot{x} = -f(x)$, where

$$f(x) = \mu g + \frac{k}{m}(\mu b + x)\left(1 - \frac{b}{\sqrt{b^2 + x^2}}\right)$$

If the mass is released from rest at $x = b$, its speed at $x = 0$ is given by

$$v_0 = \sqrt{2 \int_0^b f(x)\,dx}$$

Compute v_0 by numerical integration using the data $m = 0.8$ kg, $b = 0.4$ m, $\mu = 0.3$, $k = 80$ N/m and $g = 9.81$ m/s^2.

14. ■ Debye's formula for the heat capacity C_V of a solid is $C_V = 9Nkg(u)$, where

$$g(u) = u^3 \int_0^{1/u} \frac{x^4 e^x}{(e^x - 1)^2}\,dx$$

The terms in this equation are

$$N = \text{number of particles in the solid}$$
$$k = \text{Boltzmann constant}$$
$$u = T/\Theta_D$$
$$T = \text{absolute temperature}$$
$$\Theta_D = \text{Debye temperature}$$

Compute $g(u)$ from $u = 0$ to 1.0 in intervals of 0.05 and plot the results.

15. ■ A power spike in an electric circuit results in the current

$$i(t) = i_0 e^{-t/t_0} \sin(2t/t_0)$$

across a resistor. The energy E dissipated by the resistor is

$$E = \int_0^\infty R\,[i(t)]^2\,dt$$

Find E using the data $i_0 = 100\,\text{A}$, $R = 0.5\,\Omega$ and $t_0 = 0.01\,\text{s}$.

6.4   Gaussian Integration

Gaussian Integration Formulas

We found that Newton–Cotes formulas for approximating $\int_a^b f(x)dx$ work best if $f(x)$ is a smooth function, such as a polynomial. This is also true for Gaussian quadrature. However, Gaussian formulas are also good at estimating integrals of the form

$$\int_a^b w(x)\,f(x)dx \tag{6.15}$$

where $w(x)$, called the *weighting function*, can contain singularities, as long as they are integrable. An example of such an integral is $\int_0^1 (1 + x^2)\ln x\,dx$. Sometimes infinite limits, as in $\int_0^\infty e^{-x} \sin x\,dx$, can also be accommodated.

Gaussian integration formulas have the same form as Newton–Cotes rules:

$$I = \sum_{i=1}^n A_i\,f(x_i) \tag{6.16}$$

where, as before, I represents the approximation to the integral in Eq. (6.15). The difference lies in the way that the weights A_i and nodal abscissas x_i are determined. In Newton–Cotes integration the nodes were evenly spaced in (a, b), i.e., their locations were predetermined. In Gaussian quadrature the nodes and weights are chosen so that Eq. (6.16) yields the exact integral if $f(x)$ is a polynomial of degree $2n - 1$ or less; that is,

$$\int_a^b w(x)\,P_m(x)dx = \sum_{i=1}^n A_i\,P_m(x_i), \quad m \le 2n - 1 \tag{6.17}$$

One way of determining the weights and abscissas is to substitute $P_1(x) = 1$, $P_2(x) = x, \dots, P_{2n-1}(x) = x^{2n-1}$ in Eq. (6.17) and solve the resulting $2n$ equations

$$\int_a^b w(x)x^j dx = \sum_{i=1}^n A_i x_i^j, \quad j = 0, 1, \dots, 2n - 1$$

for the unknowns A_i and x_i, $i = 1, 2, \dots, n$.

As an illustration, let $w(x) = e^{-x}$, $a = 0$, $b = \infty$ and $n = 2$. The four equations determining x_1, x_2, A_1 and A_2 are

$$\int_0^\infty e^{-x}dx = A_1 + A_2$$

$$\int_0^\infty e^{-x}x\,dx = A_1x_1 + A_2x_2$$

$$\int_0^\infty e^{-x}x^2dx = A_1x_1^2 + A_2x_2^2$$

$$\int_0^\infty e^{-x}x^3dx = A_1x_1^3 + A_2x_2^3$$

After evaluating the integrals, we get

$$A_1 + A_2 = 1$$
$$A_1x_1 + A_2x_2 = 1$$
$$A_1x_1^2 + A_2x_2^2 = 2$$
$$A_1x_1^3 + A_2x_2^3 = 6$$

The solution is

$$x_1 = 2 - \sqrt{2} \qquad A_1 = \frac{\sqrt{2}+1}{2\sqrt{2}}$$

$$x_2 = 2 + \sqrt{2} \qquad A_2 = \frac{\sqrt{2}-1}{2\sqrt{2}}$$

so that the quadrature formula becomes

$$\int_0^\infty e^{-x}f(x)dx \approx \frac{1}{2\sqrt{2}}\left[(\sqrt{2}+1)\,f\left(2-\sqrt{2}\right) + (\sqrt{2}-1)\,f\left(2+\sqrt{2}\right)\right]$$

Due to the nonlinearity of the equations, this approach will not work well for large n. Practical methods of finding x_i and A_i require some knowledge of orthogonal polynomials and their relationship to Gaussian quadrature. There are, however, several "classical" Gaussian integration formulas for which the abscissas and weights have been computed with great precision and tabulated. These formulas can used without knowing the theory behind them, since all one needs for Gaussian integration are the values of x_i and A_i. If you do not intend to venture outside the classical formulas, you can skip the next two topics.

*Orthogonal Polynomials

Orthogonal polynomials are employed in many areas of mathematics and numerical analysis. They have been studied thoroughly and many of their properties are known. What follows is a very small compendium of a large topic.

The polynomials $\varphi_n(x)$, $n = 0, 1, 2, \ldots$ (n is the degree of the polynomial) are said to form an *orthogonal set* in the interval (a, b) with respect to the *weighting function* $w(x)$ if

$$\int_a^b w(x)\varphi_m(x)\varphi_n(x)dx = 0, \quad m \neq n \tag{6.18}$$

The set is determined, except for a constant factor, by the choice of the weighting function and the limits of integration. That is, each set of orthogonal polynomials is associated with certain $w(x)$, a and b. The constant factor is specified by standardization. Some of the classical orthogonal polynomials, named after well-known mathematicians, are listed in Table 6.1. The last column in the table shows the standardization used.

Name	Symbol	a	b	$w(x)$	$\int_a^b w(x)\left[\varphi_n(x)\right]^2 dx$
Legendre	$p_n(x)$	-1	1	1	$2/(2n+1)$
Chebyshev	$T_n(x)$	-1	1	$(1-x^2)^{-1/2}$	$\pi/2 \quad (n > 0)$
Laguerre	$L_n(x)$	0	∞	e^{-x}	1
Hermite	$H_n(x)$	$-\infty$	∞	e^{-x^2}	$\sqrt{\pi}2^n n!$

Table 6.1

Orthogonal polynomials obey recurrence relations of the form

$$a_n\varphi_{n+1}(x) = (b_n + c_n x)\varphi_n(x) - d_n\varphi_{n-1}(x) \tag{6.19}$$

If the first two polynomials of the set are known, the other members of the set can be computed from Eq. (6.19). The coefficients in the recurrence formula, together with $\varphi_0(x)$ and $\varphi_1(x)$, are given in Table 6.2.

Name	$\varphi_0(x)$	$\varphi_1(x)$	a_n	b_n	c_n	d_n
Legendre	1	x	$n+1$	0	$2n+1$	n
Chebyshev	1	x	1	0	2	1
Laguerre	1	$1-x$	$n+1$	$2n+1$	-1	n
Hermite	1	$2x$	1	0	2	2

Table 6.2

The classical orthogonal polynomials are also obtainable from the formulas

$$p_n(x) = \frac{(-1)^n}{2^n n!} \frac{d^n}{dx^n}\left[(1 - x^2)^n\right]$$

$$T_n(x) = \cos(n\cos^{-1} x), \quad n > 0$$

$$L_n(x) = \frac{e^x}{n!} \frac{d^n}{dx^n}(x^n e^{-x}) \tag{6.20}$$

$$H_n(x) = (-1)^n e^{x^2} \frac{d^n}{dx^n}(e^{-x^2})$$

and their derivatives can be calculated from

$$(1 - x^2)p_n'(x) = n[-xp_n(x) + p_{n-1}(x)]$$

$$(1 - x^2)T_n'(x) = n[-xT_n(x) + nT_{n-1}(x)]$$

$$xL_n'(x) = n[L_n(x) - L_{n-1}(x)] \tag{6.21}$$

$$H_n'(x) = 2nH_{n-1}(x)$$

Other properties of orthogonal polynomials that have relevance to Gaussian integration are:

- $\varphi_n(x)$ has n real, distinct zeroes in the interval (a, b).
- The zeroes of $\varphi_n(x)$ lie between the zeroes of $\varphi_{n+1}(x)$.
- Any polynomial $P_n(x)$ of degree n can be expressed in the form

$$P_n(x) = \sum_{i=0}^{n} c_i\varphi_i(x) \tag{6.22}$$

- It follows from Eq. (6.22) and the orthogonality property in Eq. (6.18) that

$$\int_a^b w(x)P_n(x)\varphi_{n+m}(x)dx = 0, \quad m \geq 0 \tag{6.23}$$

*Determination of Nodal Abscissas and Weights

Theorem The nodal abscissas $x_1, x_2, \ldots, x_n$ are the zeros of the polynomial $\varphi_n(x)$ that belongs to the orthogonal set defined in Eq. (6.18).

Proof We start the proof by letting $f(x) = P_{2n-1}(x)$ be a polynomial of degree $2n - 1$. Since the Gaussian integration with n nodes is exact for this polynomial, we have

$$\int_a^b w(x)P_{2n-1}(x)dx = \sum_{i=1}^{n} A_i P_{2n-1}(x_i) \tag{a}$$

A polynomial of degree $2n - 1$ can always written in the form

$$P_{2n-1}(x) = Q_{n-1}(x) + R_{n-1}(x)\varphi_n(x) \tag{b}$$

where $Q_{n-1}(x)$, $R_{n-1}(x)$ and $\varphi_n(x)$ are polynomials of the degree indicated by the subscripts.[11] Therefore,

$$\int_a^b w(x) P_{2n-1}(x)dx = \int_a^b w(x) Q_{n-1}(x)dx + \int_a^b w(x) R_{n-1}(x)\varphi_n(x)dx$$

But according to Eq. (6.23) the second integral on the right hand-side vanishes, so that

$$\int_a^b w(x) P_{2n-1}(x)dx = \int_a^b w(x) Q_{n-1}(x)dx \qquad (c)$$

Because a polynomial of degree $n-1$ is uniquely defined by n points, it is always possible to find A_i such that

$$\int_a^b w(x) Q_{n-1}(x)dx = \sum_{i=1}^n A_i Q_{n-1}(x_i) \qquad (d)$$

In order to arrive at Eq. (a), we must choose for the nodal abscissas x_i the roots of $\varphi_n(x) = 0$. According to Eq. (b) we then have

$$P_{2n-1}(x_i) = Q_{n-1}(x_i), \quad i = 1, 2, \dots, n \qquad (e)$$

which together with Eqs. (c) and (d) leads to

$$\int_a^b w(x) P_{2n-1}(x)dx = \int_a^b w(x) Q_{n-1}(x)dx = \sum_{i=1}^n A_i P_{2n-1}(x_i)$$

This completes the proof.

Theorem

$$A_i = \int_a^b w(x)\ell_i(x)dx, \quad i = 1, 2, \dots, n \qquad (6.24)$$

where $\ell_i(x)$ are the Lagrange's cardinal functions spanning the nodes at $x_1, x_2, \dots x_n$. These functions were defined in Eq. (3.2).

Proof Applying Lagrange's formula, Eq. (3.1a), to $Q_{n-1}(x)$ yields

$$Q_{n-1}(x) = \sum_{i=1}^n Q_{n-1}(x_i)\ell_i(x)$$

which upon substitution in Eq. (d) gives us

$$\sum_{i=1}^n \left[Q_{n-1}(x_i) \int_a^b w(x)\ell_i(x)dx \right] = \sum_{i=1}^n A_i Q_{n-1}(x_i)$$

or

$$\sum_{i=1}^n Q_{n-1}(x_i) \left[A_i - \int_a^b w(x)\ell_i(x)dx \right] = 0$$

---

[11] It can be shown that $Q_{n-1}(x)$ and $R_{n-1}(x)$ are unique for given $P_{2n-1}(x)$ and $\varphi_n(x)$.

This equation can be satisfied for arbitrary Q_{n-1} only if

$$A_i - \int_a^b w(x)\ell_i(x)dx = 0, \quad i = 1, 2, \ldots, n$$

which is equivalent to Eq. (6.24).

It is not difficult to compute the zeros x_i, $i = 1, 2, \ldots, n$ of a polynomial $\varphi_n(x)$ belonging to an orthogonal set by one of the methods discussed in Chapter 4. Once the zeros are known, the weights A_i, $i = 1, 2, \ldots, n$ could be found from Eq. (6.24). However the following formulas (given without proof) are easier to compute

$$\text{Gauss–Legendre} \quad A_i = \frac{2}{(1 - x_i^2)\left[p_n'(x_i)\right]^2}$$

$$\text{Gauss–Laguerre} \quad A_i = \frac{1}{x_i\left[L_n'(x_i)\right]^2} \qquad (6.25)$$

$$\text{Gauss–Hermite} \quad A_i = \frac{2^{n+1}n!\sqrt{\pi}}{\left[H_n'(x_i)\right]^2}$$

Abscissas and Weights for Gaussian Quadratures

We list here some classical Gaussian integration formulas. The tables of nodal abscissas and weights, covering $n = 2$ to 6, have been rounded off to six decimal places. These tables should be adequate for hand computation, but in programming you may need more precision or a larger number of nodes. In that case you should consult other references,[12] or use a subroutine to compute the abscissas and weights within the integration program.[13]

The truncation error in Gaussian quadrature

$$E = \int_a^b w(x)f(x)dx - \sum_{i=1}^n A_i f(x_i)$$

has the form $E = K(n)f^{(2n)}(c)$, where $a < c < b$ (the value of c is unknown; only its bounds are given). The expression for $K(n)$ depends on the particular quadrature being used. If the derivatives of $f(x)$ can be evaluated, the error formulas are useful is estimating the error bounds.

---

[12] *Handbook of Mathematical Functions*, M. Abramowitz and I.A. Stegun, Dover Publications (1965); A.H. Stroud and D. Secrest, *Gaussian Quadrature Formulas*, Prentice-Hall (1966).
[13] Several such subroutines are listed in *Numerical Recipes in Fortran 90*, W.H. Press et al., Cambridge University Press (1996).

Gauss–Legendre quadrature

$$\int_{-1}^{1} f(\xi)d\xi \approx \sum_{i=1}^{n} A_i f(\xi_i) \tag{6.26}$$

$\pm\xi_i$	A_i	$\pm\xi_i$	A_i
	$n = 2$		$n = 5$
0.577 350	1.000 000	0.000 000	0.568 889
	$n = 3$	0.538 469	0.478 629
0.000 000	0.888 889	0.906 180	0.236 927
0.774 597	0.555 556		$n = 6$
	$n = 4$	0.238 619	0.467 914
0.339 981	0.652 145	0.661 209	0.360 762
0.861 136	0.347 855	0.932 470	0.171 324

Table 6.3

This is the most often used Gaussian integration formula. The nodes are arranged symmetrically about $\xi = 0$, and the weights associated with a symmetric pair of nodes are equal. For example, for $n = 2$ we have $\xi_1 = -\xi_2$ and $A_1 = A_2$. The truncation error in Eq. (6.26) is

$$E = \frac{2^{2n+1}(n!)^4}{(2n+1)[(2n)!]^3} f^{(2n)}(c), \quad -1 < c < 1 \tag{6.27}$$

To apply Gauss–Legendre quadrature to the integral $\int_a^b f(x)dx$, we must first map the integration range (a, b) into the "standard" range $(-1, 1)$. We can accomplish this by the transformation

$$x = \frac{b+a}{2} + \frac{b-a}{2}\xi \tag{6.28}$$

Now $dx = d\xi (b-a)/2$, and the quadrature becomes

$$\int_a^b f(x)dx \approx \frac{b-a}{2}\sum_{i=1}^{n} A_i f(x_i) \tag{6.29}$$

where the abscissas x_i must be computed from Eq. (6.28). The truncation error here is

$$E = \frac{(b-a)^{2n+1}(n!)^4}{(2n+1)[(2n)!]^3} f^{(2n)}(c), \quad a < c < b \tag{6.30}$$

Gauss–Chebyshev quadrature

$$\int_{-1}^{1} \left(1 - x^2\right)^{-1/2} f(x)\,dx \approx \frac{\pi}{n} \sum_{i=1}^{n} f(x_i) \tag{6.31}$$

Note that all the weights are equal: $A_i = \pi/n$. The abscissas of the nodes, which are symmetric about $x = 0$, are given by

$$x_i = \cos \frac{(2i - 1)\pi}{2n} \tag{6.32}$$

The truncation error is

$$E = \frac{2\pi}{2^{2n}(2n)!} f^{(2n)}(c), \quad -1 < c < 1 \tag{6.33}$$

Gauss–Laguerre quadrature

$$\int_{0}^{\infty} e^{-x} f(x)\,dx \approx \sum_{i=1}^{n} A_i f(x_i) \tag{6.34}$$

x_i	A_i	x_i	A_i
n = 2		**n = 5**	
0.585 786	0.853 554	0.263 560	0.521 756
3.414 214	0.146 447	1.413 403	0.398 667
n = 3		3.596 426	(−1)0.759 424
0.415 775	0.711 093	7.085 810	(−2)0.361 175
2.294 280	0.278 517	12.640 801	(−4)0.233 670
6.289 945	(−1)0.103 892	**n = 6**	
n = 4		0.222 847	0.458 964
0.322 548	0.603 154	1.188 932	0.417 000
1.745 761	0.357 418	2.992 736	0.113 373
4.536 620	(−1)0.388 791	5.775 144	(−1)0.103 992
9.395 071	(−3)0.539 295	9.837 467	(−3)0.261 017
		15.982 874	(−6)0.898 548

Table 6.4. Multiply numbers by 10^k, where k is given in parentheses

$$E = \frac{(n!)^2}{(2n)!} f^{(2n)}(c), \quad 0 < c < \infty \tag{6.35}$$

Gauss–Hermite quadrature:

$$\int_{-\infty}^{\infty} e^{-x^2} f(x)\,dx \approx \sum_{i=1}^{n} A_i f(x_i) \tag{6.36}$$

The nodes are placed symmetrically about $x = 0$, each symmetric pair having the same weight.

$\pm x_i$	A_i	$\pm x_i$	A_i
	$n = 2$		$n = 5$
0.707 107	0.886 227	0.000 000	0.945 308
	$n = 3$	0.958 572	0.393 619
0.000 000	1.181 636	2.020 183	$(-1)\,0.199\,532$
1.224745	0.295 409		$n = 6$
	$n = 4$	0.436 077	0.724 629
0.524 648	0.804 914	1.335 849	0.157 067
1.650 680	$(-1)0.813\,128$	2.350 605	$(-2)0.453\,001$

Table 6.5. Multiply numbers by 10^k, where k is given in parentheses

$$E = \frac{\sqrt{\pi}\,n!}{2^2(2n)!}\,f^{(2n)}(c), \quad 0 < c < \infty \tag{6.37}$$

Gauss quadrature with logarithmic singularity

$$\int_0^1 f(x)\ln(x)\,dx \approx -\sum_{i=1}^{n} A_i\,f(x_i) \tag{6.38}$$

x_i	A_i	x_i	A_i
	$n = 2$		$n = 5$
0.112 009	0.718 539	$(-1)0.291\,345$	0.297 893
0.602 277	0.281 461	0.173 977	0.349 776
	$n = 3$	0.411 703	0.234 488
$(-1)0.638\,907$	0.513 405	0.677314	$(-1)0.989\,305$
0.368 997	0.391 980	0.894 771	$(-1)0.189\,116$
0.766 880	$(-1)0.946\,154$		$n = 6$
	$n = 4$	$(-1)0.216\,344$	0.238 764
$(-1)0.414\,485$	0.383 464	0.129 583	0.308 287
0.245 275	0.386 875	0.314 020	0.245 317
0.556 165	0.190 435	0.538 657	0.142 009
0.848 982	$(-1)0.392\,255$	0.756 916	$(-1)0.554\,546$
		0.922 669	$(-1)0.101\,690$

Table 6.6. Multiply numbers by 10^k, where k is given in parentheses

$$E = \frac{k(n)}{(2n)!} f^{(2n)}(c), \quad 0 < c < 1 \tag{6.39}$$

where $k(2) = 0.00\,285$, $k(3) = 0.000\,17$, $k(4) = 0.000\,01$.

■ gaussNodes

The function gaussNodes computes the nodal abscissas x_i and the corresponding weights A_i used in Gauss–Legendre quadrature.[14] It can be shown that the approximate values of the abscissas are

$$x_i = \cos \frac{\pi (i - 0.25)}{n + 0.5}$$

Using these approximations as the starting values, we compute the nodal abscissas by finding the nonnegative zeros of the Legendre polynomial $p_n(x)$ with the Newton–Raphson method (the negative zeros are obtained from symmetry). Note that gaussNodes calls the subfunction legendre, which returns $p_n(t)$ and its derivative.

```
function [x,A] = gaussNodes(n,tol)
% Computes nodal abscissas x and weights A of
% Gauss-Legendre n-point quadrature.
% USAGE: [x,A] = gaussNodes(n,epsilon,maxIter)
% tol = error tolerance (default is 1.0e4*eps).

if nargin < 2; tol = 1.0e4*eps; end
A = zeros(n,1); x = zeros(n,1);
nRoots = fix(n + 1)/2;                  % Number of non-neg. roots
for i = 1:nRoots
    t = cos(pi*(i - 0.25)/(n + 0.5)); % Approx. roots
    for j = i:30
        [p,dp] = legendre(t,n);        % Newton's
        dt = -p/dp; t = t + dt;        % root finding
        if abs(dt) < tol               % method
            x(i) = t; x(n-i+1) = -t;
            A(i) = 2/(1-t^2)/dp^2;      % Eq. (6.25)
            A(n-i+1) = A(i);
            break
```

---

[14] This function is an adaptation of a routine in *Numerical Recipes in Fortran 90*, W.H. Press et al., Cambridge University Press (1996).

```
            end
        end
end


function [p,dp] = legendre(t,n)
% Evaluates Legendre polynomial p of degree n
% and its derivative dp at x = t.
p0 = 1.0; p1 = t;
for k = 1:n-1
    p = ((2*k + 1)*t*p1 - k*p0)/(k + 1); % Eq. (6.19)
    p0 = p1;p1 = p;
end
dp = n *(p0 - t*p1)/(1 - t^2);            % Eq. (6.21)
```

■ gaussQuad

The function `gaussQuad` evaluates $\int_a^b f(x)\,dx$ with Gauss–Legendre quadrature using n nodes. The function defining $f(x)$ must be supplied by the user. The nodal abscissas and the weights are obtained by calling `gaussNodes`.

```
function I = gaussQuad(func,a,b,n)
% Gauss-Legendre quadrature.
% USAGE: I = gaussQuad(func,a,b,n)
% INPUT:
% func = handle of function to be integrated.
% a,b  = integration limits.
% n    = order of integration.
% OUTPUT:
% I = integral

c1 = (b + a)/2; c2 = (b - a)/2;   % Mapping constants
[x,A] = gaussNodes(n);            % Nodal abscissas & weights
sum = 0;
for i = 1:length(x)
    y = feval(func,c1 + c2*x(i)); % Function at node i
    sum = sum + A(i)*y;
end
I = c2*sum;
```

EXAMPLE 6.8
Evaluate $\int_{-1}^{1}(1 - x^2)^{3/2}dx$ as accurately as possible with Gaussian integration.

Solution As the integrand is smooth and free of singularities, we could use Gauss–Legendre quadrature. However, the exact integral can obtained with the Gauss–Chebyshev formula. We write

$$\int_{-1}^{1} \left(1 - x^2\right)^{3/2} dx = \int_{-1}^{1} \frac{\left(1 - x^2\right)^2}{\sqrt{1 - x^2}} dx$$

The numerator $f(x) = (1 - x^2)^2$ is a polynomial of degree four, so that Gauss–Chebyshev quadrature is exact with three nodes.

The abscissas of the nodes are obtained from Eq. (6.32). Substituting $n = 3$, we get

$$x_i = \cos \frac{(2i - 1)\pi}{2(3)}, \quad i = 1, 2, 3$$

Therefore,

$$x_1 = \cos \frac{\pi}{6} = \frac{\sqrt{3}}{2}$$

$$x_2 = \cos \frac{\pi}{2} = 0$$

$$x_2 = \cos \frac{5\pi}{6} = \frac{\sqrt{3}}{2}$$

and Eq. (6.31) yields

$$\int_{-1}^{1} \left(1 - x^2\right)^{3/2} dx = \frac{\pi}{3} \sum_{i=1}^{3} \left(1 - x_i^2\right)^2$$

$$= \frac{\pi}{3} \left[\left(1 - \frac{3}{4}\right)^2 + (1 - 0)^2 + \left(1 - \frac{3}{4}\right)^2 \right] = \frac{3\pi}{8}$$

EXAMPLE 6.9

Use Gaussian integration to evaluate $\int_{0}^{0.5} \cos \pi x \ln x \, dx$.

Solution We split the integral into two parts:

$$\int_{0}^{0.5} \cos \pi x \ln x \, dx = \int_{0}^{1} \cos \pi x \ln x \, dx - \int_{0.5}^{1} \cos \pi x \ln x \, dx$$

The first integral on the right-hand side, which contains a logarithmic singularity at $x = 0$, can be computed with the special Gaussian quadrature in Eq. (6.38). Choosing $n = 4$, we have

$$\int_{0}^{1} \cos \pi x \ln x \, dx \approx - \sum_{i=1}^{4} A_i \cos \pi x_i$$

where x_i and A_i are given in Table 6.6. The sum is evaluated in the following table:

x_i	$\cos \pi x_i$	A_i	$A_i \cos \pi x_i$
0.041 448	0.991 534	0.383 464	0.380 218
0.245 275	0.717 525	0.386 875	0.277 592
0.556 165	−0.175 533	0.190 435	−0.033 428
0.848 982	−0.889 550	0.039 225	−0.034 892
			$\Sigma = 0.589\,490$

Thus

$$\int_0^1 \cos \pi x \ln x \, dx \approx -0.589\,490$$

The second integral is free of singularities, so that it can be evaluated with Gauss–Legendre quadrature. Choosing again $n = 4$, we have

$$\int_{0.5}^1 \cos \pi x \ln x \, dx \approx 0.25 \sum_{i=1}^4 A_i \cos \pi x_i \ln x_i$$

where the nodal abscissas are (see Eq. (6.28))

$$x_i = \frac{1 + 0.5}{2} + \frac{1 - 0.5}{2}\xi_i = 0.75 + 0.25\xi_i$$

Looking up ξ_i and A_i in Table 6.3 leads to the following computations:

ξ_i	x_i	$\cos \pi x_i \ln x_i$	A_i	$A_i \cos \pi x_i \ln x_i$
−0.861 136	0.534 716	0.068 141	0.347 855	0.023 703
−0.339 981	0.665 005	0.202 133	0.652 145	0.131 820
0.339 981	0.834 995	0.156 638	0.652 145	0.102 151
0.861 136	0.965 284	0.035 123	0.347 855	0.012 218
				$\Sigma = 0.269\,892$

from which

$$\int_{0.5}^1 \cos \pi x \ln x \, dx \approx 0.25(0.269\,892) = 0.067\,473$$

Therefore,

$$\int_0^1 \cos \pi x \ln x \, dx \approx -0.589\,490 - 0.067\,473 = -0.656\,963$$

which is correct to six decimal places.

EXAMPLE 6.10
Evaluate as accurately as possible

$$F = \int_0^\infty \frac{x+3}{\sqrt{x}} e^{-x} dx$$

Solution In its present form, the integral is not suited to any of the Gaussian quadratures listed in this article. But using the transformation

$$x = t^2 \qquad dx = 2t \, dt$$

we have

$$F = 2\int_0^\infty (t^2 + 3)e^{-t^2} dt = \int_{-\infty}^\infty (t^2 + 3)e^{-t^2} dt$$

which can be evaluated exactly with Gauss–Hermite formula using only two nodes ($n = 2$). Thus

$$F = A_1(t_1^2 + 3) + A_2(t_2^2 + 3)$$
$$= 0.886\,227\left[(0.707\,107)^2 + 3\right] + 0.886\,227\left[(-0.707\,107)^2 + 3\right]$$
$$= 6.203\,59$$

EXAMPLE 6.11
Determine how many nodes are required to evaluate

$$\int_0^\pi \left(\frac{\sin x}{x}\right)^2 dx$$

with Gauss–Legendre quadrature to six decimal places. The exact integral, rounded to six places, is 1.418 15.

Solution The integrand is a smooth function; hence it is suited for Gauss–Legendre integration. There is an indeterminacy at $x = 0$, but this does not bother the quadrature since the integrand is never evaluated at that point. We used the following program that computes the quadrature with 2, 3, ... nodes until the desired accuracy is reached:

```
% Example 6.11 (Gauss-Legendre quadrature)
a = 0; b = pi; Iexact = 1.41815;
for n = 2:12
    I = gaussQuad(@fex6_11,a,b,n);
    if abs(I - Iexact) < 0.00001
        I
        n
        break
    end
end
```

The M-file of the function integrated is

```
function y = fex6_11(x)
% Function used in Example 6.11
y = (sin(x)/x)^2;
```

The program produced the following output:

```
I =

    1.41815026780139

n =

    5
```

EXAMPLE 6.12

Evaluate numerically $\int_{1.5}^{3} f(x)\,dx$, where $f(x)$ is represented by the unevenly spaced data

x	1.2	1.7	2.0	2.4	2.9	3.3
$f(x)$	−0.362 36	0.128 84	0.416 15	0.737 39	0.970 96	0.987 48

Knowing that the data points lie on the curve $f(x) = -\cos x$, evaluate the accuracy of the solution.

Solution We approximate $f(x)$ by the polynomial $P_5(x)$ that intersects all the data points, and then evaluate $\int_{1.5}^{3} f(x)\,dx \approx \int_{1.5}^{3} P_5(x)\,dx$ with the Gauss–Legendre formula. Since the polynomial is of degree five, only three nodes ($n = 3$) are required in the quadrature.

From Eq. (6.28) and Table 6.3, we obtain for the abscissas of the nodes

$$x_1 = \frac{3 + 1.5}{2} + \frac{3 - 1.5}{2}(-0.774597) = 1.6691$$

$$x_2 = \frac{3 + 1.5}{2} = 2.25$$

$$x_3 = \frac{3 + 1.5}{2} + \frac{3 - 1.5}{2}(0.774597) = 2.8309$$

We now compute the values of the interpolant $P_5(x)$ at the nodes. This can be done using the functions `newtonPoly` or `neville` listed in Art. 3.2. The results are

$$P_5(x_1) = 0.098\,08 \qquad P_5(x_2) = 0.628\,16 \qquad P_5(x_3) = 0.952\,16$$

Using Gauss–Legendre quadrature

$$I = \int_{1.5}^{3} P_5(x)\,dx = \frac{3 - 1.5}{2} \sum_{i=1}^{3} A_i P_5(x_i)$$

we get

$$I = 0.75\,[0.555\,556(0.098\,08) + 0.888\,889(0.628\,16) + 0.555\,556(0.952\,16)]$$
$$= 0.856\,37$$

Comparison with $-\int_{1.5}^{3} \cos x \, dx = 0.856\,38$ shows that the discrepancy is within the roundoff error.

PROBLEM SET 6.2

1. Evaluate

$$\int_{1}^{\pi} \frac{\ln x}{x^2 - 2x + 2} dx$$

   with Gauss–Legendre quadrature. Use (a) two nodes and (b) four nodes.

2. Use Gauss–Laguerre quadrature to evaluate $\int_{0}^{\infty} (1 - x^2)^3 e^{-x} \, dx$.

3. Use Gauss–Chebyshev quadrature with six nodes to evaluate

$$\int_{0}^{\pi/2} \frac{dx}{\sqrt{\sin x}}$$

   Compare the result with the "exact" value 2.62206. *Hint:* substitute $\sin x = t^2$.

4. The integral $\int_{0}^{\pi} \sin x \, dx$ is evaluated with Gauss–Legendre quadrature using four nodes. What are the bounds on the truncation error resulting from the quadrature?

5. How many nodes are required in Gauss–Laguerre quadrature to evaluate $\int_{0}^{\infty} e^{-x} \sin x \, dx$ to six decimal places?

6. Evaluate as accurately as possible

$$\int_{0}^{1} \frac{2x + 1}{\sqrt{x(1 - x)}} dx$$

   Hint: substitute $x = (1 + t)/2$.

7. Compute $\int_{0}^{\pi} \sin x \ln x \, dx$ to four decimal places.

8. Calculate the bounds on the truncation error if $\int_{0}^{\pi} x \sin x \, dx$ is evaluated with Gauss–Legendre quadrature using three nodes. What is the actual error?

9. Evaluate $\int_{0}^{2} (\sinh x/x) \, dx$ to four decimal places.

10. Evaluate the integral

$$\int_{0}^{\infty} \frac{x \, dx}{e^x + 1}$$

   to six decimal places. *Hint:* substitute $e^x = 1/t$.

11. ■ The equation of an ellipse is $x^2/a^2 + y^2/b^2 = 1$. Write a program that computes the length

$$S = 2 \int_{-a}^{a} \sqrt{1 + (dy/dx)^2} \, dx$$

of the circumference to four decimal places for given a and b. Test the program with $a = 2$ and $b = 1$.

12. ■ The error function, which is of importance in statistics, is defined as

$$\text{erf}(x) = \frac{2}{\sqrt{\pi}} \int_{0}^{x} e^{-t^2} \, dt$$

Write a program that uses Gauss–Legendre quadrature to evaluate $\text{erf}(x)$ for a given x to six decimal places. Note that $\text{erf}(x) = 1.000\,000$ (correct to 6 decimal places) when $x > 5$. Test the program by verifying that $\text{erf}(1.0) = 0.842\,701$.

13. ■

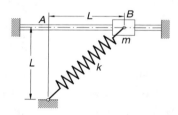

The sliding weight of mass m is attached to a spring of stiffness k that has an undeformed length L. When the mass is released from rest at B, the time it takes to reach A can be shown to be $t = C\sqrt{m/k}$, where

$$C = \int_{0}^{1} \left[\left(\sqrt{2} - 1 \right)^2 - \left(\sqrt{1 + z^2} - 1 \right)^2 \right]^{-1/2} dz$$

Compute C to six decimal places. *Hint:* the integrand has a singularity at $z = 1$ that behaves as $(1 - z^2)^{-1/2}$.

14. ■

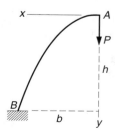

A uniform beam forms the semiparabolic cantilever arch AB. The vertical displacement of A due to the force P can be shown to be

$$\delta_A = \frac{Pb^3}{EI} C\left(\frac{h}{b}\right)$$

where EI is the bending rigidity of the beam and

$$C\left(\frac{h}{b}\right) = \int_0^1 z^2 \sqrt{1 + \left(\frac{2h}{b}z\right)^2}\, dz$$

Write a program that computes $C(h/b)$ for any given value of h/b to four decimal places. Use the program to compute $C(0.5)$, $C(1.0)$ and $C(2.0)$.

15. ■ There is no elegant way to compute $I = \int_0^{\pi/2} \ln(\sin x)\, dx$. A "brute force" method that works is to split the integral into several parts: from $x = 0$ to 0.01, from 0.01 to 0.2 and from $x = 0.2$ to $\pi/2$. In the first part we can use the approximation $\sin x \approx x$, which allows us to obtain the integral analytically. The other two parts can be evaluated with Gauss–Legendre quadrature. Use this method to evaluate I to six decimal places.

16. ■

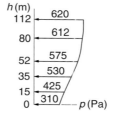

The pressure of wind was measured at various heights on a vertical wall, as shown on the diagram. Find the height of the pressure center, which is defined as

$$\bar{h} = \frac{\int_0^{112\,\text{m}} h\, p(h)\, dh}{\int_0^{112\,\text{m}} p(h)\, dh}$$

Hint: fit a cubic polynomial to the data and then apply Gauss–Legendre quadrature.

*6.5 Multiple Integrals

Multiple integrals, such as the area integral $\int \int_A f(x, y)\, dx\, dy$, can also be evaluated by quadrature. The computations are straightforward if the region of integration has a simple geometric shape, such as a triangle or a quadrilateral. Due to complications in specifying the limits of integration on x and y, quadrature is not a practical means of evaluating integrals over irregular regions. However, an irregular region A can always be approximated as an assembly of triangular or quadrilateral subregions $A_1, A_2, \ldots,$ called *finite elements*, as illustrated in Fig. 6.6. The integral over A can then be evaluated by summing the integrals over the finite elements:

$$\int \int_A f(x, y)\, dx\, dy \approx \sum_i \int \int_{A_i} f(x, y)\, dx\, dy$$

Volume integrals can computed in a similar manner, using tetrahedra or rectangular prisms for the finite elements.

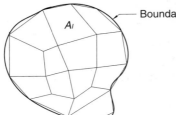

Boundary of region A

A_I

Figure 6.6. Finite element model of an irregular region.

Gauss–Legendre Quadrature over a Quadrilateral Element

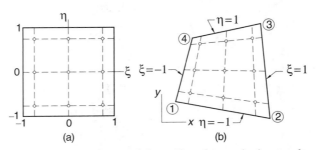

Figure 6.7. Mapping a quadrilateral into the standard rectangle.

Consider the double integral

$$I = \int_{-1}^{1} \int_{-1}^{1} f(\xi, \eta)\, d\xi\, d\eta$$

over the rectangular element shown in Fig. 6.7(a). Evaluating each integral in turn by Gauss–Legendre quadrature using n nodes in each coordinate direction, we obtain

$$I = \int_{-1}^{1} \sum_{i=1}^{n} A_i f(\xi_i, \eta) \, d\eta = \sum_{j=1}^{n} A_j \left[\sum_{i=1}^{n} A_i f(\xi_i, \eta_i) \right]$$

or

$$I = \sum_{i=1}^{n} \sum_{j=1}^{n} A_i A_j f(\xi_i, \eta_j) \tag{6.40}$$

The number of integration points n in each coordinate direction is called the *integration order*. Figure 6.7(a) shows the locations of the integration points used in third-order integration ($n = 3$). Because the integration limits were the "standard" limits $(-1, 1)$ of Gauss–Legendre quadrature, the weights and the coordinates of the integration points are as listed Table 6.3.

In order to apply quadrature to the quadrilateral element in Fig. 6.7(b), we must first map the quadrilateral into the "standard" rectangle in Fig. 6.7(a). By mapping we mean a coordinate transformation $x = x(\xi, \eta)$, $y = y(\xi, \eta)$ that results in one-to-one correspondence between points in the quadrilateral and in the rectangle. The transformation that does the job is

$$x(\xi, \eta) = \sum_{k=1}^{4} N_k(\xi, \eta) x_k \quad y(\xi, \eta) = \sum_{k=1}^{4} N_k(\xi, \eta) y_k \tag{6.41}$$

where (x_k, y_k) are the coordinates of corner k of the quadrilateral and

$$N_1(\xi, \eta) = \frac{1}{4}(1 - \xi)(1 - \eta)$$

$$N_2(\xi, \eta) = \frac{1}{4}(1 + \xi)(1 - \eta)$$

$$N_3(\xi, \eta) = \frac{1}{4}(1 + \xi)(1 + \eta) \tag{6.42}$$

$$N_4(\xi, \eta) = \frac{1}{4}(1 - \xi)(1 + \eta)$$

The functions $N_k(\xi, \eta)$, known as the *shape functions*, are bilinear (linear in each coordinate). Consequently, straight lines remain straight upon mapping. In particular, note that the sides of the quadrilateral are mapped into the lines $\xi = \pm 1$ and $\eta = \pm 1$.

Because mapping distorts areas, an infinitesimal area element $dA = dx \, dy$ of the quadrilateral is not equal to its counterpart $d\xi \, d\eta$ of the rectangle. It can be shown that the relationship between the areas is

$$dx \, dy = |\mathbf{J}(\xi, \eta)| \, d\xi \, d\eta \tag{6.43}$$

where

$$J(\xi, \eta) = \begin{bmatrix} \dfrac{\partial x}{\partial \xi} & \dfrac{\partial y}{\partial \xi} \\[2mm] \dfrac{\partial x}{\partial \eta} & \dfrac{\partial y}{\partial \eta} \end{bmatrix} \tag{6.44a}$$

is known as the *Jacobian matrix* of the mapping. Substituting from Eqs. (6.41) and (6.42) and differentiating, we find that the components of the Jacobian matrix are

$$J_{11} = \frac{1}{4}[-(1-\eta)x_1 + (1-\eta)x_2 + (1+\eta)x_3 - (1+\eta)x_4]$$

$$J_{12} = \frac{1}{4}[-(1-\eta)y_1 + (1-\eta)y_2 + (1+\eta)y_3 - (1+\eta)y_4] \tag{6.44b}$$

$$J_{21} = \frac{1}{4}[-(1-\xi)x_1 - (1+\xi)x_2 + (1+\xi)x_3 + (1-\xi)x_4]$$

$$J_{22} = \frac{1}{4}[-(1-\xi)y_1 - (1+\xi)y_2 + (1+\xi)y_3 + (1-\xi)y_4]$$

We can now write

$$\int\int_A f(x, y)\, dx\, dy = \int_{-1}^{1}\int_{-1}^{1} f[x(\xi, \eta), y(\xi, \eta)]\, |J(\xi, \eta)|\, d\xi\, d\eta \tag{6.45}$$

Since the right-hand side integral is taken over the "standard" rectangle, it can be evaluated using Eq. (6.40). Replacing $f(\xi, \eta)$ in Eq. (6.40) by the integrand in Eq. (6.45), we get the following formula for Gauss–Legendre quadrature over a quadrilateral region:

$$I = \sum_{i=1}^{n}\sum_{j=1}^{n} A_i A_j\, f\left[x(\xi_i, \eta_j), y(\xi_i, \eta_j)\right] |J(\xi_i, \eta_j)| \tag{6.46}$$

The ξ and η coordinates of the integration points and the weights can again be obtained from Table 6.3.

■ gaussQuad2

The function gaussQuad2 computes $\int\int_A f(x, y)\, dx\, dy$ over a quadrilateral element with Gauss–Legendre quadrature of integration order n. The quadrilateral is defined by the arrays x and y, which contain the coordinates of the four corners ordered in a *counterclockwise direction* around the element. The determinant of the Jacobian matrix is obtained by calling `detJ`; mapping is performed by `map`. The weights and the values of ξ and η at the integration points are computed by `gaussNodes` listed in the previous article (note that ξ and η appear as `s` and `t` in listing).

```
function I = gaussQuad2(func,x,y,n)
% Gauss-Legendre quadrature over a quadrilateral.
% USAGE: I = gaussQuad2(func,x,y,n)
% INPUT:
% func = handle of function to be integrated.
% x    = [x1;x2;x3;x4] = x-coordinates of corners.
% y    = [y1;y2;y3;y4] = y-coordinates of corners.
% n    = order of integration
% OUTPUT:
% I    = integral

[t,A] = gaussNodes(n);  I = 0;
for i = 1:n
    for j = 1:n
        [xNode,yNode] = map(x,y,t(i),t(j));
        z = feval(func,xNode,yNode);
        detJ = jac(x,y,t(i),t(j));
        I = I + A(i)*A(j)*detJ*z;
    end
end

function detJ = jac(x,y,s,t)
% Computes determinant of Jacobian matrix.
J = zeros(2);
J(1,1) = - (1 - t)*x(1) + (1 - t)*x(2)...
         + (1 + t)*x(3) - (1 + t)*x(4);
J(1,2) = - (1 - t)*y(1) + (1 - t)*y(2)...
         + (1 + t)*y(3) - (1 + t)*y(4);
J(2,1) = - (1 - s)*x(1) - (1 + s)*x(2)...
         + (1 + s)*x(3) + (1 - s)*x(4);
J(2,2) = - (1 - s)*y(1) - (1 + s)*y(2)...
         + (1 + s)*y(3) + (1 - s)*y(4);
detJ = (J(1,1)*J(2,2) - J(1,2)*J(2,1))/16;
```

```
function [xNode,yNode] = map(x,y,s,t)
% Computes x and y-coordinates of nodes.
N = zeros(4,1);
N(1) = (1 - s)*(1 - t)/4;
N(2) = (1 + s)*(1 - t)/4;
N(3) = (1 + s)*(1 + t)/4;
N(4) = (1 - s)*(1 + t)/4;
xNode = dot(N,x); yNode = dot(N,y);
```

EXAMPLE 6.13

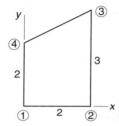

Evaluate the integral

$$I = \int \int_A \left(x^2 + y \right) dx\,dy$$

over the quadrilateral shown.

Solution The corner coordinates of the quadrilateral are

$$\mathbf{x}^T = \begin{bmatrix} 0 & 2 & 2 & 0 \end{bmatrix} \qquad \mathbf{y}^T = \begin{bmatrix} 0 & 0 & 3 & 2 \end{bmatrix}$$

The mapping is

$$
\begin{aligned}
x(\xi, \eta) &= \sum_{k=1}^{4} N_k(\xi, \eta) x_k \\
&= 0 + \frac{(1 + \xi)(1 - \eta)}{4} (2) + \frac{(1 + \xi)(1 + \eta)}{4} (2) + 0 \\
&= 1 + \xi \\
y(\xi, \eta) &= \sum_{k=1}^{4} N_k(\xi, \eta) y_k \\
&= 0 + 0 + \frac{(1 + \xi)(1 + \eta)}{4} (3) + \frac{(1 - \xi)(1 + \eta)}{4} (2) \\
&= \frac{(5 + \xi)(1 + \eta)}{4}
\end{aligned}
$$

which yields for the Jacobian matrix

$$J(\xi, \eta) = \begin{bmatrix} \dfrac{\partial x}{\partial \xi} & \dfrac{\partial y}{\partial \xi} \\[2mm] \dfrac{\partial x}{\partial \eta} & \dfrac{\partial y}{\partial \eta} \end{bmatrix} = \begin{bmatrix} 1 & \dfrac{1+\eta}{4} \\[2mm] 0 & \dfrac{5+\xi}{4} \end{bmatrix}$$

Thus the area scale factor is

$$|J(\xi, \eta)| = \frac{5+\xi}{4}$$

Now we can map the integral from the quadrilateral to the standard rectangle. Referring to Eq. (6.45), we obtain

$$I = \int_{-1}^{1} \int_{-1}^{1} \left[(1+\xi)^2 + \frac{(5+\xi)(1+\eta)}{4} \right] \frac{5+\xi}{4} \, d\xi \, d\eta$$

$$= \int_{-1}^{1} \int_{-1}^{1} \left(\frac{45}{16} + \frac{21}{8}\xi + \frac{29}{16}\xi^2 + \frac{1}{4}\xi^3 + \frac{25}{16}\eta + \frac{5}{8}\xi\eta + \frac{1}{16}\xi^2\eta \right) d\xi \, d\eta$$

Noting that only even powers of ξ and η contribute to the integral, we can simplify the integral to

$$I = \int_{-1}^{1} \int_{-1}^{1} \left(\frac{45}{16} + \frac{29}{16}\xi^2 \right) d\xi \, d\eta = \frac{41}{3}$$

EXAMPLE 6.14

Evaluate the integral

$$\int_{-1}^{1} \int_{-1}^{1} \cos \frac{\pi x}{2} \cos \frac{\pi y}{2} \, dx \, dy$$

by Gauss–Legendre quadrature of order three.

Solution From the quadrature formula in Eq. (6.40), we have

$$I = \sum_{i=1}^{3} \sum_{j=1}^{3} A_i A_j \cos \frac{\pi x_i}{2} \cos \frac{\pi y_j}{2}$$

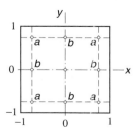

The integration points are shown in the figure; their coordinates and the corresponding weights are listed in Table 6.3. Note that the integrand, the integration points and the weights are all symmetric about the coordinate axes. It follows that the points labeled a contribute equal amounts to I; the same is true for the points labeled b. Therefore,

$$I = 4(0.555\,556)^2 \cos^2 \frac{\pi\,(0.774\,597)}{2}$$

$$+ 4(0.555\,556)(0.888\,889) \cos \frac{\pi\,(0.774\,597)}{2} \cos \frac{\pi\,(0)}{2}$$

$$+ (0.888\,889)^2 \cos^2 \frac{\pi\,(0)}{2}$$

$$= 1.623\,391$$

The exact value of the integral is $16/\pi^2 \approx 1.621\,139$.

EXAMPLE 6.15

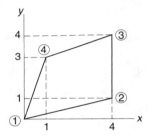

Utilize $\texttt{gaussQuad2}$ to evaluate $I = \int \int_A f(x, y)\,dx\,dy$ over the quadrilateral shown, where

$$f(x, y) = (x - 2)^2(y - 2)^2$$

Use enough integration points for an "exact" answer.

Solution The required integration order is determined by the integrand in Eq. (6.45):

$$I = \int_{-1}^{1} \int_{-1}^{1} f\,[x(\xi, \eta),\, y(\xi, \eta)]\,\,|\boldsymbol{J}(\xi, \eta)|\,d\xi\,d\eta \qquad\qquad \text{(a)}$$

We note that $|\boldsymbol{J}(\xi, \eta)|$, defined in Eqs. (6.44), is biquadratic. Since the specified $f(x, y)$ is also biquadratic, the integrand in Eq. (a) is a polynomial of degree 4 in both ξ and η. Thus third-order integration ($n = 3$) is sufficient for an "exact" result. Here is the MATLAB command that performs the integration:

```
>> I = gaussQuad2(@fex6_15,[0;4;4;1],[0;1;4;3],3)
I =
    11.3778
>>
```

The M-file that returns the function to be integrated is

```
function z = fex6_15(x,y)
% Function used in Example 6.15.
z = ((x - 2)*(y - 2))^2;
```

Quadrature over a Triangular Element

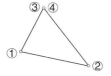

Figure 6.8. Quadrilateral with two coincident corners.

A triangle may be viewed as a degenerate quadrilateral with two of its corners occupying the same location, as illustrated in Fig. 6.8. Therefore, the integration formulas over a quadrilateral region can also be used for a triangular element. However, it is computationally advantageous to use integration formulas specially developed for triangles, which we present without derivation.[15]

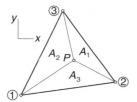

Figure 6.9. Triangular element.

Consider the triangular element in Fig. 6.9. Drawing straight lines from the point P in the triangle to each of the corners divides the triangle into three parts with areas A_1, A_2 and A_3. The so-called *area coordinates* of P are defined as

$$\alpha_i = \frac{A_i}{A}, \quad i = 1, 2, 3 \tag{6.47}$$

where A is the area of the element. Since $A_1 + A_2 + A_3 = A$, the area coordinates are related by

$$\alpha_1 + \alpha_2 + \alpha_3 = 1 \tag{6.48}$$

Note that α_i ranges from 0 (when P lies on the side opposite to corner i) to 1 (when P is at corner i).

---

[15] The triangle formulas are extensively used in the finite method analysis. See, for example, O.C. Zienkiewicz and R.L Taylor, *The Finite Element Method*, Vol. 1, 4th ed., McGraw-Hill (1989).

A convenient formula of computing A from the corner coordinates (x_i, y_i) is

$$A = \frac{1}{2} \begin{vmatrix} 1 & 1 & 1 \\ x_1 & x_2 & x_3 \\ y_1 & y_2 & y_3 \end{vmatrix} \tag{6.49}$$

The area coordinates are mapped into the Cartesian coordinates by

$$x(\alpha_1, \alpha_2, \alpha_3) = \sum_{i=1}^{3} \alpha_i x_i \qquad y(\alpha_1, \alpha_2, \alpha_3) = \sum_{i=1}^{3} \alpha_i y_i \tag{6.50}$$

The integration formula over the element is

$$\int \int_A f[x(\alpha), y(\alpha)] \, dA = A \sum_k W_k f[x(\alpha_k), y(\alpha_k)] \tag{6.51}$$

where α_k represents the area coordinates of the integration point k, and W_k are the weights. The locations of the integration points are shown in Fig. 6.10, and the corresponding values of α_k and W_k are listed in Table 6.7. The quadrature in Eq. (6.51) is exact if $f(x, y)$ is a polynomial of the degree indicated.

(a) Linear

(b) Quadratic

(c) Cubic

Figure 6.10. Integration points of triangular elements.

Degree of $f(x, y)$	Point	α_k	W_k
(a) Linear	a	1/3, 1/3, 1/3	1
(b) Quadratic	a	1/2, 0 , 1/2	1/3
	b	1/2, 1/2, 0	1/3
	c	0, 1/2 , 1/2	1/3
(c) Cubic	a	1/3, 1/3, 1/3	−27/48
	b	1/5, 1/5, 3/5	25/48
	c	3/5. 1/5 , 1/5	25/48
	d	1/5, 3/5, 1/5	25/48

Table 6.7

■ triangleQuad

The function `triangleQuad` computes $\int \int_A f(x, y) \, dx \, dy$ over a triangular region using the cubic formula—case (c) in Fig. 6.10. The triangle is defined by its corner coordinate arrays x and y, where the coordinates must be listed in a counterclockwise direction around the triangle.

```
function I = triangleQuad(func,x,y)
% Cubic quadrature over a triangle.
% USAGE: I = triangleQuad(func,x,y)
% INPUT:
% func = handle of function to be integrated.
% x    = [x1;x2;x3] x-coordinates of corners.
% y    = [y1;y2;y3] y-coordinates of corners.
% OUTPUT:
% I    = integral

alpha = [1/3 1/3 1/3; 1/5 1/5 3/5;...
         3/5 1/5 1/5; 1/5 3/5 1/5];
W = [-27/48; 25/48; 25/48; 25/48];
xNode = alpha*x; yNode = alpha*y;
A = (x(2)*y(3) - x(3)*y(2)...
   - x(1)*y(3) + x(3)*y(1)...
   + x(1)*y(2) - x(2)*y(1))/2;
sum = 0;
for i = 1:4
    z = feval(func,xNode(i),yNode(i));
    sum = sum + W(i)*z;
end
I = A*sum
```

EXAMPLE 6.16

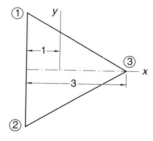

Evaluate $I = \int \int_A f(x, y)\, dx\, dy$ over the equilateral triangle shown, where[16]

$$f(x, y) = \frac{1}{2}(x^2 + y^2) - \frac{1}{6}(x^3 - 3xy^2) - \frac{2}{3}$$

Use the quadrature formulas for (1) a quadrilateral and (2) a triangle.

[16] This function is identical to the Prandtl stress function for torsion of a bar with the cross section shown; the integral is related to the torsional stiffness of the bar. See, for example, S.P. Timoshenko and J.N. Goodier, *Theory of Elasticity*, 3rd ed., McGraw-Hill (1970).

Solution of Part (1) Let the triangle be formed by collapsing corners 3 and 4 of a quadrilateral. The corner coordinates of this quadrilateral are $\mathbf{x} = [-1, -1, 2, 2]^T$ and $\mathbf{y} = [\sqrt{3}, -\sqrt{3}, 0, 0]^T$. To determine the minimum required integration order for an exact result, we must examine $f[x(\xi, \eta), y(\xi, \eta)] \, |\mathbf{J}(\xi, \eta)|$, the integrand in Eq. (6.45). Since $|\mathbf{J}(\xi, \eta)|$ is biquadratic, and $f(x, y)$ is cubic in x, the integrand is a polynomial of degree 5 in x. Therefore, third-order integration will suffice. The command used for the computations is similar to the one in Example 6.15:

```
>> I = gaussQuad2(@fex6_16,[-1;-1;2;2],...
        [sqrt(3);-sqrt(3);0;0],3)
I =
   -1.5588
```

The function that returns $z = f(x, y)$ is

```
function z = fex6_16(x,y)
% Function used in Example 6.16
z = (x^2 + y^2)/2 - (x^3 - 3*x*y^2)/6 - 2/3;
```

Solution of Part (2) The following command executes quadrature over the triangular element:

```
>> I = triangleQuad(@fex6_16,[-1; -1; 2],[sqrt(3);-sqrt(3); 0])
I =
   -1.5588
```

Since the integrand is a cubic, this result is also exact.

Note that only four function evaluations were required when using the triangle formulas. In contrast, the function had to be evaluated at nine points in Part (1).

EXAMPLE 6.17
The corner coordinates of a triangle are $(0, 0)$, $(16, 10)$ and $(12, 20)$. Compute $\int \int_A (x^2 - y^2) \, dx \, dy$ over this triangle.

Solution

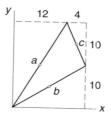

Because $f(x, y)$ is quadratic, quadrature over the three integration points shown in Fig. 6.10(b) will be sufficient for an "exact" result. Note that the integration points lie in the middle of each side; their coordinates are (6, 10), (8, 5) and (14, 15). The area of the triangle is obtained from Eq. (6.49):

$$A = \frac{1}{2} \begin{vmatrix} 1 & 1 & 1 \\ x_1 & x_2 & x_3 \\ y_1 & y_2 & y_3 \end{vmatrix} = \frac{1}{2} \begin{vmatrix} 1 & 1 & 1 \\ 0 & 16 & 12 \\ 0 & 10 & 20 \end{vmatrix} = 100$$

From Eq. (6.51) we get

$$I = A \sum_{k=a}^{c} W_k f(x_k, y_k)$$

$$= 100 \left[\frac{1}{3} f(6, 10) + \frac{1}{3} f(8, 5) + \frac{1}{3} f(14, 15) \right]$$

$$= \frac{100}{3} \left[(6^2 - 10^2) + (8^2 - 5^2) + (14^2 - 15^2) \right] = 1800$$

PROBLEM SET 6.3

1. Use Gauss–Legendre quadrature to compute

$$\int_{-1}^{1} \int_{-1}^{1} (1 - x^2)(1 - y^2) \, dx \, dy$$

2. Evaluate the following integral with Gauss–Legendre quadrature:

$$\int_{y=0}^{2} \int_{x=0}^{3} x^2 y^2 \, dx \, dy$$

3. Compute the approximate value of

$$\int_{-1}^{1} \int_{-1}^{1} e^{-(x^2+y^2)} \, dx \, dy$$

with Gauss–Legendre quadrature. Use integration order (a) two and (b) three. (The true value of the integral is 2.230 985.)

4. Use third-order Gauss–Legendre quadrature to obtain an approximate value of

$$\int_{-1}^{1}\int_{-1}^{1} \cos\frac{\pi(x-y)}{2}\, dx\, dy$$

(The exact value of the integral is 1.621 139.)

5.

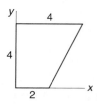

Map the integral $\int\int_A xy\, dx\, dy$ from the quadrilateral region shown to the "standard" rectangle and then evaluate it analytically.

6.

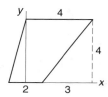

Compute $\int\int_A x\, dx\, dy$ over the quadrilateral region shown by first mapping it into the "standard" rectangle and then integrating analytically.

7.

Use quadrature to compute $\int\int_A x^2\, dx\, dy$ over the triangle shown.

8. Evaluate $\int\int_A x^3\, dx\, dy$ over the triangle shown in Prob. 7.

9.

Evaluate $\int\int_A (3-x)y\, dx\, dy$ over the region shown.

10. Evaluate $\int\int_A x^2 y\, dx\, dy$ over the triangle shown in Prob. 9.

11. ■

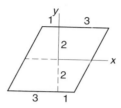

Evaluate $\int\int_A xy(2 - x^2)(2 - xy)\, dx\, dy$ over the region shown.

12. ■ Compute $\int\int_A xy\exp(-x^2)\, dx\, dy$ over the region shown in Prob. 11 to four decimal places.

13. ■

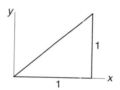

Evaluate $\int\int_A(1 - x)(y - x)y\, dx\, dy$ over the triangle shown.

14. ■ Estimate $\int\int_A \sin\pi x\, dx\, dy$ over the region shown in Prob. 13. Use the cubic integration formula for a triangle. (The exact integral is $1/\pi$.)

15. ■ Compute $\int\int_A \sin\pi x \sin\pi(y - x)\, dx\, dy$ to six decimal places, where A is the triangular region shown in Prob. 13. Consider the triangle as a degenerate quadrilateral.

16. ■

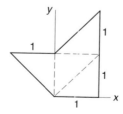

Write a program to evaluate $\int\int_A f(x, y)\, dx\, dy$ over an irregular region that has been divided into several triangular elements. Use the program to compute $\int\int_A xy(y - x)\, dx\, dy$ over the region shown.

MATLAB Functions

$\texttt{I = quad(func,a,b,tol)}$ uses adaptive Simpson's rule for evaluating $I = \int_a^b f(x)\,dx$ with an error tolerance $\texttt{tol}$ (default is 1.0e-6). To speed up execution, vectorize the computation of $\texttt{func}$ by using array operators $\texttt{.*}$, $\texttt{./}$ and $\texttt{.^}$ in the definition of $\texttt{func}$. For example, if $f(x) = x^3 \sin x + 1/x$, specify the function as

```
function y = func(x)
y = (x.^3).*sin(x) + 1./x
```

$\texttt{I = dblquad(func,xMin,xMax,yMin,yMax,tol)}$ uses $\texttt{quad}$ to integrate over a rectangle:

$$I = \int_{yMin}^{yMax} \int_{xMin}^{xMax} f(x, y)\,dx\,dy$$

$\texttt{I = quadl(func,a,b,tol)}$ employs adaptive Lobatto quadrature (this method is not discussed in this book). It is recommended if very high accuracy is desired and the integrand is smooth.

There are no functions for Gaussian quadrature.

7 Initial Value Problems

$$\text{Solve } \mathbf{y}' = \mathbf{F}(x, \mathbf{y}),\ \mathbf{y}(a) = \alpha$$

7.1 Introduction

The general form of a first-order differential equation is

$$y' = f(x, y) \tag{7.1a}$$

where $y' = dy/dx$ and $f(x, y)$ is a given function. The solution of this equation contains an arbitrary constant (the constant of integration). To find this constant, we must know a point on the solution curve; that is, y must be specified at some value of x, say at $x = a$. We write this auxiliary condition as

$$y(a) = \alpha \tag{7.1b}$$

An ordinary differential equation of order n

$$y^{(n)} = f\left(x, y, y', \dots, y^{(n-1)}\right) \tag{7.2}$$

can always be transformed into n first-order equations. Using the notation

$$y_1 = y \qquad y_2 = y' \qquad y_3 = y'' \quad \dots \quad y_n = y^{(n-1)} \tag{7.3}$$

the equivalent first-order equations are

$$y_1' = y_2 \qquad y_2' = y_3 \qquad y_3' = y_4 \quad \dots \quad y_n' = f(x, y_1, y_2, \dots, y_n) \tag{7.4a}$$

The solution now requires the knowledge n auxiliary conditions. If these conditions are specified at the same value of x, the problem is said to be an *initial value problem*. Then the auxiliary conditions, called *initial conditions*, have the form

$$y_1(a) = \alpha_1 \qquad y_2(a) = \alpha_2 \qquad y_3(a) = \alpha_3 \quad \dots \quad y_n(a) = \alpha_n \tag{7.4b}$$

If y_i are specified at different values of x, the problem is called a *boundary value problem*.

For example,

$$y'' = -y \qquad y(0) = 1 \qquad y'(0) = 0$$

is an initial value problem since both auxiliary conditions imposed on the solution are given at $x = 0$. On the other hand,

$$y'' = -y \qquad y(0) = 1 \qquad y(\pi) = 0$$

is a boundary value problem because the two conditions are specified at different values of x.

In this chapter we consider only initial value problems. The more difficult boundary value problems are discussed in the next chapter. We also make extensive use of vector notation, which allows us manipulate sets of first-order equations in a concise form. For example, Eqs. (7.4) are written as

$$\mathbf{y}' = \mathbf{F}(x, \mathbf{y}) \qquad \mathbf{y}(a) = \boldsymbol{\alpha} \tag{7.5a}$$

where

$$\mathbf{F}(x, \mathbf{y}) = \begin{bmatrix} y_2 \\ y_3 \\ \vdots \\ y_n \\ f(x, \mathbf{y}) \end{bmatrix} \tag{7.5b}$$

A numerical solution of differential equations is essentially a table of x- and $\mathbf{y}$-values listed at discrete intervals of x.

7.2   Taylor Series Method

The Taylor series method is conceptually simple and capable of high accuracy. Its basis is the truncated Taylor series for $\mathbf{y}$ about x:

$$\mathbf{y}(x + h) \approx \mathbf{y}(x) + \mathbf{y}'(x)h + \frac{1}{2!}\mathbf{y}''(x)h^2 + \frac{1}{3!}\mathbf{y}'''(x)h^3 + \cdots + \frac{1}{n!}\mathbf{y}^{(m)}(x)h^m \tag{7.6}$$

Because Eq. (7.6) predicts $\mathbf{y}$ at $x + h$ from the information available at x, it is also a formula for numerical integration. The last term kept in the series determines the *order of integration*. For the series in Eq. (7.6) the integration order is m.

The truncation error, due to the terms omitted from the series, is

$$\mathbf{E} = \frac{1}{(m+1)!}\mathbf{y}^{(m+1)}(\xi)h^{m+1}, \quad x < \xi < x + h$$

Using the finite difference approximation

$$\mathbf{y}^{(m+1)}(\xi) \approx \frac{\mathbf{y}^{(m)}(x+h) - \mathbf{y}^{(m)}(x)}{h}$$

we obtain the more usable form

$$\mathbf{E} \approx \frac{h^m}{(m+1)!} \left[\mathbf{y}^{(m)}(x+h) - \mathbf{y}^{(m)}(x) \right] \tag{7.7}$$

which could be incorporated in the algorithm to monitor the error in each integration step.

■ taylor

The function `taylor` implements the Taylor series method of integration of order four. It can handle any number of first-order differential equations $y_i' = f_i(x, y_1, y_2, \ldots, y_n)$, $i = 1, 2, \ldots, n$. The user is required to supply the function `deriv` that returns the $4 \times n$ array

$$\mathbf{d} = \begin{bmatrix} (\mathbf{y}')^T \\ (\mathbf{y}'')^T \\ (\mathbf{y}''')^T \\ (\mathbf{y}^{(4)})^T \end{bmatrix} = \begin{bmatrix} y_1' & y_2' & \cdots & y_n' \\ y_1'' & y_2'' & \cdots & y_n'' \\ y_1''' & y_2''' & \cdots & y_n''' \\ y_1^{(4)} & y_2^{(4)} & \cdots & y_n^{(4)} \end{bmatrix}$$

The function returns the arrays `xSol` and `ySol` that contain the values of x and $\mathbf{y}$ at intervals h.

```
function [xSol,ySol] = taylor(deriv,x,y,xStop,h)
% 4th-order Taylor series method of integration.
% USAGE: [xSol,ySol] = taylor(deriv,x,y,xStop,h)
% INPUT:
% deriv = handle of function that returns the matrix
%         d = [dy/dx d^2y/dx^2 d^3y/dx^3 d^4y/dx^4].
% x,y   = initial values; y must be a row vector.
% xStop = terminal value of x
% h     = increment of x used in integration (h > 0).
% OUTPUT:
% xSol  = x-values at which solution is computed.
% ySol  = values of y corresponding to the x-values.

if size(y,1) > 1; y = y'; end  % y must be row vector
xSol = zeros(2,1); ySol = zeros(2,length(y));
xSol(1) = x; ySol(1,:) = y;
```

```
k = 1;
while x < xStop
    h = min(h,xStop - x);
    d = feval(deriv,x,y);          % Derivatives of [y]
    hh = 1;
    for j = 1:4                    % Build Taylor series
        hh = hh*h/j;               % hh = h^j/j!
        y = y + d(j,:)*hh;
    end
    x = x + h; k = k + 1;
    xSol(k) = x; ySol(k,:) = y; % Store current soln.
end
```

■ printSol

This function prints the results xSol and ySol in tabular form. The amount of data is controlled by the printout frequency freq. For example, if freq = 5, every fifth integration step would be displayed. If freq = 0, only the initial and final values will be shown.

```
function printSol(xSol,ySol,freq)
% Prints xSol and ySoln arrays in tabular format.
% USAGE: printSol(xSol,ySol,freq)
% freq = printout frequency (prints every freq-th
%        line of xSol and ySol).

[m,n] = size(ySol);
if freq == 0;freq = m; end
head = '    x';
for i = 1:n
    head = strcat(head,'              y',num2str(i));
end
fprintf(head); fprintf('\n')
for i = 1:freq:m
    fprintf('%14.4e',xSol(i),ySol(i,:)); fprintf('\n')
end
if i ~= m; fprintf('%14.4e',xSol(m),ySol(m,:)); end
```

EXAMPLE 7.1
Given that

$$y' + 4y = x^2 \qquad y(0) = 1$$

determine $y(0.2)$ with the fourth-order Taylor series method using a single integration step. Also compute the estimated error from Eq. (7.7) and compare it with the actual error. The analytical solution of the differential equation is

$$y = \frac{31}{32}e^{-4x} + \frac{1}{4}x^2 - \frac{1}{8}x + \frac{1}{32}$$

Solution The Taylor series up to and including the term with h^4 is

$$y(h) = y(0) + y'(0)h + \frac{1}{2!}y''(0)h^2 + \frac{1}{3!}y'''(0)h^3 + \frac{1}{4!}y^{(4)}(0)h^4 \qquad \text{(a)}$$

Differentiation of the differential equation yields

$$y' = -4y + x^2$$
$$y'' = -4y' + 2x = 16y - 4x^2 + 2x$$
$$y''' = 16y' - 8x + 2 = -64y + 16x^2 - 8x + 2$$
$$y^{(4)} = -64y' + 32x - 8 = 256y - 64x^2 + 32x - 8$$

Thus

$$y'(0) = -4(1) = -4$$
$$y''(0) = 16(1) = 16$$
$$y'''(0) = -64(1) + 2 = -62$$
$$y^{(4)}(0) = 256(1) - 8 = 248$$

With $h = 0.2$ Eq. (a) becomes

$$y(0.2) = 1 + (-4)(0.2) + \frac{1}{2!}(16)(0.2)^2 + \frac{1}{3!}(-62)(0.2)^3 + \frac{1}{4!}(248)(0.2)^4$$
$$= 0.4539$$

According to Eq. (7.7) the approximate truncation error is

$$E = \frac{h^4}{5!}\left[y^{(4)}(0.2) - y^{(4)}(0)\right]$$

where

$$y^{(4)}(0) = 248$$
$$y^{(4)}(0.2) = 256(0.4539) - 64(0.2)^2 + 32(0.2) - 8 = 112.04$$

Therefore,

$$E = \frac{(0.2)^4}{5!}(112.04 - 248) = -0.0018$$

The analytical solution yields

$$y(0.2) = \frac{31}{32}e^{-4(0.2)} + \frac{1}{4}(0.2)^2 - \frac{1}{8}(0.2) + \frac{1}{32} = 0.4515$$

so that the actual error is $0.4515 - 0.4539 = -0.0024$.

EXAMPLE 7.2

Solve

$$y'' = -0.1y' - x \quad y(0) = 0 \quad y'(0) = 1$$

from $x = 0$ to 2 with the Taylor series method of order four using $h = 0.25$.

Solution With $y_1 = y$ and $y_2 = y'$ the equivalent first-order equations and initial conditions are

$$\mathbf{y}' = \begin{bmatrix} y_1' \\ y_2' \end{bmatrix} = \begin{bmatrix} y_2 \\ -0.1y_2 - x \end{bmatrix} \quad \mathbf{y}(0) = \begin{bmatrix} 0 \\ 1 \end{bmatrix}$$

Repeated differentiation of the differential equations yields

$$\mathbf{y}'' = \begin{bmatrix} y_2' \\ -0.1y_2' - 1 \end{bmatrix} = \begin{bmatrix} -0.1y_2 - x \\ 0.01y_2 + 0.1x - 1 \end{bmatrix}$$

$$\mathbf{y}''' = \begin{bmatrix} -0.1y_2' - 1 \\ 0.01y_2' + 0.1 \end{bmatrix} = \begin{bmatrix} 0.01y_2 + 0.1x - 1 \\ -0.001y_2 - 0.01x + 0.1 \end{bmatrix}$$

$$\mathbf{y}^{(4)} = \begin{bmatrix} 0.01y_2' + 0.1 \\ -0.001y_2' - 0.01 \end{bmatrix} = \begin{bmatrix} -0.001y_2 - 0.01x + 0.1 \\ 0.0001y_2 + 0.001x - 0.01 \end{bmatrix}$$

Thus the derivative array required by `taylor` is

$$\mathbf{d} = \begin{bmatrix} y_2 & -0.1y_2 - x \\ -0.1y_2 - x & 0.01y_2 + 0.1x - 1 \\ 0.01y_2 + 0.1x - 1 & -0.001y_2 - 0.01x + 0.1 \\ -0.001y_2 - 0.01x + 0.1 & 0.0001y_2 + 0.001x - 0.01 \end{bmatrix}$$

which is computed by

```
function d = fex7_2(x,y)
% Derivatives used in Example 7.2

d = zeros(4,2);
d(1,1) = y(2);
d(1,2) = -0.1*y(2) - x;
d(2,1) = d(1,2);
d(2,2) = 0.01*y(2) + 0.1*x -1;
```

```
d(3,1) = d(2,2);
d(3,2) = -0.001*y(2) - 0.01*x + 0.1;
d(4,1) = d(3,2);
d(4,2) = 0.0001*y(2) + 0.001*x - 0.01;
```

Here is the solution:

```
>> [x,y] = taylor(@fex7_2, 0, [0 1], 2, 0.25);
>> printSol(x,y,1)
       x              y1              y2
   0.0000e+000    0.0000e+000    1.0000e+000
   2.5000e-001    2.4431e-001    9.4432e-001
   5.0000e-001    4.6713e-001    8.2829e-001
   7.5000e-001    6.5355e-001    6.5339e-001
   1.0000e+000    7.8904e-001    4.2110e-001
   1.2500e+000    8.5943e-001    1.3281e-001
   1.5000e+000    8.5090e-001   -2.1009e-001
   1.7500e+000    7.4995e-001   -6.0625e-001
   2.0000e+000    5.4345e-001   -1.0543e+000
```

The analytical solution of the problem is

$$y = 100x - 5x^2 + 990(e^{-0.1x} - 1)$$

from which we obtain $y(2) = 0.543\,45$ and $y'(2) = -1.0543$, which agree with the numerical solution.

The main drawback of the Taylor series method is that it requires repeated differentiation of the dependent variables. These expressions may become very long and thus error-prone and tedious to compute. Moreover, there is the extra work of coding each of the derivatives.

7.3   Runge–Kutta Methods

The aim of Runge–Kutta methods is to eliminate the need for repeated differentiation of the differential equations. Since no such differentiation is involved in the first-order Taylor series integration formula

$$\mathbf{y}(x + h) = \mathbf{y}(x) + \mathbf{y}'(x)h = \mathbf{y}(x) + \mathbf{F}(x, \mathbf{y})h \tag{7.8}$$

it can be considered as the first-order Runge–Kutta method; it is also called *Euler's method*. Due to excessive truncation error, this method is rarely used in practice.

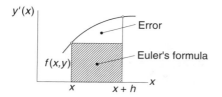

Figure 7.1. Graphical representation of Euler's formula.

Let us now take a look at the graphical interpretation of Euler's formula. For the sake of simplicity, we assume that there is a single dependent variable y, so that the differential equation is $y' = f(x, y)$. The change in the solution y between x and $x + h$ is

$$y(x + h) - y(h) = \int_x^{x+h} y'\, dx = \int_x^{x+h} f(x, y)dx$$

which is the area of the panel under the $y'(x)$ plot, shown in Fig. 7.1. Euler's formula approximates this area by the area of the cross-hatched rectangle. The area between the rectangle and the plot represents the truncation error. Clearly, the truncation error is proportional to the slope of the plot; that is, proportional to $y''(x)$.

Second-Order Runge–Kutta Method

To arrive at the second-order method, we assume an integration formula of the form

$$\mathbf{y}(x + h) = \mathbf{y}(x) + c_0\mathbf{F}(x, \mathbf{y})h + c_1\mathbf{F}\left[x + ph, \mathbf{y} + qh\mathbf{F}(x, \mathbf{y})\right]h \tag{a}$$

and attempt to find the parameters c_0, c_1, p and q by matching Eq. (a) to the Taylor series

$$\mathbf{y}(x + h) = \mathbf{y}(x) + \mathbf{y}'(x)h + \frac{1}{2!}\mathbf{y}''(x)h^2 + \mathcal{O}(h^3)$$

$$= \mathbf{y}(x) + \mathbf{F}(x, \mathbf{y})h + \frac{1}{2}\mathbf{F}'(x, \mathbf{y})h^2 + \mathcal{O}(h^3) \tag{b}$$

Noting that

$$\mathbf{F}'(x, \mathbf{y}) = \frac{\partial \mathbf{F}}{\partial x} + \sum_{i=1}^{n} \frac{\partial \mathbf{F}}{\partial y_i} y_i' = \frac{\partial \mathbf{F}}{\partial x} + \sum_{i=1}^{n} \frac{\partial \mathbf{F}}{\partial y_i} F_i(x, \mathbf{y})$$

where n is the number of first-order equations, we can write Eq. (b) as

$$\mathbf{y}(x + h) = \mathbf{y}(x) + \mathbf{F}(x, \mathbf{y})h + \frac{1}{2}\left(\frac{\partial \mathbf{F}}{\partial x} + \sum_{i=1}^{n} \frac{\partial \mathbf{F}}{\partial y_i} F_i(x, \mathbf{y})\right)h^2 + \mathcal{O}(h^3) \tag{c}$$

Returning to Eq. (a), we can rewrite the last term by applying a Taylor series in several variables:

$$\mathbf{F}\left[x + ph, \mathbf{y} + qh\mathbf{F}(x, \mathbf{y})\right] = \mathbf{F}(x, \mathbf{y}) + \frac{\partial \mathbf{F}}{\partial x}ph + qh\sum_{i=1}^{n} \frac{\partial \mathbf{F}}{\partial y_i} F_i(x, \mathbf{y}) + \mathcal{O}(h^2)$$

so that Eq. (a) becomes

$$\mathbf{y}(x+h) = \mathbf{y}(x) + (c_0 + c_1)\,\mathbf{F}(x,\mathbf{y})h + c_1\left[\frac{\partial \mathbf{F}}{\partial x}\,ph + qh\sum_{i=1}^{n}\frac{\partial \mathbf{F}}{\partial y_i}F_i(x,\mathbf{y})\right]h + \mathcal{O}(h^3) \quad \text{(d)}$$

Comparing Eqs. (c) and (d), we find that they are identical if

$$c_0 + c_1 = 1 \quad c_1 p = \frac{1}{2} \quad c_1 q = \frac{1}{2} \tag{e}$$

Because Eqs. (e) represent three equations in four unknown parameters, we can assign any value to one of the parameters. Some of the popular choices and the names associated with the resulting formulas are:

$$
\begin{array}{llll}
c_0 = 0 & c_1 = 1 & p = 1/2 & q = 1/2 \quad \text{Modified Euler's method} \\
c_0 = 1/2 & c_1 = 1/2 & p = 1 & q = 1 \quad\ \ \text{Heun's method} \\
c_0 = 1/3 & c_1 = 2/3 & p = 3/4 & q = 3/4 \quad \text{Ralston's method}
\end{array}
$$

All these formulas are classified as second-order Runge–Kutta methods, with no formula having a numerical superiority over the others. Choosing the *modified Euler's method*, we substitute the corresponding parameters into Eq. (a) to yield

$$\mathbf{y}(x+h) = \mathbf{y}(x) + \mathbf{F}\left[x + \frac{h}{2}, \mathbf{y} + \frac{h}{2}\mathbf{F}(x,\mathbf{y})\right]h \tag{f}$$

This integration formula can be conveniently evaluated by the following sequence of operations

$$\mathbf{K}_1 = h\mathbf{F}(x,\mathbf{y})$$
$$\mathbf{K}_2 = h\mathbf{F}\left(x + \frac{h}{2}, \mathbf{y} + \frac{1}{2}\mathbf{K}_1\right) \tag{7.9}$$
$$\mathbf{y}(x+h) = \mathbf{y}(x) + \mathbf{K}_2$$

Second-order methods are seldom used in computer application. Most programmers prefer integration formulas of order four, which achieve a given accuracy with less computational effort.

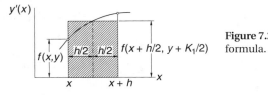

Figure 7.2. Graphical representation of modified Euler formula.

Figure 7.2 displays the graphical interpretation of modified Euler's formula for a single differential equation $y' = f(x, y)$. The first of Eqs. (7.9) yields an estimate of y at the midpoint of the panel by Euler's formula: $y(x + h/2) = y(x) + f(x, y)h/2 =$

$y(x) + K_1/2$. The second equation then approximates the area of the panel by the area K_2 of the cross-hatched rectangle. The error here is proportional to the curvature y''' of the plot.

Fourth-Order Runge–Kutta Method

The fourth-order Runge–Kutta method is obtained from the Taylor series along the same lines as the second-order method. Since the derivation is rather long and not very instructive, we skip it. The final form of the integration formula again depends on the choice of the parameters; that is, there is no unique Runge–Kutta fourth-order formula. The most popular version, which is known simply as the *Runge–Kutta method*, entails the following sequence of operations:

$$\mathbf{K}_1 = h\mathbf{F}(x, \mathbf{y})$$

$$\mathbf{K}_2 = h\mathbf{F}\left(x + \frac{h}{2}, \mathbf{y} + \frac{\mathbf{K}_1}{2}\right)$$

$$\mathbf{K}_3 = h\mathbf{F}\left(x + \frac{h}{2}, y + \frac{\mathbf{K}_2}{2}\right) \qquad (7.10)$$

$$\mathbf{K}_4 = h\mathbf{F}(x + h, \mathbf{y} + \mathbf{K}_3)$$

$$\mathbf{y}(x + h) = \mathbf{y}(x) + \frac{1}{6}(\mathbf{K}_1 + 2\mathbf{K}_2 + 2\mathbf{K}_3 + \mathbf{K}_4)$$

The main drawback of this method is that it does not lend itself to an estimate of the truncation error. Therefore, we must guess the integration step size h, or determine it by trial and error. In contrast, the so-called *adaptive methods* can evaluate the truncation error in each integration step and adjust the value of h accordingly (but at a higher cost of computation). One such adaptive method is introduced in the next article.

■ runKut4

The function `runKut4` implements the Runge–Kutta method of order four. The user must provide `runKut4` with the function `dEqs` that defines the first-order differential equations $\mathbf{y}' = \mathbf{F}(x, \mathbf{y})$.

```
function [xSol,ySol] = runKut4(dEqs,x,y,xStop,h)
% 4th-order Runge--Kutta integration.
% USAGE: [xSol,ySol] = runKut4(dEqs,x,y,xStop,h)
% INPUT:
% dEqs  = handle of function that specifies the
```

```
%          1st-order differential equations
%          F(x,y) = [dy1/dx dy2/dx dy3/dx ...].
% x,y   = initial values; y must be row vector.
% xStop = terminal value of x.
% h     = increment of x used in integration.
% OUTPUT:
% xSol = x-values at which solution is computed.
% ySol = values of y corresponding to the x-values.

if size(y,1) > 1 ; y = y'; end  % y must be row vector
xSol = zeros(2,1); ySol = zeros(2,length(y));
xSol(1) = x; ySol(1,:) = y;
i = 1;
while x < xStop
    i = i + 1;
    h = min(h,xStop - x);
    K1 = h*feval(dEqs,x,y);
    K2 = h*feval(dEqs,x + h/2,y + K1/2);
    K3 = h*feval(dEqs,x + h/2,y + K2/2);
    K4 = h*feval(dEqs,x+h,y + K3);
    y = y + (K1 + 2*K2 + 2*K3 + K4)/6;
    x = x + h;
    xSol(i) = x; ySol(i,:) = y;  % Store current soln.
end
```

EXAMPLE 7.3

Use the second-order Runge–Kutta method to integrate

$$y' = \sin y \quad y(0) = 1$$

from $x = 0$ to 0.5 in steps of $h = 0.1$. Keep four decimal places in the computations.

Solution In this problem we have

$$f(x, y) = \sin y$$

so that the integration formulas in Eqs. (7.9) are

$$K_1 = hf(x, y) = 0.1 \sin y$$

$$K_2 = hf\left(x + \frac{h}{2}, y + \frac{1}{2}K_1\right) = 0.1 \sin\left(y + \frac{1}{2}K_1\right)$$

$$y(x + h) = y(x) + K_2$$

Noting that $y(0) = 1$, we may proceed with the integration as follows:

$$K_1 = 0.1 \sin 1.0000 = 0.0841$$

$$K_2 = 0.1 \sin \left(1.0000 + \frac{0.0841}{2}\right) = 0.0863$$

$$y(0.1) = 1.0 + 0.0863 = 1.0863$$

$$K_1 = 0.1 \sin 1.0863 = 0.0885$$

$$K_2 = 0.1 \sin \left(1.0863 + \frac{0.0885}{2}\right) = 0.0905$$

$$y(0.2) = 1.0863 + 0.0905 = 1.1768$$

and so on. A summary of the computations is shown in the table below.

x	y	K_1	K_2
0.0	1.0000	0.0841	0.0863
0.1	1.0863	0.0885	0.0905
0.2	1.1768	0.0923	0.0940
0.3	1.2708	0.0955	0.0968
0.4	1.3676	0.0979	0.0988
0.5	1.4664		

The exact solution can be shown to be

$$x(y) = \ln(\csc y - \cot y) + 0.604582$$

which yields $x(1.4664) = 0.5000$. Therefore, up to this point the numerical solution is accurate to four decimal places. However, it is unlikely that this precision would be maintained if we were to continue the integration. Since the errors (due to truncation and roundoff) tend to accumulate, longer integration ranges require better integration formulas and more significant figures in the computations.

EXAMPLE 7.4

Solve

$$y'' = -0.1y' - x \qquad y(0) = 0 \qquad y'(0) = 1$$

from $x = 0$ to 2 in increments of $h = 0.25$ with the fourth-order Runge–Kutta method. (This problem was solved by the Taylor series method in Example 7.2.)

Solution Letting $y_1 = y$ and $y_2 = y'$, we write the equivalent first-order equations as

$$\mathbf{F}(x, \mathbf{y}) = \mathbf{y}' = \begin{bmatrix} y_1' \\ y_2' \end{bmatrix} = \begin{bmatrix} y_2 \\ -0.1y_2 - x \end{bmatrix}$$

which are coded in the following function:

```
function F = fex7_4(x,y)
% Differential. eqs. used in Example 7.4
F = zeros(1,2);
F(1) = y(2); F(2) = -0.1*y(2) - x;
```

Comparing the function `fex7_4` here with `fex7_2` in Example 7.2 we note that it is much simpler to input the differential equations for the Runge–Kutta method than for the Taylor series method. Here are the results of integration:

```
>> [x,y] = runKut4(@fex7_4,0,[0 1],2,0.25);
>> printSol(x,y,1)
       x              y1              y2
  0.0000e+000    0.0000e+000    1.0000e+000
  2.5000e-001    2.4431e-001    9.4432e-001
  5.0000e-001    4.6713e-001    8.2829e-001
  7.5000e-001    6.5355e-001    6.5339e-001
  1.0000e+000    7.8904e-001    4.2110e-001
  1.2500e+000    8.5943e-001    1.3281e-001
  1.5000e+000    8.5090e-001   -2.1009e-001
  1.7500e+000    7.4995e-001   -6.0625e-001
  2.0000e+000    5.4345e-001   -1.0543e+000
```

These results are the same as obtained by the Taylor series method in Example 7.2. This was expected, since both methods are of the same order.

EXAMPLE 7.5
Use the fourth-order Runge–Kutta method to integrate

$$y' = 3y - 4e^{-x} \qquad y(0) = 1$$

from $x = 0$ to 10 in steps of $h = 0.1$. Compare the result with the analytical solution $y = e^{-x}$.

Solution The function specifying the differential equation is

```
function F = fex7_5(x,y)
% Differential eq. used in Example 7.5.
F = 3*y - 4*exp(-x);
```

The solution is (every 20th line was printed):

```
>> [x,y] = runKut4(@fex7_5,0,1,10,0.1);
>> printSol(x,y,20)
      x               y1
   0.0000e+000    1.0000e+000
   2.0000e+000    1.3250e-001
   4.0000e+000   -1.1237e+000
   6.0000e+000   -4.6056e+002
   8.0000e+000   -1.8575e+005
   1.0000e+001   -7.4912e+007
```

It is clear that something went wrong. According to the analytical solution, y should decrease to zero with increasing x, but the output shows the opposite trend: after an initial decrease, the magnitude of y increases dramatically. The explanation is found by taking a closer look at the analytical solution. The general solution of the given differential equation is

$$y = Ce^{3x} + e^{-x}$$

which can be verified by substitution. The initial condition $y(0) = 1$ yields $C = 0$, so that the solution to the problem is indeed $y = e^{-x}$.

The cause of trouble in the numerical solution is the dormant term Ce^{3x}. Suppose that the initial condition contains a small error ε, so that we have $y(0) = 1 + \varepsilon$. This changes the analytical solution to

$$y = \varepsilon e^{3x} + e^{-x}$$

We now see that the term containing the error ε becomes dominant as x is increased. Since errors inherent in the numerical solution have the same effect as small changes in initial conditions, we conclude that our numerical solution is the victim of *numerical instability* due to sensitivity of the solution to initial conditions. The lesson here is: do not always trust the results of numerical integration.

EXAMPLE 7.6

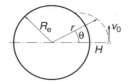

A spacecraft is launched at an altitude $H = 772$ km above sea level with the speed $v_0 = 6700$ m/s in the direction shown. The differential equations describing the motion of the spacecraft are

$$\ddot{r} = r\dot{\theta}^2 - \frac{GM_e}{r^2} \qquad \ddot{\theta} = -\frac{2\dot{r}\dot{\theta}}{r}$$

where r and θ are the polar coordinates of the spacecraft. The constants involved in the motion are

$$G = 6.672 \times 10^{-11}\ \text{m}^3\text{kg}^{-1}\text{s}^{-2} = \text{universal gravitational constant}$$

$$M_e = 5.9742 \times 10^{24}\ \text{kg} = \text{mass of the earth}$$

$$R_e = 6378.14\ \text{km} = \text{radius of the earth at sea level}$$

(1) Derive the first-order differential equations and the initial conditions of the form $\dot{\mathbf{y}} = \mathbf{F}(t, \mathbf{y})$, $\mathbf{y}(0) = \mathbf{b}$. (2) Use the fourth-order Runge–Kutta method to integrate the equations from the time of launch until the spacecraft hits the earth. Determine θ at the impact site.

Solution of Part (1) We have

$$GM_e = \left(6.672 \times 10^{-11}\right)\left(5.9742 \times 10^{24}\right) = 3.9860 \times 10^{14}\ \text{m}^3\text{s}^{-2}$$

Letting

$$\mathbf{y} = \begin{bmatrix} y_1 \\ y_2 \\ y_3 \\ y_4 \end{bmatrix} = \begin{bmatrix} r \\ \dot{r} \\ \theta \\ \dot{\theta} \end{bmatrix}$$

the equivalent first-order equations become

$$\dot{\mathbf{y}} = \begin{bmatrix} \dot{y}_1 \\ \dot{y}_2 \\ \dot{y}_3 \\ \dot{y}_4 \end{bmatrix} = \begin{bmatrix} y_1 \\ y_0 y_3^2 - 3.9860 \times 10^{14}/y_0^2 \\ y_3 \\ -2y_1 y_3/y_0 \end{bmatrix}$$

with the initial conditions

$$r(0) = R_e + H = R_e = (6378.14 + 772) \times 10^3 = 7.15014 \times 10^6 \text{ m}$$

$$\dot{r}(0) = 0$$

$$\theta(0) = 0$$

$$\dot{\theta}(0) = v_0/r(0) = (6700)/(7.15014 \times 10^6) = 0.937045 \times 10^{-3} \text{ rad/s}$$

Therefore,

$$\mathbf{y}(0) = \begin{bmatrix} 7.15014 \times 10^6 \\ 0 \\ 0 \\ 0.937045 \times 10^{-3} \end{bmatrix}$$

Solution of Part (2) The function that returns the differential equations is

```
function F = fex7_6(x,y)
% Differential eqs. used in Example 7.6.
F = zeros(1,4);
F(1) = y(2);
F(2) = y(1)*y(4)^2 - 3.9860e14/y(1)^2;
F(3) = y(4);
F(4) = -2*y(2)*y(4)/y(1);
```

The program used for numerical integration is listed below. Note that the independent variable t is denoted by x.

```
% Example 7.6 (Runge-Kutta integration)
x = 0; y = [7.15014e6 0 0 0.937045e-3];
xStop = 1200; h = 50; freq = 2;
[xSol,ySol] = runKut4(@fex7_6,x,y,xStop,h);
printSol(xSol,ySol,freq)
```

Here is the output:

```
>>    x              y1             y2             y3             y4
0.0000e+000   7.1501e+006   0.0000e+000   0.0000e+000   9.3704e-004
1.0000e+002   7.1426e+006  -1.5173e+002   9.3771e-002   9.3904e-004
2.0000e+002   7.1198e+006  -3.0276e+002   1.8794e-001   9.4504e-004
3.0000e+002   7.0820e+006  -4.5236e+002   2.8292e-001   9.5515e-004
```

```
4.0000e+002   7.0294e+006  -5.9973e+002   3.7911e-001   9.6951e-004
5.0000e+002   6.9622e+006  -7.4393e+002   4.7697e-001   9.8832e-004
6.0000e+002   6.8808e+006  -8.8389e+002   5.7693e-001   1.0118e-003
7.0000e+002   6.7856e+006  -1.0183e+003   6.7950e-001   1.0404e-003
8.0000e+002   6.6773e+006  -1.1456e+003   7.8520e-001   1.0744e-003
9.0000e+002   6.5568e+006  -1.2639e+003   8.9459e-001   1.1143e-003
1.0000e+003   6.4250e+006  -1.3708e+003   1.0083e+000   1.1605e-003
1.1000e+003   6.2831e+006  -1.4634e+003   1.1269e+000   1.2135e-003
1.2000e+003   6.1329e+006  -1.5384e+003   1.2512e+000   1.2737e-003
```

The spacecraft hits the earth when r equals $R_e = 6.378\,14 \times 10^6$ m. This occurs between $t = 1000$ and 1100 s. A more accurate value of t can be obtained by polynomial interpolation. If no great precision is needed, linear interpolation will do. Letting $1000 + \Delta t$ be the time of impact, we can write

$$r(1000 + \Delta t) = R_e$$

Expanding r in a two-term Taylor series, we get

$$r(1000) + \dot{r}(1000)\Delta t = R_e$$

$$6.4250 \times 10^6 + \left(-1.3708 \times 10^3\right)\Delta t = 6378.14 \times 10^3$$

from which

$$\Delta t = 34.184 \text{ s}$$

Thus the time of impact is 1034.2 s.

The coordinate θ of the impact site can be estimated in a similar manner. Using again two terms of the Taylor series, we have

$$\theta(1000 + \Delta t) = \theta(1000) + \dot{\theta}(1000)\Delta t$$

$$= 1.0083 + \left(1.1605 \times 10^{-3}\right)(34.184)$$

$$= 1.0480 \text{ rad} = 60.00°$$

PROBLEM SET 7.1

1. Given

$$y' + 4y = x^2 \qquad y(0) = 1$$

compute $y(0.1)$ using one step of the Taylor series method of order (a) two and (b) four. Compare the result with the analytical solution

$$y(x) = \frac{31}{32}e^{-4x} + \frac{1}{4}x^2 - \frac{1}{8}x + \frac{1}{32}$$

2. Solve Prob. 1 with one step of the Runge–Kutta method of order (a) two and (b) four.

3. Integrate

$$y' = \sin y \qquad y(0) = 1$$

from $x = 0$ to 0.5 with the second-order Taylor series method using $h = 0.1$. Compare the result with Example 7.3.

4. Verify that the problem

$$y' = y^{1/3} \qquad y(0) = 0$$

has two solutions: $y = 0$ and $y = (2x/3)^{3/2}$. Which of the solutions would be reproduced by numerical integration if the initial condition is set at (a) $y = 0$ and (b) $y = 10^{-16}$? Verify your conclusions by integrating with any numerical method.

5. Convert the following differential equations into first-order equations of the form $\mathbf{y}' = \mathbf{F}(x, \mathbf{y})$:

$$
\begin{array}{ll}
\text{(a)} & \ln y' + y = \sin x \\
\text{(b)} & y''y - xy' - 2y^2 = 0 \\
\text{(c)} & y^{(4)} - 4y''\sqrt{1 - y^2} = 0 \\
\text{(d)} & \left(y''\right)^2 = |32y'x - y^2|
\end{array}
$$

6. In the following sets of coupled differential equations t is the independent variable. Convert these equations into first-order equations of the form $\dot{\mathbf{y}} = \mathbf{F}(t, \mathbf{y})$:

$$
\begin{array}{lll}
\text{(a)} & \ddot{y} = x - 2y & \ddot{x} = y - x \\
\text{(b)} & \ddot{y} = -y\left(\dot{y}^2 + \dot{x}^2\right)^{1/4} & \ddot{x} = -x\left(\dot{y}^2 + \dot{x}\right)^{1/4} - 32 \\
\text{(c)} & \dot{y}^2 + t\sin y = 4\dot{x} & x\ddot{x} + t\cos y = 4\dot{y}
\end{array}
$$

7. ■ The differential equation for the motion of a simple pendulum is

$$\frac{d^2\theta}{dt^2} = -\frac{g}{L}\sin\theta$$

where

$$\theta = \text{angular displacement from the vertical}$$

$$g = \text{gravitational acceleration}$$

$$L = \text{length of the pendulum}$$

With the transformation $\tau = t\sqrt{g/L}$ the equation becomes

$$\frac{d^2\theta}{d\tau^2} = -\sin\theta$$

Use numerical integration to determine the period of the pendulum if the amplitude is $\theta_0 = 1$ rad. Note that for small amplitudes ($\sin\theta \approx \theta$) the period is $2\pi\sqrt{L/g}$.

8. ■ A skydiver of mass m in a vertical free fall experiences an aerodynamic drag force $F_D = c_D\dot{y}^2$, where y is measured downward from the start of the fall. The differential equation describing the fall is

$$\ddot{y} = g - \frac{c_D}{m}\dot{y}^2$$

Determine the time of a 500 m fall. Use $g = 9.80665$ m/s^2, $c_D = 0.2028$ kg/m and $m = 80$ kg.

9. ■

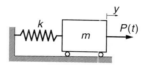

The spring–mass system is at rest when the force $P(t)$ is applied, where

$$P(t) = \begin{cases} 10t \text{ N} & \text{when } t < 2 \text{ s} \\ 20 \text{ N} & \text{when } t \geq 2 \text{ s} \end{cases}$$

The differential equation for the ensuing motion is

$$\ddot{y} = \frac{P(t)}{m} - \frac{k}{m}y$$

Determine the maximum displacement of the mass. Use $m = 2.5$ kg and $k = 75$ N/m.

10. ■

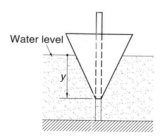

The conical float is free to slide on a vertical rod. When the float is disturbed from its equilibrium position, it undergoes oscillating motion described by the differential equation

$$\ddot{y} = g\left(1 - ay^3\right)$$

where $a = 16$ m^{-3} (determined by the density and dimensions of the float) and $g = 9.80665$ m/s^2. If the float is raised to the position $y = 0.1$ m and released, determine the period and the amplitude of the oscillations.

11. ■

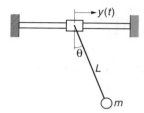

The pendulum is suspended from a sliding collar. The system is at rest when the oscillating motion $y(t) = Y \sin \omega t$ is imposed on the collar, starting at $t = 0$. The differential equation describing the motion of the pendulum is

$$\ddot{\theta} = -\frac{g}{L} \sin \theta + \frac{\omega^2}{L} Y \cos \theta \sin \omega t$$

Plot θ vs. t from $t = 0$ to 10 s and determine the largest θ during this period. Use $g = 9.80665$ m/s^2, $L = 1.0$ m, $Y = 0.25$ m and $\omega = 2.5$ rad/s.

12. ■

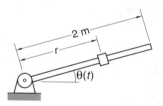

The system consisting of a sliding mass and a guide rod is at rest with the mass at $r = 0.75$ m. At time $t = 0$ a motor is turned on that imposes the motion $\theta(t) = (\pi/12) \cos \pi t$ on the rod. The differential equation describing the resulting motion of the slider is

$$\ddot{r} = \left(\frac{\pi^2}{12}\right)^2 r \sin^2 \pi t - g \sin \left(\frac{\pi}{12} \cos \pi t\right)$$

Determine the time when the slider reaches the tip of the rod. Use $g = 9.80665$ m/s^2.

13. ■

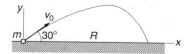

A ball of mass $m = 0.25$ kg is launched with the velocity $v_0 = 50$ m/s in the direction shown. If the aerodynamic drag force acting on the ball is $F_D = C_D v^{3/2}$, the differential equations describing the motion are

$$\ddot{x} = -\frac{C_D}{m}\dot{x}v^{1/2} \qquad \ddot{y} = -\frac{C_D}{m}\dot{y}v^{1/2} - g$$

where $v = \sqrt{\dot{x}^2 + \dot{y}^2}$. Determine the time of flight and the range R. Use $C_D = 0.03$ kg/(m·s)$^{1/2}$ and $g = 9.80665$ m/s^2.

14. ■ The differential equation describing the angular position θ of a mechanical arm is

$$\ddot{\theta} = \frac{a(b - \theta) - \theta\dot{\theta}^2}{1 + \theta^2}$$

where $a = 100$ s^{-2} and $b = 15$. If $\theta(0) = 2\pi$ and $\dot{\theta}(0) = 0$, compute θ and $\dot{\theta}$ when $t = 0.5$ s.

15. ■

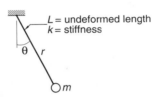

$L =$ undeformed length
$k =$ stiffness

The mass m is suspended from an elastic cord with an extensional stiffness k and undeformed length L. If the mass is released from rest at $\theta = 60°$ with the cord unstretched, find the length r of the cord when the position $\theta = 0$ is reached for the first time. The differential equations describing the motion are

$$\ddot{r} = r\dot{\theta}^2 + g\cos\theta - \frac{k}{m}(r - L)$$

$$\ddot{\theta} = \frac{-2\dot{r}\dot{\theta} - g\sin\theta}{r}$$

Use $g = 9.80665$ m/s^2, $k = 40$ N/m, $L = 0.5$ m and $m = 0.25$ kg.

16. ■ Solve Prob. 15 if the pendulum is released from the position $\theta = 60°$ with the cord stretched by 0.075 m.

17. ■

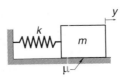

Consider the mass–spring system where dry friction is present between the block and the horizontal surface. The frictional force has a constant magnitude μmg (μ is the coefficient of friction) and always opposes the motion. The differential equation for the motion of the block can be expressed as

$$\ddot{y} = -\frac{k}{m}y - \mu g\frac{\dot{y}}{|\dot{y}|}$$

where y is measured from the position where the spring is unstretched. If the block is released from rest at $y = y_0$, verify by numerical integration that the next positive peak value of y is $y_0 - 4\mu mg/k$ (this relationship can be derived analytically). Use $k = 3000$ N/m, $m = 6$ kg, $\mu = 0.5$, $g = 9.80665$ m/s^2 and $y_0 = 0.1$ m.

18. ■ Integrate the following problems from $x = 0$ to 20 and plot y vs. x:

(a) $y'' + 0.5(y^2 - 1)y' + y = 0$ $y(0) = 1$ $y'(0) = 0$
(b) $y'' = y\cos 2x$ $y(0) = 0$ $y'(0) = 1$

These differential equations arise in nonlinear vibration analysis.

19. ■ The solution of the problem

$$y'' + \frac{1}{x}y' + y \qquad y(0) = 1 \qquad y'(0) = 0$$

is the Bessel function $J_0(x)$. Use numerical integration to compute $J_0(5)$ and compare the result with -0.17760, the value listed in mathematical tables. *Hint*: to avoid singularity at $x = 0$, start the integration at $x = 10^{-12}$.

20. ■ Consider the initial value problem

$$y'' = 16.81y \qquad y(0) = 1.0 \qquad y'(0) = -4.1$$

(a) Derive the analytical solution. (b) Do you anticipate difficulties in numerical solution of this problem? (c) Try numerical integration from $x = 0$ to 8 to see if your concerns were justified.

21. ■

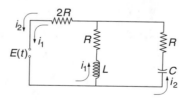

Kirchoff's equations for the circuit shown are

$$L\frac{di_1}{dt} + Ri_1 + 2R(i_1 + i_2) = E(t) \tag{a}$$

$$\frac{q_2}{C} + Ri_2 + 2R(i_2 + i_1) = E(t) \tag{b}$$

Differentiating Eq. (b) and substituting the charge–current relationship $dq_2/dt = i_2$, we get

$$\frac{di_1}{dt} = \frac{-3Ri_1 - 2Ri_2 + E(t)}{L} \tag{c}$$

$$\frac{di_2}{dt} = -\frac{2}{3}\frac{di_1}{dt} - \frac{i_2}{3RC} + \frac{1}{3R}\frac{dE}{dt} \tag{d}$$

We could substitute di_1/dt from Eq. (c) into Eq. (d), so that the latter would assume the usual form $di_2/dt = f(t, i_1, i_2)$, but it is more convenient to leave the equations as they are. Assuming that the voltage source is turned on at time $t = 0$, plot the loop currents i_1 and i_2 from $t = 0$ to 0.05 s. Use $E(t) = 240 \sin(120\pi t)$ V, $R = 1.0\,\Omega$, $L = 0.2 \times 10^{-3}$ H and $C = 3.5 \times 10^{-3}$ F.

22. ∎

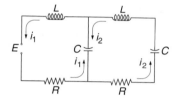

The constant voltage source E of the circuit shown is turned on at $t = 0$, causing transient currents i_1 and i_2 in the two loops that last about 0.05 s. Plot these currents from $t = 0$ to 0.05 s, using the following data: $E = 9$ V, $R = 0.25\,\Omega$, $L = 1.2 \times 10^{-3}$ H and $C = 5 \times 10^{-3}$ F. Kirchoff's equations for the two loops are

$$L\frac{di_1}{dt} + Ri_1 + \frac{q_1 - q_2}{C} = E$$

$$L\frac{di_2}{dt} + Ri_2 + \frac{q_2 - q_1}{C} + \frac{q_2}{C} = 0$$

Additional two equations are the current–charge relationships

$$\frac{dq_1}{dt} = i_1 \qquad \frac{di_2}{dt} = i_2$$

7.4   Stability and Stiffness

Loosely speaking, a method of numerical integration is said to be stable if the effects of local errors do not accumulate catastrophically; that is, if the global error remains bounded. If the method is unstable, the global error will increase exponentially, eventually causing numerical overflow. Stability has nothing to do with accuracy; in fact, an inaccurate method can be very stable.

Stability is determined by three factors: the differential equations, the method of solution and the value of the increment h. Unfortunately, it is not easy to determine stability beforehand, unless the differential equation is linear.

Stability of Euler's Method

As a simple illustration of stability, consider the problem

$$y' = -\lambda y \qquad y(0) = \beta \tag{7.11}$$

where λ is a positive constant. The exact solution of this problem is

$$y(x) = \beta e^{-\lambda x}$$

Let us now investigate what happens when we attempt to solve Eq. (7.11) numerically with Euler's formula

$$y(x + h) = y(x) + hy'(x) \tag{7.12}$$

Substituting $y'(x) = -\lambda y(x)$, we get

$$y(x + h) = (1 - \lambda h)y(x)$$

If $|1 - \lambda h| > 1$, the method is clearly unstable since $|y|$ increases in every integration step. Thus Euler's method is stable only if $|1 - \lambda h| \leq 1$, or

$$h \leq 2/\lambda \tag{7.13}$$

The results can be extended to a system of n differential equations of the form

$$\mathbf{y}' = -\Lambda \mathbf{y} \tag{7.14}$$

where Λ is a constant matrix with the positive eigenvalues $\lambda_i, i = 1, 2, \ldots, n$. It can be shown that Euler's implicit method of integration formula is stable only if

$$h < 2/\lambda_{max} \tag{7.15}$$

where λ_{max} is the largest eigenvalue of Λ.

Stiffness

An initial value problem is called *stiff* if some terms in the solution vector $\mathbf{y}(x)$ vary much more rapidly with x than others. Stiffness can be easily predicted for the differential equations $\mathbf{y}' = -\Lambda \mathbf{y}$ with constant coefficient matrix Λ. The solution of these equations is $\mathbf{y}(x) = \sum_i C_i \mathbf{v}_i \exp(-\lambda_i x)$, where λ_i are the eigenvalues of Λ and $\mathbf{v}_i$ are the corresponding eigenvectors. It is evident that the problem is stiff if there is a large disparity in the magnitudes of the positive eigenvalues.

Numerical integration of stiff equations requires special care. The step size h needed for stability is determined by the largest eigenvalue λ_{max}, even if the terms $\exp(-\lambda_{max}x)$ in the solution decay very rapidly and becomes insignificant as we move away from the origin.

For example, consider the differential equation[17]

$$y'' + 1001y' + 1000y = 0 \qquad (7.16)$$

Using $y_1 = y$ and $y_2 = y'$, the equivalent first-order equations are

$$\mathbf{y}' = \begin{bmatrix} y_2 \\ -1000y_1 - 1001y_2 \end{bmatrix}$$

In this case

$$\Lambda = \begin{bmatrix} 0 & -1 \\ 1000 & 1001 \end{bmatrix}$$

The eigenvalues of Λ are the roots of

$$|\Lambda - \lambda \mathbf{I}| = \begin{vmatrix} -\lambda & -1 \\ 1000 & 1001 - \lambda \end{vmatrix} = 0$$

Expanding the determinant we get

$$-\lambda(1001 - \lambda) + 1000 = 0$$

which has the solutions $\lambda_1 = 1$ and $\lambda_2 = 1000$. These equations are clearly stiff. According to Eq. (7.15) we would need $h < 2/\lambda_2 = 0.002$ for Euler's method to be stable. The Runge–Kutta method would have approximately the same limitation on the step size.

When the problem is very stiff, the usual methods of solution, such as the Runge–Kutta formulas, become impractical due to the very small h required for stability. These problems are best solved with methods that are specially designed for stiff equations. Stiff problem solvers, which are outside the scope of this text, have much better stability characteristics; some of them are even unconditionally stable. However, the higher degree of stability comes at a cost—the general rule is that stability can be improved only by reducing the order of the method (and thus increasing the truncation error).

EXAMPLE 7.7

(1) Show that the problem

$$y'' = -\frac{19}{4}y - 10y' \qquad y(0) = -9 \qquad y'(0) = 0$$

---

[17] This example is taken from C.E. Pearson, *Numerical Methods in Engineering and Science*, van Nostrand and Reinhold (1986).

is moderately stiff and estimate h_{max}, the largest value of h for which the Runge–Kutta method would be stable. (2) Confirm the estimate by computing $y(10)$ with $h \approx h_{max}/2$ and $h \approx 2h_{max}$.

Solution of Part (1) With the notation $y = y_1$ and $y' = y_2$ the equivalent first-order differential equations are

$$\mathbf{y}' = \begin{bmatrix} y_2 \\ -\dfrac{19}{4}y_1 - 10y_2 \end{bmatrix} = -\Lambda \begin{bmatrix} y_1 \\ y_2 \end{bmatrix}$$

where

$$\Lambda = \begin{bmatrix} 0 & -1 \\ \dfrac{19}{4} & 10 \end{bmatrix}$$

The eigenvalues of Λ are given by

$$|\Lambda - \lambda \mathbf{I}| = \begin{vmatrix} -\lambda & -1 \\ \dfrac{19}{4} & 10 - \lambda \end{vmatrix} = 0$$

which yields $\lambda_1 = 1/2$ and $\lambda_2 = 19/2$. Because λ_2 is quite a bit larger than λ_1, the equations are moderately stiff.

Solution of Part (2) An estimate for the upper limit of the stable range of h can be obtained from Eq. (7.15):

$$h_{max} = \frac{2}{\lambda_{max}} = \frac{2}{19/2} = 0.2153$$

Although this formula is strictly valid for Euler's method, it is usually not too far off for higher-order integration formulas.

Here are the results from the Runge–Kutta method with $h = 0.1$ (by specifying `freq = 0` in `printSol`, only the initial and final values were printed):

```
>>      x               y1               y2
    0.0000e+000    -9.0000e+000     0.0000e+000
    1.0000e+001    -6.4011e-002     3.2005e-002
```

The analytical solution is

$$y(x) = -\frac{19}{2}e^{-x/2} + \frac{1}{2}e^{-19x/2}$$

yielding $y(10) = -0.0640\,11$, which agrees with the value obtained numerically.

With $h = 0.5$ we encountered instability, as expected:

```
>>      x              y1              y2
    0.0000e+000    -9.0000e+000    0.0000e+000
    1.0000e+001    2.7030e+020    -2.5678e+021
```

7.5 Adaptive Runge–Kutta Method

Determination of a suitable step size h can be a major headache in numerical integration. If h is too large, the truncation error may be unacceptable; if h is too small, we are squandering computational resources. Moreover, a constant step size may not be appropriate for the entire range of integration. For example, if the solution curve starts off with rapid changes before becoming smooth (as in a stiff problem), we should use a small h at the beginning and increase it as we reach the smooth region. This is where *adaptive methods* come in. They estimate the truncation error at each integration step and automatically adjust the step size to keep the error within prescribed limits.

The adaptive Runge–Kutta methods use so-called *embedded integration formulas*. These formulas come in pairs: one formula has the integration order m, the other one is of order $m+1$. The idea is to use both formulas to advance the solution from x to $x + h$. Denoting the results by $\mathbf{y}_m(x + h)$ and $\mathbf{y}_{m+1}(x + h)$, we may estimate the truncation error in the formula of order m as

$$\mathbf{E}(h) = \mathbf{y}_{m+1}(x + h) - \mathbf{y}_m(x + h) \tag{7.17}$$

What makes the embedded formulas attractive is that they share the points where $\mathbf{F}(x, \mathbf{y})$ is evaluated. This means that once $\mathbf{y}_m(x + h)$ has been computed, relatively small additional effort is required to calculate $\mathbf{y}_{m+1}(x + h)$.

Here are the Runge–Kutta embedded formulas of orders 5 and 4 that were originally derived by Fehlberg; hence they are known as *Runge–Kutta–Fehlberg formulas*:

$$\mathbf{K}_1 = h\mathbf{F}(x, \mathbf{y})$$

$$\mathbf{K}_i = h\mathbf{F}\left(x + A_i h, \mathbf{y} + \sum_{j=0}^{i-1} B_{ij}\mathbf{K}_j\right), \quad i = 2, 3, \ldots, 6 \tag{7.1}$$

$$\mathbf{y}_5(x + h) = \mathbf{y}(x) + \sum_{i=1}^{6} C_i \mathbf{K}_i \quad \text{(5th-order formula)} \tag{7.19a}$$

$$\mathbf{y}_4(x + h) = \mathbf{y}(x) + \sum_{i=1}^{6} D_i \mathbf{K}_i \quad \text{(4th-order formula)} \tag{7.19b}$$

The coefficients appearing in these formulas are not unique. The tables below give the coefficients proposed by Cash and Karp[18] which are claimed to be an improvement over Fehlberg's original values.

i	A_i	B_{ij}					C_i	D_i
1	—	—	—	—	—	—	$\dfrac{37}{378}$	$\dfrac{2825}{27\,648}$
2	$\dfrac{1}{5}$	$\dfrac{1}{5}$	—	—	—	—	0	0
3	$\dfrac{3}{10}$	$\dfrac{3}{40}$	$\dfrac{9}{40}$	—	—	—	$\dfrac{250}{621}$	$\dfrac{18\,575}{48\,384}$
4	$\dfrac{3}{5}$	$\dfrac{3}{10}$	$-\dfrac{9}{10}$	$\dfrac{6}{5}$	—	—	$\dfrac{125}{594}$	$\dfrac{13\,525}{55\,296}$
5	1	$-\dfrac{11}{54}$	$\dfrac{5}{2}$	$-\dfrac{70}{27}$	$\dfrac{35}{27}$	—	0	$\dfrac{277}{14\,336}$
6	$\dfrac{7}{8}$	$\dfrac{1631}{55296}$	$\dfrac{175}{512}$	$\dfrac{575}{13824}$	$\dfrac{44275}{110592}$	$\dfrac{253}{4096}$	$\dfrac{512}{1771}$	$\dfrac{1}{4}$

Table 7.1. Cash–Karp coefficients for Runge–Kutta–Fehlberg formulas

The solution is advanced with the fifth-order formula in Eq. (7.19a). The fourth-order formula is used only implicitly in estimating the truncation error

$$\mathbf{E}(h) = \mathbf{y}_5(x+h) - \mathbf{y}_4(x+h) = \sum_{i=1}^{6}(C_i - D_i)\mathbf{K}_i \tag{7.20}$$

Since Eq. (7.20) actually applies to the fourth-order formula, it tends to overestimate the error in the fifth-order formula.

Note that $\mathbf{E}(h)$ is a vector, its components $E_i(h)$ representing the errors in the dependent variables y_i. This brings up the question: what is the error measure $e(h)$ that we wish to control? There is no single choice that works well in all problems. If we want to control the largest component of $\mathbf{E}(h)$, the error measure would be

$$e(h) = \max_i |E_i(h)| \tag{7.21}$$

[18] J.R. Cash and A.H. Carp, *ACM Transactions on Mathematical Software* **16**, 201–222 (1990).

We could also control some gross measure of the error, such as the root-mean-square error defined by

$$\bar{E}(h) = \sqrt{\frac{1}{n} \sum_{i=1}^{n} E_i^2(h)} \qquad (7.22)$$

where n is the number of first-order equations. Then we would use

$$e(h) = \bar{E}(h) \qquad (7.23)$$

for the error measure. Since the root-mean-square error is easier to handle, we adopt it for our program.

Error control is achieved by adjusting the increment h so that the per-step error e is approximately equal to a prescribed tolerance ε. Noting that the truncation error in the fourth-order formula is $\mathcal{O}(h^5)$, we conclude that

$$\frac{e(h_1)}{e(h_2)} \approx \left(\frac{h_1}{h_2}\right)^5 \qquad (a)$$

Let us now suppose that we performed an integration step with h_1 that resulted in the error $e(h_1)$. The step size h_2 that we should have used can now be obtained from Eq. (a) by setting $e(h_2) = \varepsilon$:

$$h_2 = h_1 \left[\frac{\varepsilon}{e(h_1)}\right]^{1/5} \qquad (b)$$

If $h_2 \geq h_1$, we could repeat the integration step with h_2, but since the error associated with h_1 was below the tolerance, that would be a waste of a perfectly good result. So we accept the current step and try h_2 in the next step. On the other hand, if $h_2 < h_1$, we must scrap the current step and repeat it with h_2. As Eq. (b) is only an approximation, it is prudent to incorporate a small margin of safety. In our program we use the formula

$$h_2 = 0.9 h_1 \left[\frac{\varepsilon}{e(h_1)}\right]^{1/5} \qquad (7.24)$$

Recall that $e(h)$ applies to a single integration step; that is, it is a measure of the local truncation error. The all-important global truncation error is due to the accumulation of the local errors. What should ε be set at in order to achieve a global error no greater than $\varepsilon_{\text{global}}$? Since $e(h)$ is a conservative estimate of the actual error, setting $\varepsilon = \varepsilon_{\text{global}}$ will usually be adequate. If the number integration steps is large, it is advisable to decrease ε accordingly.

Is there any reason to use the nonadaptive methods at all? Usually no; however, there are special cases where adaptive methods break down. For example, adaptive methods generally do not work if $\mathbf{F}(x, \mathbf{y})$ contains discontinuous functions. Because the error behaves erratically at the point of discontinuity, the program can get stuck in an infinite loop trying to find the appropriate value of h. We would also use a nonadaptive method if the output is to have evenly spaced values of x.

■ runKut5

The adaptive Runge–Kutta method is implemented in the function runKut5 listed below. The input argument h is the trial value of the increment for the first integration step.

```
function [xSol,ySol] = runKut5(dEqs,x,y,xStop,h,eTol)
% 5th-order Runge-Kutta integration.
% USAGE: [xSol,ySol] = runKut5(dEqs,x,y,xStop,h,eTol)
% INPUT:
% dEqs  = handle of function that specifyies the
%         1st-order differential equations
%         F(x,y) = [dy1/dx dy2/dx dy3/dx ...].
% x,y   = initial values; y must be row vector.
% xStop = terminal value of x.
% h     = trial value of increment of x.
% eTol  = per-step error tolerance (default = 1.0e-6).
% OUTPUT:
% xSol = x-values at which solution is computed.
% ySol = values of y corresponding to the x-values.

if size(y,1) > 1 ; y = y'; end  % y must be row vector
if nargin < 6; eTol = 1.0e-6; end
n = length(y);
A = [0 1/5 3/10 3/5 1 7/8];
B = [    0           0        0         0          0
         1/5         0        0         0          0
         3/40        9/40     0         0          0
         3/10       -9/10     6/5       0          0
        -11/54       5/2     -70/27     35/27      0
      1631/55296 175/512 575/13824 44275/110592 253/4096];
C = [37/378 0 250/621 125/594 0 512/1771];
D = [2825/27648 0 18575/48384 13525/55296 277/14336 1/4];
% Initialize solution
xSol = zeros(2,1); ySol = zeros(2,n);
xSol(1) = x; ySol(1,:) = y;
stopper = 0; k = 1;
```

```
for p = 2:5000
    % Compute K's from Eq. (7.18)
    K = zeros(6,n);
    K(1,:) = h*feval(dEqs,x,y);
    for i = 2:6
        BK = zeros(1,n);
        for j = 1:i-1
            BK = BK + B(i,j)*K(j,:);
        end
        K(i,:) = h*feval(dEqs, x + A(i)*h, y + BK);
    end
    % Compute change in y and per-step error from
    % Eqs.(7.19) & (7.20)
    dy = zeros(1,n); E = zeros(1,n);
    for i = 1:6
        dy = dy + C(i)*K(i,:);
        E = E + (C(i) - D(i))*K(i,:);
    end
    e = sqrt(sum(E.*E)/n);
    % If error within tolerance, accept results and
    % check for termination
    if e <= eTol
        y = y + dy; x = x + h;
        k = k + 1;
        xSol(k) = x; ySol(k,:) = y;
        if stopper == 1;
            break
        end
    end
    % Size of next integration step from Eq. (7.24)
    if e~= 0; hNext = 0.9*h*(eTol/e)^0.2;
    else; hNext=h;
    end
    % Check if next step is the last one (works
    % with positive and negative h)
    if (h > 0) == (x + hNext >= xStop )
        hNext = xStop - x; stopper = 1;
    end
    h = hNext;
end
```

EXAMPLE 7.8

The aerodynamic drag force acting on a certain object in free fall can be approximated by

$$F_D = av^2 e^{-by}$$

where

$$v = \text{velocity of the object in m/s}$$
$$y = \text{elevation of the object in meters}$$
$$a = 7.45 \text{ kg/m}$$
$$b = 10.53 \times 10^{-5} \text{ m}^{-1}$$

The exponential term accounts for the change of air density with elevation. The differential equation describing the fall is

$$m\ddot{y} = -mg + F_D$$

where $g = 9.80665$ m/s^2 and $m = 114$ kg is the mass of the object. If the object is released at an elevation of 9 km, determine its elevation and speed after a 10s fall with the adaptive Runge–Kutta method.

Solution The differential equation and the initial conditions are

$$\ddot{y} = -g + \frac{a}{m}\dot{y}^2 \exp(-by)$$

$$= -9.80665 + \frac{7.45}{114}\dot{y}^2 \exp(-10.53 \times 10^{-5} y)$$

$$y(0) = 9000 \text{ m} \qquad \dot{y}(0) = 0$$

Letting $y_1 = y$ and $y_2 = \dot{y}$, we obtain the equivalent first-order equations and the initial conditions as

$$\dot{\mathbf{y}} = \begin{bmatrix} \dot{y}_1 \\ \dot{y}_2 \end{bmatrix} = \begin{bmatrix} y_2 \\ -9.80665 + (65.351 \times 10^{-3}) y_2^2 \exp(-10.53 \times 10^{-5} y_1) \end{bmatrix}$$

$$\mathbf{y}(0) = \begin{bmatrix} 9000 \text{ m} \\ 0 \end{bmatrix}$$

The function describing the differential equations is

```
function F = fex7_8(x,y)
% Diff. eqs. used in Example 7.8
F = zeros(1,2);
F(1) = y(2);
F(2) = -9.80665...
       + 65.351e-3 * y(2)^2 * exp(-10.53e-5 * y(1));
```

The commands for performing the integration and displaying the results are shown below. We specified a per-step error tolerance of 10^{-2} in runKut5. Considering the magnitude of **y**, this should be enough for five decimal point accuracy in the solution.

```
>> [x,y] = runKut5(@fex7_8,0,[9000 0],10,0.5,1.0e-2);
>> printSol(x,y,1)
```

Execution of the commands resulted in the following output:

```
>>      x              y1              y2
    0.0000e+000    9.0000e+003     0.0000e+000
    5.0000e-001    8.9988e+003    -4.8043e+000
    1.9246e+000    8.9841e+003    -1.4632e+001
    3.2080e+000    8.9627e+003    -1.8111e+001
    4.5031e+000    8.9384e+003    -1.9195e+001
    5.9732e+000    8.9099e+003    -1.9501e+001
    7.7786e+000    8.8746e+003    -1.9549e+001
    1.0000e+001    8.8312e+003    -1.9519e+001
```

The first integration step was carried out with the prescribed trial value $h = 0.5$ s. Apparently the error was well within the tolerance, so that the step was accepted. Subsequent step sizes, determined from Eq. (7.24), were considerably larger.

Inspecting the output, we see that at $t = 10$ s the object is moving with the speed $v = -\dot{y} = 19.52$ m/s at an elevation of $y = 8831$ m.

EXAMPLE 7.9

Integrate the moderately stiff problem

$$y'' = -\frac{19}{4}y - 10y' \qquad y(0) = -9 \qquad y'(0) = 0$$

from $x = 0$ to 10 with the adaptive Runge–Kutta method and plot the results (this problem also appeared in Example 7.7).

Solution Since we use an adaptive method, there is no need to worry about the stable range of h, as we did in Example 7.7. As long as we specify a reasonable tolerance for the per-step error, the algorithm will find the appropriate step size. Here are the commands and the resulting output:

```
>> [x,y] = runKut5(@fex7_7,0,[-9 0],10,0.1);
>> printSol(x,y,4)
```

```
>>         x               y1              y2
    0.0000e+000   -9.0000e+000    0.0000e+000
    9.8941e-002   -8.8461e+000    2.6651e+000
    2.1932e-001   -8.4511e+000    3.6653e+000
    3.7058e-001   -7.8784e+000    3.8061e+000
    5.7229e-001   -7.1338e+000    3.5473e+000
    8.6922e-001   -6.1513e+000    3.0745e+000
    1.4009e+000   -4.7153e+000    2.3577e+000
    2.8558e+000   -2.2783e+000    1.1391e+000
    4.3990e+000   -1.0531e+000    5.2656e-001
    5.9545e+000   -4.8385e-001    2.4193e-001
    7.5596e+000   -2.1685e-001    1.0843e-001
    9.1159e+000   -9.9591e-002    4.9794e-002
    1.0000e+001   -6.4010e-002    3.2005e-002
```

The results are in agreement with the analytical solution.

The plots of y and y' show every fourth integration step. Note the high density of points near $x = 0$ where y' changes rapidly. As the y'-curve becomes smoother, the distance between the points increases.

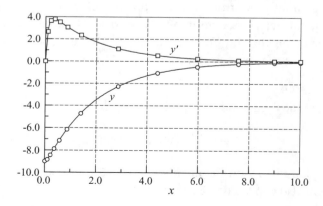

7.6   Bulirsch–Stoer Method

Midpoint Method

The midpoint formula of numerical integration of $\mathbf{y}' = \mathbf{F}(x, \mathbf{y})$ is

$$\mathbf{y}(x + h) = \mathbf{y}(x - h) + 2h\mathbf{F}\left[x, \mathbf{y}(x)\right] \tag{7.25}$$

It is a second-order formula, like the modified Euler's formula. We discuss it here because it is the basis of the powerful *Bulirsch–Stoer method,* which is the technique of choice in problems where high accuracy is required.

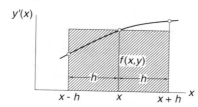

Figure 7.3. Graphical repesentation of the midpoint formula.

Figure 7.3 illustrates the midpoint formula for a single differential equation $y' = f(x, y)$. The change in y over the two panels shown is

$$y(x + h) - y(x - h) = \int_{x-h}^{x+h} y'(x)dx$$

which equals the area under the $y'(x)$ curve. The midpoint method approximates this area by the area $2hf(x, y)$ of the cross-hatched rectangle.

Figure 7.4. Mesh used in the midpoint method.

Consider now advancing the solution of $\mathbf{y}'(x) = \mathbf{F}(x, \mathbf{y})$ from $x = x_0$ to $x_0 + H$ with the midpoint formula. We divide the interval of integration into n steps of length $h = H/n$ each, as shown in Fig. 7.4, and carry out the computations

$$\mathbf{y}_1 = \mathbf{y}_0 + h\mathbf{F}_0$$

$$\mathbf{y}_2 = \mathbf{y}_0 + 2h\mathbf{F}_1$$

$$\mathbf{y}_3 = \mathbf{y}_1 + 2h\mathbf{F}_2 \tag{7.26}$$

$$\vdots$$

$$\mathbf{y}_n = \mathbf{y}_{n-2} + 2h\mathbf{F}_{n-1}$$

Here we used the notation $\mathbf{y}_i = \mathbf{y}(x_i)$ and $\mathbf{F}_i = \mathbf{F}(x_i, \mathbf{y}_i)$. The first of Eqs. (7.26) uses the Euler formula to "seed" the midpoint method; the other equations are midpoint formulas. The final result is obtained by averaging $\mathbf{y}_n$ in Eq. (7.26) and the estimate $\mathbf{y}_n \approx \mathbf{y}_{n-1} + h\mathbf{F}_n$ available from Euler formula:

$$\mathbf{y}(x_0 + H) = \frac{1}{2}\left[(\mathbf{y}_n + (\mathbf{y}_{n-1} + h\mathbf{F}_n)\right] \tag{7.27}$$

Richardson Extrapolation

It can be shown that the error in Eq. (7.27) is

$$\mathbf{E} = \mathbf{c}_1 h^2 + \mathbf{c}_2 h^4 + \mathbf{c}_3 h^6 + \cdots$$

Herein lies the great utility of the midpoint method: we can eliminate as many of the leading error terms as we wish by Richardson's extrapolation. For example, we could compute $\mathbf{y}(x_0 + H)$ with a certain value of h and then repeat the process with $h/2$. Denoting the corresponding results by $\mathbf{g}(h)$ and $\mathbf{g}(h/2)$, Richardson's extrapolation—see Eq. (5.9)—then yields the improved result

$$\mathbf{y}_{\text{better}}(x_0 + H) = \frac{4\mathbf{g}(h/2) - \mathbf{g}(h)}{3}$$

which is fourth-order accurate. Another round of integration with $h/4$ followed by Richardson's extrapolation get us sixth-order accuracy, etc.

The $\mathbf{y}$'s in Eqs. (7.26) should be viewed as a intermediate variables, because unlike $\mathbf{y}(x_0 + H)$, they cannot be refined by Richardson's extrapolation.

■ midpoint

The function `midpoint` in this module combines the midpoint method with Richardson extrapolation. The first application of the midpoint method uses two integration steps. The number of steps is doubled in successive integrations, each integration being followed by Richardson extrapolation. The procedure is stopped when two successive solutions differ (in the root-mean-square sense) by less than a prescribed tolerance.

```
function y = midpoint(dEqs,x,y,xStop,tol)
% Modified midpoint method for intergration of y' = F(x,y).
% USAGE: y = midpoint(dEqs,xStart,yStart,xStop,tol)
% INPUT:
% dEqs  = handle of function that returns the first-order
%         differential equations F(x,y) = [dy1/dx,dy2/dx,...].
% x, y  = initial values; y must be a row vector.
% xStop = terminal value of x.
% tol   = per-step error tolerance (default = 1.0e-6).
```

```
% OUTPUT:
% y = y(xStop).

if size(y,1) > 1 ; y = y'; end  % y must be row vector
if nargin <5; tol = 1.0e-6; end
kMax = 51;
n = length(y);
r = zeros(kMax,n);  % Storage for Richardson extrapolation.
% Start with two integration steps.
nSteps = 2;
r(1,1:n) = mid(dEqs,x,y,xStop,nSteps);
rOld = r(1,1:n);
for k = 2:kMax
    % Double the number of steps & refine results by
    % Richardson extrapolation.
    nSteps = 2*k;
    r(k,1:n) = mid(dEqs,x,y,xStop,nSteps);
    r = richardson(r,k,n);
    % Check for convergence.
    dr = r(1,1:n) - rOld;
    e = sqrt(dot(dr,dr)/n);
    if e < tol; y = r(1,1:n); return; end
    rOld = r(1,1:n);
end
error('Midpoint method did not converge')

function r = richardson(r,k,n)
% Richardson extrapolation.
for j = k-1:-1:1
    c =(k/(k-1))^(2*(k-j));
    r(j,1:n) =(c*r(j+1,1:n) - r(j,1:n))/(c - 1.0);
end
return

function y = mid(dEqs,x,y,xStop,nSteps)
% Midpoint formulas.
h = (xStop - x)/nSteps;
y0 = y;
y1 = y0 + h*feval(dEqs,x,y0);
for i = 1:nSteps-1
```

```
    x = x + h;
    y2 = y0 + 2.0*h*feval(dEqs,x,y1);
    y0 = y1;
    y1 = y2;
end
y = 0.5*(y1 + y0 + h*feval(dEqs,x,y2));
```

Bulirsch–Stoer Algorithm

When used on its own, the module `midpoint` has a major shortcoming: the solution at points between the initial and final values of x cannot be refined by Richardson extrapolation, so that **y** is usable only at the last point. This deficiency is rectified in the Bulirsch–Stoer method. The fundamental idea behind the method is simple: apply the midpoint method in a piecewise fashion. That is, advance the solution in stages of length H, using the midpoint method with Richardson extrapolation to perform the integration in each stage. The value of H can be quite large, since the precision of the result is determined mainly by the step length h in the midpoint method, not by H.

The original Bulirsch and Stoer technique[19] is a complex procedure that incorporates many refinements missing in our algorithm. However, the function `bulStoer` given below retains the essential ideas of Bulirsch and Stoer.

What are the relative merits of adaptive Runge–Kutta and Bulirsch–Stoer methods? The Runge–Kutta method is more robust, having higher tolerance for nonsmooth functions and stiff problems. In most applications where high precision is not required, it also tends to be more efficient. However, this is not the case in the computation of high-accuracy solutions involving smooth functions, where the Bulirsch–Stoer algorithm shines.

■ bulStoer

This function contains a simplified algorithm for the Bulirsch–Stoer method.

```
function [xSol,ySol] = bulStoer(dEqs,x,y,xStop,H,tol)
% Simplified Bulirsch-Stoer method for integration of y' = F(x,y).
% USAGE: [xSol,ySol] = bulStoer(dEqs,x,y,xStop,H,tol)
% INPUT:
% dEqs   = handle of function that returns the first-order
%            differential equations F(x,y) = [dy1/dx,dy2/dx,...].
```

---

[19] Stoer, J., and Bulirsch, R., *Introduction to Numerical Analysis*, Springer, 1980.

```
% x, y   = initial values; y must be a row vector.
% xStop = terminal value of x.
% H       = increment of x at which solution is stored.
% tol    = per-step error tolerance (default = 1.0e-6).
% OUTPUT:
% xSol, ySol = solution at increments H.

if size(y,1) > 1 ; y = y'; end  % y must be row vector
if nargin < 6; tol = 1.0e-6; end
n = length(y);
xSol = zeros(2,1); ySol = zeros(2,n);
xSol(1) = x; ySol(1,:) = y;
k = 1;
while x < xStop
    k = k + 1;
    H = min(H,xStop - x);
    y = midpoint(dEqs,x,y,x + H,tol);
    x = x + H;
    xSol(k) = x; ySol(k,:) = y;
end
```

EXAMPLE 7.10

Compute the solution of the initial value problem

$$y' = \sin y \qquad y(0) = 1$$

at $x = 0.5$ with the midpoint formulas using $n = 2$ and $n = 4$, followed by Richardson extrapolation (this problem was solved with the second-order Runge–Kutta method in Example 7.3).

Solution With $n = 2$ the step length is $h = 0.25$. The midpoint formulas, Eqs. (7.26) and (7.27), yield

$$y_1 = y_0 + hf_0 = 1 + 0.25 \sin 1.0 = 1.210\,368$$

$$y_2 = y_0 + 2hf_1 = 1 + 2(0.25) \sin 1.210\,368 = 1.467\,873$$

$$y_h(0.5) = \frac{1}{2}(y_1 + y_0 + hf_2)$$

$$= \frac{1}{2}(1.210\,368 + 1.467\,873 + 0.25 \sin 1.467\,873)$$

$$= 1.463\,459$$

Using $n = 4$ we have $h = 0.125$ and the midpoint formulas become

$$y_1 = y_0 + hf_0 = 1 + 0.125 \sin 1.0 = 1.105\,184$$

$$y_2 = y_0 + 2hf_1 = 1 + 2(0.125) \sin 1.105\,184 = 1.223\,387$$

$$y_3 = y_1 + 2hf_2 = 1.105\,184 + 2(0.125) \sin 1.223\,387 = 1.340\,248$$

$$y_4 = y_2 + 2hf_3 = 1.223\,387 + 2(0.125) \sin 1.340\,248 = 1.466\,772$$

$$y_{h/2}(0.5) = \frac{1}{2}(y_4 + y_3 + hf_4)$$

$$= \frac{1}{2}(1.466\,772 + 1.340\,248 + 0.125 \sin 1.466\,772)$$

$$= 1.465\,672$$

Richardson extrapolation results in

$$y(0.5) = \frac{4(1.465\,672) - 1.463\,459}{3} = 1.466\,410$$

which compares favorably with the "true" solution $y(0.5) = 1.466\,404$.

EXAMPLE 7.11

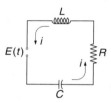

The differential equations governing the loop current i and the charge q on the capacitor of the electric circuit shown are

$$L\frac{di}{dt} + Ri + \frac{q}{C} = E(t) \qquad \frac{dq}{dt} = i$$

If the applied voltage E is suddenly increased from zero to 9 V, plot the resulting loop current during the first ten seconds. Use $R = 1.0\,\Omega$, $L = 2$ H and $C = 0.45$ F.

Solution Letting

$$\mathbf{y} = \begin{bmatrix} y_1 \\ y_2 \end{bmatrix} = \begin{bmatrix} q \\ i \end{bmatrix}$$

and substituting the given data, the differential equations become

$$\dot{\mathbf{y}} = \begin{bmatrix} \dot{y}_1 \\ \dot{y}_2 \end{bmatrix} = \begin{bmatrix} y_2 \\ (-Ry_2 - y_1/C + E)/L \end{bmatrix}$$

The initial conditions are

$$\mathbf{y}(0) = \begin{bmatrix} 0 \\ 0 \end{bmatrix}$$

We solved the problem with the function `bulStoer` using the increment $H = 0.5$ s. The following program utilizes the plotting facilities of MATLAB:

```
% Example 7.11 (Bulirsch-Stoer integration)
[xSol,ySol] = bulStoer(@fex7_11,0,[0 0],10,0.5);
plot(xSol,ySol(:,2),'k:o')
grid on
xlabel('Time (s)')
ylabel('Current (A)')
```

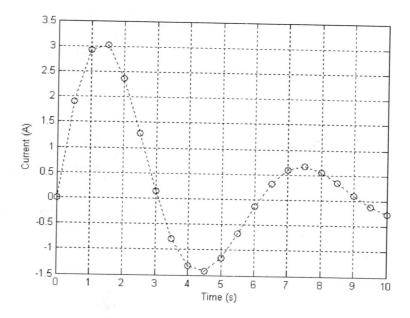

Recall that in each interval H (the spacing of open circles) the integration was performed by the modified midpoint method and refined by Richardson's extrapolation.

PROBLEM SET 7.2

1. Derive the analytical solution of the problem

$$y'' + y' - 380y = 0 \qquad y(0) = 1 \qquad y'(0) = -20$$

Would you expect difficulties in solving this problem numerically?

2. Consider the problem

$$y' = x - 10y \qquad y(0) = 10$$

(a) Verify that the analytical solution is $y(x) = 0.1x - 0.01 + 10.01e^{-10x}$. (b) Determine the step size h that you would use in numerical solution with the (nonadaptive) Runge–Kutta method.

3. ■ Integrate the initial value problem in Prob. 2 from $x = 0$ to 5 with the Runge–Kutta method using (a) $h = 0.1$; (b) $h = 0.25$; and (c) $h = 0.5$. Comment on the results.

4. ■ Integrate the initial value problem in Prob. 2 from $x = 0$ to 10 with the adaptive Runge–Kutta method.

5. ■

The differential equation describing the motion of the mass–spring–dashpot system is

$$\ddot{y} + \frac{c}{m}\dot{y} + \frac{k}{m}y = 0$$

where $m = 2$ kg, $c = 460$ N·s/m and $k = 450$ N/m. The initial conditions are $y(0) = 0.01$ m and $\dot{y}(0) = 0$. (a) Show that this is a stiff problem and determine a value of h that you would use in numerical integration with the nonadaptive Runge–Kutta method. (b) Carry out the integration from $t = 0$ to 0.2 s with the chosen h and plot $\dot{y}$ vs. t.

6. ■ Integrate the initial value problem specified in Prob. 5 with the adaptive Runge–Kutta method from $t = 0$ to 0.2 s, and plot $\dot{y}$ vs. t.

7. ■ Compute the numerical solution of the differential equation

$$y'' = 16.81y$$

from $x = 0$ to 2 with the adaptive Runge–Kutta method. Use the initial conditions (a) $y(0) = 1.0$, $y'(0) = -4.1$; and (b) $y(0) = 1.0$, $y'(0) = -4.11$. Explain the large difference in the two solutions. *Hint:* derive the analytical solutions.

8. ■ Integrate

$$y'' + y' - y^2 = 0 \qquad y(0) = 1 \qquad y'(0) = 0$$

from $x = 0$ to 3.5. Investigate whether the sudden increase in y near the upper limit is real or an artifact caused by instability. *Hint:* experiment with different values of h.

9. ■ Solve the stiff problem—see Eq. (7.16)

$$y'' + 1001y' + 1000y = 0 \qquad y(0) = 1 \qquad y'(0) = 0$$

from $x = 0$ to 0.2 with the adaptive Runge–Kutta method and plot y' vs. x.

10. ■ Solve

$$y'' + 2y' + 3y = 0 \qquad y(0) = 0 \qquad y'(0) = \sqrt{2}$$

with the adaptive Runge–Kutta method from $x = 0$ to 5 (the analytical solution is $y = e^{-x} \sin \sqrt{2}x$).

11. ■ Use the adaptive Runge–Kutta method to solve the differential equation

$$y'' = 2yy'$$

from $x = 0$ to 10 with the initial conditions $y(0) = 1$, $y'(0) = -1$. Plot y vs. x.

12. ■ Repeat Prob. 11 with the initial conditions $y(0) = 0$, $y'(0) = 1$ and the integration range $x = 0$ to 1.5.

13. ■ Use the adaptive Runge–Kutta method to integrate

$$y' = \left(\frac{9}{y} - y\right) x \qquad y(0) = 5$$

from $x = 0$ to 5 and plot y vs. x.

14. Solve Prob. 13 with the Bulirsch–Stoer method using $H = 0.5$.

15. ■ Integrate

$$x^2 y'' + xy' + y = 0 \qquad y(1) = 0 \qquad y'(1) = -2$$

from $x = 1$ to 20, and plot y and y' vs. x. Use the Bulirsch–Stoer method.

16. ■

The magnetized iron block of mass m is attached to a spring of stiffness k and free length L. The block is at rest at $x = L$ when the electromagnet is turned on, exerting the repulsive force $F = c/x^2$ on the block. The differential equation of the resulting motion is

$$m\ddot{x} = \frac{c}{x^2} - k(x - L)$$

Determine the amplitude and the period of the motion by numerical integration with the adaptive Runge–Kutta method. Use $c = 5\ \text{N·m}^2$, $k = 120\ \text{N/m}$, $L = 0.2\ \text{m}$ and $m = 1.0\ \text{kg}$.

17. ■

The bar ABC is attached to the vertical rod with a horizontal pin. The assembly is free to rotate about the axis of the rod. In the absence of friction, the equations of motion of the system are

$$\ddot{\theta} = \dot{\phi}^2 \sin\theta \cos\theta \qquad \ddot{\phi} = -2\dot{\theta}\dot{\phi}\cot\theta$$

If the system is set into motion with the initial conditions $\theta(0) = \pi/12$ rad, $\dot{\theta}(0) = 0$, $\phi(0) = 0$ and $\dot{\phi}(0) = 20$ rad/s, obtain a numerical solution with the adaptive Runge–Kutta method from $t = 0$ to 1.5 s and plot $\dot{\phi}$ vs. t.

18. ■ Solve the circuit problem in Example 7.11 if $R = 0$ and

$$E(t) = \begin{cases} 0 \text{ when } t < 0 \\ 9\sin\pi t \text{ when } t \geq 0 \end{cases}$$

19. ■ Solve Prob. 21 in Problem Set 1 if $E = 240$ V (constant).

20. ■

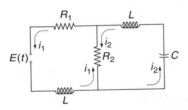

Kirchoff's equations for the circuit in the figure are

$$L\frac{di_1}{dt} + R_1 i_1 + R_2(i_1 - i_2) = E(t)$$

$$L\frac{di_2}{dt} + R_2(i_2 - i_1) + \frac{q_2}{C} = 0$$

where

$$\frac{dq_2}{dt} = i_2$$

Using the data $R_1 = 4\,\Omega$, $R_2 = 10\,\Omega$, $L = 0.032$ H, $C = 0.53$ F and

$$E(t) = \begin{cases} 20 \text{ V if } 0 < t < 0.005 \text{ s} \\ 0 \text{ otherwise} \end{cases}$$

plot the transient loop currents i_1 and i_2 from $t = 0$ to 0.05 s.

21. ■ Consider a closed biological system populated by M number of prey and N number of predators. Volterra postulated that the two populations are related by the differential equations

$$\dot{M} = aM - bMN$$

$$\dot{N} = -cN + dMN$$

where a, b, c and d are constants. The steady-state solution is $M_0 = c/d$, $N_0 = a/b$; if numbers other than these are introduced into the system, the populations undergo periodic fluctuations. Introducing the notation

$$y_1 = M/M_0 \qquad y_2 = N/N_0$$

allows us to write the differential equations as

$$\dot{y}_1 = a(y_1 - y_1 y_2)$$

$$\dot{y}_2 = b(-y_2 + y_1 y_2)$$

Using $a = 1.0$/year, $b = 0.2$/year, $y_1(0) = 0.1$ and $y_2(0) = 1.0$, plot the two populations from $t = 0$ to 50 years.

22. ■ The equations

$$\dot{u} = -au + av$$

$$\dot{v} = cu - v - uw$$

$$\dot{w} = -bw + uv$$

known as the Lorenz equations, are encountered in theory of fluid dynamics. Letting $a = 5.0$, $b = 0.9$ and $c = 8.2$, solve these equations from $t = 0$ to 10 with the initial conditions $u(0) = 0$, $v(0) = 1.0$, $w(0) = 2.0$ and plot $u(t)$. Repeat the solution with $c = 8.3$. What conclusions can you draw from the results?

MATLAB Functions

[xSol,ySol] = ode23(dEqs,[xStart,xStop],yStart) low-order (probably third order) adaptive Runge–Kutta method. The function dEqs must return the differential equations as a column vector (recall that runKut4 and runKut5 require row vectors). The range of integration is from xStart to xStop with the initial conditions yStart (also a column vector).

[xSol,ySol] = ode45(dEqs,[xStart xStop],yStart) is similar to ode23, but uses a higher-order Runge–Kutta method (probably fifth order).

These two methods, as well as all the methods described in in this book, belong to a group known as *single-step methods*. The name stems from the fact that the information at a single point on the solution curve is sufficient to compute the next point. There are also *multistep methods* that utilize several points on the curve to extrapolate the solution at the next step. These methods were popular once, but have lost some of their luster in the last few years. Multistep methods have two shortcomings that complicate their implementation:

- The methods are not self-starting, but must be provided with the solution at the first few points by a single-step method.
- The integration formulas assume equally spaced steps, which makes it makes it difficult to change the step size.

Both of these hurdles can be overcome, but the price is complexity of the algorithm that increases with sophistication of the method. The benefits of multistep methods are minimal—the best of them can outperform their single-step counterparts in certain problems, but these occasions are rare. MATLAB provides one general-purpose multistep method:

`[xSol,ySol] = ode113(dEqs,[xStart xStop],yStart)`uses the variable-order Adams–Bashforth–Moulton method.

MATLAB has also several functions for solving stiff problems. These are `ode15s` (this is the first method to try when a stiff problem is encountered), `ode23s`, `ode23t` and `ode23tb`.

8 Two-Point Boundary Value Problems

$$\text{Solve } y'' = f(x, y, y'), \quad y(a) = \alpha, \quad y(b) = \beta$$

8.1 Introduction

In two-point boundary value problems the auxiliary conditions associated with the differential equation, called the *boundary conditions*, are specified at two different values of x. This seemingly small departure from initial value problems has a major repercussion—it makes boundary value problems considerably more difficult to solve. In an initial value problem we were able to start at the point where the initial values were given and march the solution forward as far as needed. This technique does not work for boundary value problems, because there are not enough starting conditions available at either end point to produce a unique solution.

One way to overcome the lack of starting conditions is to guess the missing values. The resulting solution is very unlikely to satisfy boundary conditions at the other end, but by inspecting the discrepancy we can estimate what changes to make to the initial conditions before integrating again. This iterative procedure is known as the *shooting method*. The name is derived from analogy with target shooting—take a shot and observe where it hits the target, then correct the aim and shoot again.

Another means of solving two-point boundary value problems is the *finite difference method*, where the differential equations are approximated by finite differences at evenly spaced mesh points. As a consequence, a differential equation is transformed into set of simultaneous algebraic equations.

The two methods have a common problem: they give rise to nonlinear sets of equations if the differential equation is not linear. As we noted in Chapter 4, all methods of solving nonlinear equations are iterative procedures that can consume a lot of computational resources. Thus solution of nonlinear boundary value problems is not cheap. Another complication is that iterative methods need reasonably good starting values in order to converge. Since there is no set formula for determining these, an algorithm for solving nonlinear boundary value problems requires intelligent input; it cannot be treated as a "black box."

8.2   Shooting Method

Second-Order Differential Equation

The simplest two-point boundary value problem is a second-order differential equation with one condition specified at $x = a$ and another one at $x = b$. Here is an example of a second-order boundary value problem:

$$y'' = f(x, y, y'), \quad y(a) = \alpha, \quad y(b) = \beta \tag{8.1}$$

Let us now attempt to turn Eqs. (8.1) into the initial value problem

$$y'' = f(x, y, y'), \quad y(a) = \alpha, \quad y'(a) = u \tag{8.2}$$

The key to success is finding the correct value of u. This could be done by trial and error: guess u and solve the initial value problem by marching from $x = a$ to b. If the solution agrees with the prescribed boundary condition $y(b) = \beta$, we are done; otherwise we have to adjust u and try again. Clearly, this procedure is very tedious.

More systematic methods become available to us if we realize that the determination of u is a root-finding problem. Because the solution of the initial value problem depends on u, the computed boundary value $y(b)$ is a function of u; that is

$$y(b) = \theta(u)$$

Hence u is a root of

$$r(u) = \theta(u) - \beta = 0 \tag{8.3}$$

where $r(u)$ is the *boundary residual* (difference between the computed and specified boundary values). Equation (8.3) can be solved by any one of the root-finding methods discussed in Chapter 4. We reject the method of bisection because it involves too many evaluations of $\theta(u)$. In the Newton–Raphson method we run into the problem of having to compute $d\theta/du$, which can be done, but not easily. That leaves Brent's algorithm as our method of choice.

Here is the procedure we use in solving nonlinear boundary value problems:

1. Specify the starting values u_1 and u_2 which *must bracket the root u* of Eq. (8.3).
2. Apply Brent's method to solve Eq. (8.3) for u. Note that each iteration requires evaluation of $\theta(u)$ by solving the differential equation as an initial value problem.

3. Having determined the value of u, solve the differential equations once more and record the results.

If the differential equation is linear, any root-finding method will need only one interpolation to determine u. But since Brent's method uses quadratic interpolation, it needs three points: u_1, u_2 and u_3, the latter being provided by a bisection step. This is wasteful, since linear interpolation with u_1 and u_2 would also result in the correct value of u. Therefore, we replace Brent's method with linear interpolation whenever the differential equation is linear.

■ linInterp

Here is the algorithm for linear interpolation:

```
function root = linInterp(func,x1,x2)
% Finds the zero of the linear function f(x) by straight
% line interpolation between x1 and x2.
% func = handle of function that returns f(x).

f1 = feval(func,x1); f2 = feval(func,x2);
root = x2 - f2*(x2 - x1)/(f2 - f1);
```

EXAMPLE 8.1
Solve the nonlinear boundary value problem

$$y'' + 3yy' = 0 \quad y(0) = 0 \quad y(2) = 1$$

Solution The equivalent first-order equations are

$$\mathbf{y}' = \begin{bmatrix} y_1' \\ y_2' \end{bmatrix} = \begin{bmatrix} y_2 \\ -3y_1y_2 \end{bmatrix}$$

with the boundary conditions

$$y_1(0) = 0 \quad y_1(2) = 1$$

Now comes the daunting task of estimating the trial values of $y_2(0) = y'(0)$, the unspecified initial condition. We could always pick two numbers at random and hope for the best. However, it is possible to reduce the element of chance with a little detective work. We start by making the reasonable assumption that y is smooth (does not wiggle) in the interval $0 \le x \le 2$. Next we note that y has to increase from 0 to 1, which requires $y' > 0$. Since both y and y' are positive, we conclude that y'' must be negative in order to satisfy the differential equation. Now we are in a position to make a rough sketch of y:

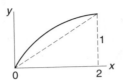

Looking at the sketch it is clear that $y'(0) > 0.5$, so that $y'(0) = 1$ and 2 appear to be reasonable values for the brackets of $y'(0)$; if they are not, Brent's method will display an error message.

In the program listed below we chose the nonadaptive Runge–Kutta method (runKut4) for integration. Note that three user-supplied functions are needed to describe the problem at hand. Apart from the function dEqs(x,y) that defines the differential equations, we also need the functions inCond(u) to specify the initial conditions for integration, and residual(u) that provides Brent's method with the boundary residual. By changing a few statements in these functions, the program can be applied to any second-order boundary value problem. It also works for third-order equations if integration is started at the end where two of the three boundary conditions are specified.

```
function shoot2
% Shooting method for 2nd-order boundary value problem
% in Example 8.1.

global XSTART XSTOP H       % Make these params. global.
XSTART = 0; XSTOP = 2;      % Range of integration.
H = 0.1;                    % Step size.
freq = 2;                   % Frequency of printout.
u1 = 1; u2 = 2;             % Trial values of unknown
                            % initial condition u.

x = XSTART;
u = brent(@residual,u1,u2);
[xSol,ySol] = runKut4(@dEqs,x,inCond(u),XSTOP,H);
printSol(xSol,ySol,freq)

function F = dEqs(x,y)      % First-order differential
F = [y(2), -3*y(1)*y(2)];  % equations.

function y = inCond(u)      % Initial conditions (u is
y = [0 u];                 % the unknown condition).
```

```
function r = residual(u)   % Boundary residual.
global XSTART XSTOP H
x = XSTART;
[xSol,ySol] = runKut4(@dEqs,x,inCond(u),XSTOP,H);
r = ySol(size(ySol,1),1) - 1;
```

Here is the solution :

```
>>         x               y1               y2
     0.0000e+000     0.0000e+000     1.5145e+000
     2.0000e-001     2.9404e-001     1.3848e+000
     4.0000e-001     5.4170e-001     1.0743e+000
     6.0000e-001     7.2187e-001     7.3287e-001
     8.0000e-001     8.3944e-001     4.5752e-001
     1.0000e+000     9.1082e-001     2.7013e-001
     1.2000e+000     9.5227e-001     1.5429e-001
     1.4000e+000     9.7572e-001     8.6471e-002
     1.6000e+000     9.8880e-001     4.7948e-002
     1.8000e+000     9.9602e-001     2.6430e-002
     2.0000e+000     1.0000e+000     1.4522e-002
```

Note that $y'(0) = 1.5145$, so that our initial guesses of 1.0 and 2.0 were on the mark.

EXAMPLE 8.2

Numerical integration of the initial value problem

$$y'' + 4y = 4x \quad y(0) = 0 \quad y'(0) = 0$$

yielded $y'(2) = 1.653\,64$. Use this information to determine the value of $y'(0)$ that would result in $y'(2) = 0$.

Solution We use linear interpolation

$$u = u_2 - \theta(u_2)\frac{u_2 - u_1}{\theta(u_2) - \theta(u_1)}$$

where in our case $u = y'(0)$ and $\theta(u) = y'(2)$. So far we are given $u_1 = 0$ and $\theta(u_1) = 1.653\,64$. To obtain the second point, we need another solution of the initial value problem. An obvious solution is $y = x$, which gives us $y(0) = 0$ and $y'(0) = y'(2) = 1$. Thus the second point is $u_2 = 1$ and $\theta(u_2) = 1$. Linear interpolation now yields

$$y'(0) = u = 1 - (1)\frac{1 - 0}{1 - 1.653\,64} = 2.529\,89$$

Since the problem is linear, no further iterations are needed.

EXAMPLE 8.3

Solve the third-order boundary value problem

$$y''' = 2y'' + 6xy \qquad y(0) = 2 \qquad y(5) = y'(5) = 0$$

and plot y vs. x.

Solution The first-order equations and the boundary conditions are

$$\mathbf{y}' = \begin{bmatrix} y_1' \\ y_2' \\ y_3' \end{bmatrix} = \begin{bmatrix} y_2 \\ y_3 \\ 2y_3 + 6xy_1 \end{bmatrix}$$

$$y_1(0) = 2 \qquad y_1(5) = y_2(5) = 0$$

The program listed below is based on `shoot2` in Example 8.1. Because two of the three boundary conditions are specified at the right end, we start the integration at $x = 5$ and proceed with negative h toward $x = 0$. Two of the three initial conditions are prescribed as $y_1(5) = y_2(5) = 0$, whereas the third condition $y_3(5)$ is unknown. Because the differential equation is linear, the two guesses for $y_3(5)$ (u_1 and u_2) are not important; we left them as they were in Example 8.1. The adaptive Runge–Kutta method (`runKut5`) was chosen for the integration.

```
function shoot3
% Shooting method for 3rd-order boundary value
% problem in Example 8.3.

global XSTART XSTOP H      % Make these params. global.
XSTART = 5; XSTOP = 0;     % Range of integration.
H = -0.1;                  % Step size.
freq = 2;                  % Frequency of printout.
u1 = 1; u2 = 2;            % Trial values of unknown
                           % initial condition u.
x = XSTART;
u = linInterp(@residual,u1,u2);
[xSol,ySol] = runKut5(@dEqs,x,inCond(u),XSTOP,H);
printSol(xSol,ySol,freq)

function F = dEqs(x,y)     % 1st-order differential eqs.
F = [y(2), y(3), 2*y(3) + 6*x*y(1)];

function y = inCond(u)     % Initial conditions.
y = [0 0 u];
```

```
function r = residual(u) % Boundary residual.
global XSTART XSTOP H
x = XSTART;
[xSol,ySol] = runKut5(@dEqs,x,inCond(u),XSTOP,H);
r = ySol(size(ySol,1),1) - 2;
```

We skip the rather long printout of the solution and show just the plot:

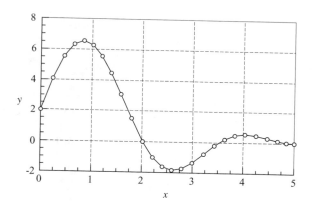

Higher-Order Equations

Consider the fourth-order differential equation

$$y^{(4)} = f(x, y, y', y'', y''') \tag{8.4a}$$

with the boundary conditions

$$y(a) = \alpha_1 \qquad y''(a) = \alpha_2 \qquad y(b) = \beta_1 \qquad y''(b) = \beta_2 \tag{8.4b}$$

To solve Eq. (8.4a) with the shooting method, we need four initial conditions at $x = a$, only two of which are specified. Denoting the two unknown initial values by u_1 and u_2, we have the set of initial conditions

$$y(a) = \alpha_1 \qquad y'(a) = u_1 \qquad y''(a) = \alpha_2 \qquad y'''(a) = u_2 \tag{8.5}$$

If Eq. (8.4a) is solved with the shooting method using the initial conditions in Eq. (8.5), the computed boundary values at $x = b$ depend on the choice of u_1 and u_2. We express this dependence as

$$y(b) = \theta_1(u_1, u_2) \qquad y''(b) = \theta_2(u_1, u_2) \tag{8.6}$$

The correct choice of u_1 and u_2 yields the given boundary conditions at $x = b$; that is, it satisfies the equations

$$\theta_1(u_1, u_2) = \beta_1 \qquad \theta_2(u_1, u_2) = \beta_2$$

or, using vector notation

$$\theta(\mathbf{u}) = \beta \tag{8.7}$$

These are simultaneous (generally nonlinear) equations that can be solved by the Newton–Raphson method discussed in Art. 4.6. It must be pointed out again that intelligent estimates of u_1 and u_2 are needed if the differential equation is not linear.

EXAMPLE 8.4

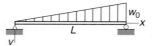

The displacement v of the simply supported beam can be obtained by solving the boundary value problem

$$\frac{d^4v}{dx^4} = \frac{w_0}{EI}\frac{x}{L} \qquad v = \frac{d^2v}{dx^2} = 0 \text{ at } x = 0 \text{ and } x = L$$

where EI is the bending rigidity. Determine by numerical integration the slopes at the two ends and the displacement at mid-span.

Solution Introducing the dimensionless variables

$$\xi = \frac{x}{L} \qquad y = \frac{EI}{w_0 L^4}v$$

the problem is transformed to

$$\frac{d^4y}{d\xi^4} = \xi \qquad y = \frac{d^2y}{d\xi^2} = 0 \text{ at } \xi = 0 \text{ and } \xi = 1$$

The equivalent first-order equations and the boundary conditions are (the prime denotes $d/d\xi$)

$$\mathbf{y}' = \begin{bmatrix} y_1' \\ y_2' \\ y_3' \\ y_4' \end{bmatrix} = \begin{bmatrix} y_2 \\ y_3 \\ y_4 \\ \xi \end{bmatrix}$$

$$y_1(0) = y_3(0) = y_1(1) = y_3(1) = 0$$

The program listed below is similar to the one in Example 8.1. With appropriate changes in functions dEqs(x,y), inCond(u) and residual(u) the program can solve boundary value problems of any order greater than two. For the problem at hand we chose the Bulirsch–Stoer algorithm to do the integration because it gives us control over the printout (we need y precisely at mid-span). The nonadaptive Runge–Kutta method could also be used here, but we would have to guess a suitable step size h.

```
function shoot4
% Shooting method for 4th-order boundary value
% problem in Example 8.4.

global XSTART XSTOP H      % Make these params. global.
XSTART = 0; XSTOP = 1;     % Range of integration.
H = 0.5;                   % Step size.
freq = 1;                  % Frequency of printout.
u = [0 1];                 % Trial values of u(1).
                           % and u(2).
x = XSTART;
u = newtonRaphson2(@residual,u);
[xSol,ySol] = bulStoer(@dEqs,x,inCond(u),XSTOP,H);
printSol(xSol,ySol,freq)

function F = dEqs(x,y)     % Differential equations.
F = [y(2) y(3) y(4) x;];

function y = inCond(u)     % Initial conditions; u(1)
y = [0 u(1) 0 u(2)];       % and u(2) are unknowns.

function r = residual(u)  % Boundary residuals.
global XSTART XSTOP H
r = zeros(length(u),1);
x = XSTART;
[xSol,ySol] = bulStoer(@dEqs,x,inCond(u),XSTOP,H);
lastRow = size(ySol,1);
r(1)= ySol(lastRow,1);
r(2) = ySol(lastRow,3);
```

Here is the output:

```
>>    x              y1              y2              y3              y4
0.0000e+000   0.0000e+000   1.9444e-002   0.0000e+000  -1.6667e-001
5.0000e-001   6.5104e-003   1.2150e-003  -6.2500e-002  -4.1667e-002
1.0000e+000  -4.8369e-017  -2.2222e-002  -5.8395e-018   3.3333e-001
```

Noting that

$$\frac{dv}{dx} = \frac{dv}{d\xi}\frac{d\xi}{dx} = \left(\frac{w_0 L^4}{EI}\frac{dy}{d\xi}\right)\frac{1}{L} = \frac{w_0 L^3}{EI}\frac{dy}{d\xi}$$

we obtain

$$\left.\frac{dv}{dx}\right|_{x=0} = 19.444 \times 10^{-3} \frac{w_0 L^3}{EI}$$

$$\left.\frac{dv}{dx}\right|_{x=L} = -22.222 \times 10^{-3} \frac{w_0 L^3}{EI}$$

$$v|_{x=0.5L} = 6.5104 \times 10^{-3} \frac{w_0 L^4}{EI}$$

which agree with the analytical solution (easily obtained by direct integration of the differential equation).

EXAMPLE 8.5
Solve the nonlinear differential equation

$$y^{(4)} + \frac{4}{x} y^3 = 0$$

with the boundary conditions

$$y(0) = y'(0) = 0 \qquad y''(1) = 0 \qquad y'''(1) = 1$$

and plot y vs. x.

Solution Our first task is to handle the indeterminacy of the differential equation at the origin, where $x = y = 0$. The problem is resolved by applying L'Hospital's rule: $4y^3/x \to 12y^2 y'$ as $x \to 0$. Thus the equivalent first-order equations and the boundary conditions that we use in the solution are

$$\mathbf{y}' = \begin{bmatrix} y_1' \\ y_2' \\ y_3' \\ y_4' \end{bmatrix} = \begin{bmatrix} y_2 \\ y_3 \\ y_4 \\ \begin{cases} -12y_1^2 y_2 & \text{near } x = 0 \\ -4y_1^3/x & \text{otherwise} \end{cases} \end{bmatrix}$$

$$y_1(0) = y_2(0) = 0 \qquad y_3(1) = 0 \qquad y_4(1) = 1$$

Because the problem is nonlinear, we need reasonable estimates for $y''(0)$ and $y'''(0)$. On the basis of the boundary conditions $y''(1) = 0$ and $y'''(1) = 1$, the plot of y'' is likely to look something like this:

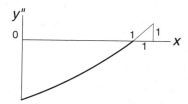

If we are right, then $y''(0) < 0$ and $y'''(0) > 0$. Based on this rather scanty information, we try $y''(0) = -1$ and $y'''(0) = 1$.

The following program uses the adaptive Runge–Kutta method (runKut5) for integration:

```
function shoot4nl
% Shooting method for nonlinear 4th-order boundary
% value problem in Example 8.5.

global XSTART XSTOP H      % Make these params. global.
XSTART = 0; XSTOP = 1;     % Range of integration.
H = 0.1;                   % Step size.
freq = 1;                  % Frequency of printout.
u = [-1 1];                % Trial values of u(1)
                           % and u(2).
x = XSTART;
u = newtonRaphson2(@residual,u);
[xSol,ySol] = runKut5(@dEqs,x,inCond(u),XSTOP,H);
printSol(xSol,ySol,freq)

function F = dEqs(x,y)     % Differential equations.
F = zeros(1,4);
F(1) = y(2); F(2) = y(3); F(3) = y(4);
if x < 10.0e-4; F(4) = -12*y(2)*y(1)^2;
else;           F(4) = -4*(y(1)^3)/x;
end

function y = inCond(u)     % Initial conditions; u(1)
y = [0 0 u(1) u(2)];       % and u(2) are unknowns.

function r = residual(u)   % Bounday residuals.
global XSTART XSTOP H
r = zeros(length(u),1);
x = XSTART;
[xSol,ySol] = runKut5(@dEqs,x,inCond(u),XSTOP,H);
lastRow = size(ySol,1);
r(1) = ySol(lastRow,3);
r(2) = ySol(lastRow,4) - 1;
```

The results are:

```
>>   x                y1               y2               y3               y4
0.0000e+000   0.0000e+000   0.0000e+000  -9.7607e-001   9.7131e-001
1.0000e-001  -4.7184e-003  -9.2750e-002  -8.7893e-001   9.7131e-001
3.9576e-001  -6.6403e-002  -3.1022e-001  -5.9165e-001   9.7152e-001
7.0683e-001  -1.8666e-001  -4.4722e-001  -2.8896e-001   9.7627e-001
9.8885e-001  -3.2061e-001  -4.8968e-001  -1.1144e-002   9.9848e-001
1.0000e+000  -3.2607e-001  -4.8975e-001   6.4879e-016   1.0000e+000
```

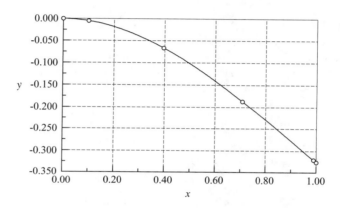

By good fortune, our initial estimates $y''(0) = -1$ and $y'''(0) = 1$ were very close to the final values.

PROBLEM SET 8.1

1. Numerical integration of the initial value problem

$$y'' + y' - y = 0 \qquad y(0) = 0 \qquad y'(0) = 1$$

yielded $y(1) = 0.741028$. What is the value of $y'(0)$ that would result in $y(1) = 1$, assuming that $y(0)$ is unchanged?

2. The solution of the differential equation

$$y''' + y'' + 2y' = 6$$

with the initial conditions $y(0) = 2$, $y'(0) = 0$ and $y''(0) = 1$, yielded $y(1) = 3.03765$. When the solution was repeated with $y''(0) = 0$ (the other conditions being unchanged), the result was $y(1) = 2.72318$. Determine the value of $y''(0)$ so that $y(1) = 0$.

3. Roughly sketch the solution of the following boundary value problems. Use the sketch to estimate $y'(0)$ for each problem.

   (a) $y'' = -e^{-y}$ $y(0) = 1$ $y(1) = 0.5$
   (b) $y'' = 4y^2$ $y(0) = 10$ $y'(1) = 0$
   (c) $y'' = \cos(xy)$ $y(0) = 1$ $y(1) = 2$

4. Using a rough sketch of the solution estimate of $y(0)$ for the following boundary value problems.

   (a) $y'' = y^2 + xy$ $y'(0) = 0$ $y(1) = 2$
   (b) $y'' = -\dfrac{2}{x}y' - y^2$ $y'(0) = 0$ $y(1) = 2$
   (c) $y'' = -x(y')^2$ $y'(0) = 2$ $y(1) = 1$

5. Obtain a rough estimate of $y''(0)$ for the boundary value problem

$$y''' + 5y''y^2 = 0$$

$$y(0) = 0 \qquad y'(0) = 1 \qquad y(1) = 0$$

6. Obtain rough estimates of $y''(0)$ and $y'''(0)$ for the boundary value problem

$$y^{(4)} + 2y'' + y'\sin y = 0$$

$$y(0) = y'(0) = 0 \qquad y(1) = 5 \qquad y'(1) = 0$$

7. Obtain rough estimates of $\dot{x}(0)$ and $\dot{y}(0)$ for the boundary value problem

$$\ddot{x} + 2x^2 - y = 0 \qquad x(0) = 1 \qquad x(1) = 0$$
$$\ddot{y} + y^2 - 2x = 1 \qquad y(0) = 0 \qquad y(1) = 1$$

8. ■ Solve the boundary value problem

$$y'' + (1 - 0.2x)\, y^2 = 0 \qquad y(0) = 0 \qquad y(\pi/2) = 1$$

9. ■ Solve the boundary value problem

$$y'' + 2y' + 3y^2 = 0 \qquad y(0) = 0 \qquad y(2) = -1$$

10. ■ Solve the boundary value problem

$$y'' + \sin y + 1 = 0 \qquad y(0) = 0 \qquad y(\pi) = 0$$

11. ■ Solve the boundary value problem

$$y'' + \frac{1}{x}y' + y = 0 \qquad y(0.01) = 1 \qquad y'(2) = 0$$

and plot y vs. x. *Warning*: y changes very rapidly near $x = 0$.

12. ■ Solve the boundary value problem

$$y'' - \left(1 - e^{-x}\right)y = 0 \qquad y(0) = 1 \qquad y(\infty) = 0$$

and plot y vs. x. *Hint*: Replace the infinity by a finite value β. Check your choice of β by repeating the solution with 1.5β. If the results change, you must increase β.

13. ■ Solve the boundary value problem

$$y''' = -\frac{1}{x}y'' + \frac{1}{x^2}y' + 0.1(y')^3$$

$$y(1) = 0 \qquad y''(1) = 0 \qquad y(2) = 1$$

14. ■ Solve the boundary value problem

$$y''' + 4y'' + 6y' = 10$$

$$y(0) = y''(0) = 0 \qquad y(3) - y'(3) = 5$$

15. ■ Solve the boundary value problem

$$y''' + 2y'' + \sin y = 0$$

$$y(-1) = 0 \qquad y'(-1) = -1 \qquad y'(1) = 1$$

16. ■ Solve the differential equation in Prob. 15 with the boundary conditions

$$y(-1) = 0 \qquad y(0) = 0 \qquad y(1) = 1$$

(this is a three-point boundary value problem).

17. ■ Solve the boundary value problem

$$y^{(4)} = -xy^2$$

$$y(0) = 5 \qquad y''(0) = 0 \qquad y'(1) = 0 \qquad y'''(1) = 2$$

18. ■ Solve the boundary value problem

$$y^{(4)} = -2yy''$$

$$y(0) = y'(0) = 0 \qquad y(4) = 0 \qquad y'(4) = 1$$

19. ■

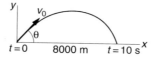

A projectile of mass m in free flight experiences the aerodynamic drag force $F_D = cv^2$, where v is the velocity. The resulting equations of motion are

$$\ddot{x} = -\frac{c}{m}v\dot{x} \qquad \ddot{y} = -\frac{c}{m}v\dot{y} - g$$

$$v = \sqrt{\dot{x}^2 + \dot{y}^2}$$

If the projectile hits a target 8 km away after a 10 s flight, determine the launch velocity v_0 and its angle of inclination θ. Use $m = 20$ kg, $c = 3.2 \times 10^{-4}$ kg/m and $g = 9.80665$ m/s^2.

20. ■

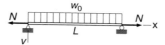

The simply supported beam carries a uniform load of intensity w_0 and the tensile force N. The differential equation for the vertical displacement v can be shown to be

$$\frac{d^4v}{dx^4} - \frac{N}{EI}\frac{d^2v}{dx^2} = \frac{w_0}{EI}$$

where EI is the bending rigidity. The boundary conditions are $v = d^2v/dx^2 = 0$ at $x = 0$ and $x = L$. Changing the variables to $\xi = \dfrac{x}{L}$ and $y = \dfrac{EI}{w_0 L^4}v$ transforms the problem to the dimensionless form

$$\frac{d^4y}{d\xi^4} - \beta\frac{d^2y}{d\xi^2} = 1 \qquad \beta = \frac{NL^2}{EI}$$

$$y|_{\xi=0} = \left.\frac{d^2y}{d\xi^2}\right|_{\xi=0} = y|_{\xi=1} = \left.\frac{d^2y}{d\xi^2}\right|_{\xi=1} = 0$$

Determine the maximum displacement if (a) $\beta = 1.65929$ and (b) $\beta = -1.65929$ (N is compressive).

21. ■ Solve the boundary value problem

$$y''' + yy'' = 0 \qquad y(0) = y'(0) = 0, \; y'(\infty) = 2$$

and plot $y(x)$ and $y'(x)$. This problem arises in determining the velocity profile of the boundary layer in incompressible flow (Blasius solution).

8.3   Finite Difference Method

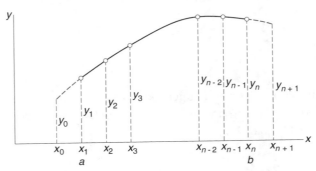

Figure 8.1. Finite difference mesh

In the finite difference method we divide the range of integration (a, b) into $n - 1$ equal subintervals of length h each, as shown in Fig. 8.1. The values of the numerical solution at the mesh points are denoted by y_i, $i = 1, 2 \ldots, n$; the two points outside (a, b) will be explained shortly. We then make two approximations:

1. The derivatives of y in the differential equation are replaced by the finite difference expressions. It is common practice to use the first central difference approximations (see Chapter 5):

$$y_i' = \frac{y_{i+1} - y_{i-1}}{2h} \qquad y_i'' = \frac{y_{i-1} - 2y_i + y_{i+1}}{h^2} \qquad \text{etc.} \qquad (8.8)$$

2. The differential equation is enforced only at the mesh points.

As a result, the differential equations are replaced by n simultaneous algebraic equations, the unknowns being y_i, $i = 1, 2, \ldots .n$. If the differential equation is nonlinear, the algebraic equations will also be nonlinear and must be solved by the Newton–Raphson method.

Since the truncation error in a first central difference approximation is $\mathcal{O}(h^2)$, the finite difference method is not as accurate as the shooting method—recall that the Runge–Kutta method has a truncation error of $O(h^5)$. Therefore, the convergence criterion in the Newton–Raphson method should not be too severe.

Second-Order Differential Equation

Consider the second-order differential equation

$$y'' = f(x, y, y')$$

with the boundary conditions

$$y(a) = \alpha \quad \text{or} \quad y'(a) = \alpha$$

$$y(b) = \beta \quad \text{or} \quad y'(b) = \beta$$

Approximating the derivatives at the mesh points by finite differences, the problem becomes

$$\frac{y_{i-1} - 2y_i + y_{i+1}}{h^2} = f\left(x_i, y_i, \frac{y_{i+1} - y_{i-1}}{2h}\right), \quad i = 1, 2, \ldots, n \tag{8.9}$$

$$y_1 = \alpha \quad \text{or} \quad \frac{y_2 - y_0}{2h} = \alpha \tag{8.10a}$$

$$y_n = \beta \quad \text{or} \quad \frac{y_{n+1} - y_{n-1}}{2h} = \beta \tag{8.10b}$$

Note the presence of y_0 and y_{n+1}, which are associated with points outside the solution domain (a, b). This "spillover" can be eliminated by using the boundary conditions. But before we do that, let us rewrite Eqs. (8.9) as

$$y_0 - 2y_1 + y_2 - h^2 f\left(x_1, y_1, \frac{y_2 - y_0}{2h}\right) = 0 \tag{a}$$

$$y_{i-1} - 2y_i + y_{i+1} - h^2 f\left(x_i, y_i, \frac{y_{i+1} - y_{i-1}}{2h}\right) = 0, \quad i = 2, 3, \ldots, n-1 \tag{b}$$

$$y_{n-1} - 2y_n + y_{n+1} - h^2 f\left(x_n, y_n, \frac{y_{n+1} - y_{n-1}}{2h}\right) = 0 \tag{c}$$

The boundary conditions on y are easily dealt with: Eq. (a) is simply replaced by $y_1 - \alpha = 0$ and Eq. (c) is replaced by $y_n - \beta = 0$. If y' are prescribed, we obtain from Eqs. (8.10) $y_0 = y_2 - 2h\alpha$ and $y_{n+1} = y_{n-1} + 2h\beta$, which are then substituted into Eqs. (a) and (c), respectively. Hence we finish up with n equations in the unknowns y_i, $i = 1, 2 \ldots, n$:

$$\left. \begin{array}{ll} y_1 - \alpha = 0 & \text{if } y(a) = \alpha \\ -2y_1 + 2y_2 - h^2 f(x_1, y_1, \alpha) - 2h\alpha = 0 & \text{if } y'(a) = \alpha \end{array} \right\} \tag{8.11a}$$

$$y_{i-1} - 2y_i + y_{i+1} - h^2 f\left(x_i, y_i, \frac{y_{i+1} - y_{i-1}}{2h}\right) = 0 \quad i = 2, 3, \ldots, n-1 \tag{8.11b}$$

$$\left. \begin{array}{ll} y_n - \beta = 0 & \text{if } y(b) = \beta \\ 2y_{n-1} - 2y_n - h^2 f(x_n, y_n, \beta) + 2h\beta = 0 & \text{if } y'(b) = \beta \end{array} \right\} \tag{8.11c}$$

EXAMPLE 8.6

Write out Eqs. (8.11) for the following linear boundary value problem using $n = 11$:

$$y'' = -4y + 4x \qquad y(0) = 0 \qquad y'(\pi/2) = 0$$

Solve these equations with a computer program.

Solution In this case $\alpha = 0$ (applicable to y), $\beta = 0$ (applicable to y') and $f(x, y, y') = -4y + 4x$. Hence Eqs. (8.11) are

$$y_1 = 0$$

$$y_{i-1} - 2y_i + y_{i+1} - h^2(-4y_i + 4x_i) = 0, \quad i = 2, 3, \ldots, 10$$

$$2y_{10} - 2y_{11} - h^2(-4y_{11} + 4x_{11}) = 0$$

or, using matrix notation

$$\begin{bmatrix} 1 & 0 & & & & \\ 1 & -2+4h^2 & 1 & & & \\ & \ddots & \ddots & \ddots & & \\ & & 1 & -2+4h^2 & 1 \\ & & & 2 & -2+4h^2 \end{bmatrix} \begin{bmatrix} y_1 \\ y_2 \\ \vdots \\ y_{10} \\ y_{11} \end{bmatrix} = \begin{bmatrix} 0 \\ 4h^2 x_2 \\ \vdots \\ 4h^2 x_{10} \\ 4h^2 x_{11} \end{bmatrix}$$

Note that the coefficient matrix is tridiagonal, so that the equations can be solved efficiently by the functions LUdec3 and LUsol3 described in Art. 2.4. Recalling that these functions store the diagonals of the coefficient matrix in vectors **c**, **d** and **e**, we arrive at the following program:

```
function fDiff6
% Finite difference method for the second-order,
% linear boundary value problem in Example 8.6.

xStart = 0; xStop = pi/2;   % Range of integration.
n = 11;                     % Number of mesh points.
freq = 1;                   % Printout frequency.

h = (xStop - xStart)/(n-1);
x = linspace(xStart,xStop,n)';
[c,d,e,b] = fDiffEqs(x,h,n);
[c,d,e] = LUdec3(c,d,e);
printSol(x,LUsol3(c,d,e,b),freq)
```

```
function [c,d,e,b] = fDiffEqs(x,h,n)
% Sets up the tridiagonal coefficient matrix and the
% constant vector of the finite difference equations.
h2 = h*h;
d = ones(n,1)*(-2 + 4*h2);
c = ones(n-1,1);
e = ones(n-1,1);
b = ones(n,1)*4*h2.*x;
d(1) = 1; e(1) = 0; b(1) = 0;c(n-1) = 2;
```

The solution is

```
>>    x                y1
0.0000e+000      0.0000e+000
1.5708e-001      3.1417e-001
3.1416e-001      6.1284e-001
4.7124e-001      8.8203e-001
6.2832e-001      1.1107e+000
7.8540e-001      1.2917e+000
9.4248e-001      1.4228e+000
1.0996e+000      1.5064e+000
1.2566e+000      1.5500e+000
1.4137e+000      1.5645e+000
1.5708e+000      1.5642e+000
```

The exact solution of the problem is

$$y = x - \sin 2x$$

which yields $y(\pi/2) = \pi/2 = 1.57080$. Thus the error in the numerical solution is about 0.4%. More accurate results can be achieved by increasing n. For example, with $n = 101$, we would get $y(\pi/2) = 1.57073$, which is in error by only 0.0002%.

EXAMPLE 8.7
Solve the boundary value problem

$$y'' = -3yy' \qquad y(0) = 0 \qquad y(2) = 1$$

with the finite difference method. (This problem was solved in Example 8.1 by the shooting method.) Use $n = 11$ and compare the results to the solution in Example 8.1.

Solution As the problem is nonlinear, Eqs. (8.11) must be solved by the Newton–Raphson method. The program listed below can be used as a model for other second-order boundary value problems. The subfunction `residual(y)` returns the residuals of the finite difference equations, which are the left-hand sides of Eqs. (8.11). The differential equation $y'' = f(x, y, y')$ is defined in the subfunction `y2Prime`. In this problem we chose for the initial solution $y_i = 0.5x_i$, which corresponds to the dashed straight line shown in the rough plot of y in Example 8.1. Note that we relaxed the convergence criterion in the Newton–Raphson method to 1.0×10^{-5}, which is more in line with the truncation error in the finite difference method.

```
function fDiff7
% Finite difference method for the second-order,
% nonlinear boundary value problem in Example 8.7.

global N H X                    % Make these params. global.
xStart = 0; xStop = 2;          % Range of integration.
N = 11;                         % Number of mesh points.
freq = 1;                       % Printout frequency.
X = linspace(xStart,xStop,N)';
y = 0.5*X;                      % Starting values of y.

H = (xStop - xStart)/(N-1);
y = newtonRaphson2(@residual,y,1.0e-5);
printSol(X,y,freq)


function r = residual(y);
% Residuals of finite difference equations (left-hand
% sides of Eqs (8.11)).
global N H X
r = zeros(N,1);
r(1) = y(1); r(N) = y(N) - 1;
for i = 2:N-1
    r(i) = y(i-1) - 2*y(i) + y(i+1)...
         - H*H*y2Prime(X(i),y(i),(y(i+1) - y(i-1))/(2*H));
end


function F = y2Prime(x,y,yPrime)
% Second-order differential equation F = y''.
F = -3*y*yPrime;
```

Here is the output from the program:

```
>>       x                y1
      0.0000e+000    0.0000e+000
      2.0000e-001    3.0240e-001
      4.0000e-001    5.5450e-001
      6.0000e-001    7.3469e-001
      8.0000e-001    8.4979e-001
      1.0000e+000    9.1813e-001
      1.2000e+000    9.5695e-001
      1.4000e+000    9.7846e-001
      1.6000e+000    9.9020e-001
      1.8000e+000    9.9657e-001
      2.0000e+000    1.0000e+000
```

The maximum discrepancy between the above solution and the one in Example 8.1 occurs at $x = 0.6$. In Example 8.1 we have $y(0.6) = 0.072187$, so that the difference between the solutions is

$$\frac{0.073469 - 0.072187}{0.072187} \times 100\% \approx 1.8\%$$

As the shooting method used in Example 8.1 is considerably more accurate than the finite difference method, the discrepancy can be attributed to truncation errors in the finite difference solution. This error would be acceptable in many engineering problems. Again, accuracy can be increased by using a finer mesh. With $n = 101$ we can reduce the error to 0.07%, but we must question whether the tenfold increase in computation time is really worth the extra precision.

Fourth-Order Differential Equation

For the sake of brevity we limit our discussion to the special case where y' and y''' do not appear explicitly in the differential equation; that is, we consider

$$y^{(4)} = f(x, y, y'')$$

We assume that two boundary conditions are prescribed at each end of the solution domain (a, b). Problems of this form are commonly encountered in beam theory.

Again we divide the solution domain into $n - 1$ intervals of length h each. Replacing the derivatives of y by finite differences at the mesh points, we get the finite difference equations

$$\frac{y_{i-2} - 4y_{i-1} + 6y_i - 4y_{i+1} + y_{i+2}}{h^4} = f\left(x_i, y_i, \frac{y_{i-1} - 2y_i + y_{i+1}}{h^2}\right) \qquad (8.12)$$

where $i = 1, 2, \ldots, n$. It is more revealing to write these equations as

$$y_{-1} - 4y_0 + 6y_1 - 4y_2 + y_3 - h^4 f\left(x_1, y_1, \frac{y_0 - 2y_1 + y_2}{h^2}\right) = 0 \qquad (8.13a)$$

$$y_0 - 4y_1 + 6y_2 - 4y_3 + y_4 - h^4 f\left(x_2, y_2, \frac{y_1 - 2y_2 + y_3}{h^2}\right) = 0 \qquad (8.13b)$$

$$y_1 - 4y_2 + 6y_3 - 4y_4 + y_5 - h^4 f\left(x_3, y_3, \frac{y_2 - 2y_3 + y_4}{h^2}\right) = 0 \qquad (8.13c)$$

$$\vdots$$

$$y_{n-3} - 4y_{n-2} + 6y_{n-1} - 4y_n + y_{n+1} - h^4 f\left(x_{n-1}, y_{n-1}, \frac{y_{n-2} - 2y_{n-1} + y_n}{h^2}\right) = 0 \qquad (8.13d)$$

$$y_{n-2} - 4y_{n-1} + 6y_n - 4y_{n+1} + y_{n+2} - h^4 f\left(x_n, y_n, \frac{y_{n-1} - 2y_n + y_{n+1}}{h^2}\right) = 0 \qquad (8.13e)$$

We now see that there are four unknowns that lie outside the solution domain: y_{-1}, y_0, y_{n+1} and y_{n+2}. This "spillover" can be eliminated by applying the boundary conditions, a task that is facilitated by Table 8.1.

Bound. cond.	Equivalent finite difference expression
$y(a) = \alpha$	$y_1 = \alpha$
$y'(a) = \alpha$	$y_0 = y_2 - 2h\alpha$
$y''(a) = \alpha$	$y_0 = 2y_1 - y_2 + h^2\alpha$
$y'''(a) = \alpha$	$y_{-1} = 2y_0 - 2y_2 + y_3 - 2h^3\alpha$
$y(b) = \beta$	$y_n = \beta$
$y'(b) = \beta$	$y_{n+1} = y_{n-1} + 2h\beta$
$y''(b) = \beta$	$y_{n+1} = 2y_n - y_{n-1} + h^2\beta$
$y'''(b) = \beta$	$y_{n+2} = 2y_{n+1} - 2y_{n-1} + y_{n-2} + 2h^3\beta$

Table 8.1

The astute observer may notice that some combinations of boundary conditions will not work in eliminating the "spillover." One such combination is clearly $y(a) = \alpha_1$ and $y'''(a) = \alpha_2$. The other one is $y'(a) = \alpha_1$ and $y''(a) = \alpha_2$. In the context of beam theory, this makes sense: we can impose either a displacement y or a shear force EIy''' at a point, but it is impossible to enforce both of them simultaneously. Similarly, it makes no physical sense to prescribe both the slope y' and the bending moment EIy'' at the same point.

EXAMPLE 8.8

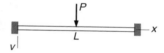

The uniform beam of length L and bending rigidity EI is attached to rigid supports at both ends. The beam carries a concentrated load P at its mid-span. If we utilize symmetry and model only the left half of the beam, the displacement v can be obtained by solving the boundary value problem

$$EI\frac{d^4v}{dx^4} = 0$$

$$v|_{x=0} = 0 \qquad \frac{dv}{dx}\Big|_{x=0} = 0 \qquad \frac{dv}{dx}\Big|_{x=L/2} = 0 \qquad EI\frac{d^3v}{dx^3}\Big|_{x=L/2} = -P/2$$

Use the finite difference method to determine the displacement and the bending moment $M = -EI\,(d^2v/dx^2)$ at the mid-span (the exact values are $v = PL^3/(192EI)$ and $M = PL/8$).

Solution By introducing the dimensionless variables

$$\xi = \frac{x}{L} \qquad y = \frac{EI}{PL^3}v$$

the problem becomes

$$\frac{d^4y}{d\xi^4} = 0$$

$$y|_{\xi=0} = 0 \qquad \frac{dy}{d\xi}\Big|_{\xi=0} = 0 \qquad \frac{dy}{d\xi}\Big|_{\xi=1/2} = 0 \qquad \frac{d^3y}{d\xi^3}\Big|_{\xi=1/2} = -\frac{1}{2}$$

We now proceed to writing Eqs. (8.13) taking into account the boundary conditions. Referring to Table 8.1, we obtain the finite difference expressions of the boundary conditions at the left end as $y_1 = 0$ and $y_0 = y_2$. Hence Eqs. (8.13a) and (8.13b) become

$$y_1 = 0 \tag{a}$$

$$-4y_1 + 7y_2 - 4y_3 + y_4 = 0 \tag{b}$$

Equation (8.13c) is

$$y_1 - 4y_2 + 6y_3 - 4y_4 + y_5 = 0 \tag{c}$$

At the mid-span the boundary conditions are equivalent to $y_{n+1} = y_{n-1}$ and

$$y_{n+2} = 2y_{n+1} - 2y_{n-1} + y_{n-2} + 2h^3(-1/2)$$

Substitution into Eqs. (8.13d) and (8.13e) yields

$$y_{n-3} - 4y_{n-2} + 7y_{n-1} - 4y_n = 0 \tag{d}$$

$$2y_{n-2} - 8y_{n-1} + 6y_n = h^3 \tag{e}$$

The coefficient matrix of Eqs. (a)–(e) can be made symmetric by dividing Eq. (e) by 2. The result is

$$
\begin{bmatrix}
1 & 0 & 0 & & & & & \\
0 & 7 & -4 & 1 & & & & \\
0 & -4 & 6 & -4 & 1 & & & \\
& \ddots & \ddots & \ddots & \ddots & \ddots & & \\
& & 1 & -4 & 6 & -4 & 1 \\
& & & 1 & -4 & 7 & -4 \\
& & & & 1 & -4 & 3
\end{bmatrix}
\begin{bmatrix}
y_1 \\ y_2 \\ y_3 \\ \vdots \\ y_{n-2} \\ y_{n-1} \\ y_n
\end{bmatrix}
=
\begin{bmatrix}
0 \\ 0 \\ 0 \\ \vdots \\ 0 \\ 0 \\ 0.5h^3
\end{bmatrix}
$$

The above system of equations can be solved with the decomposition and back substitution routines in the functions LUdec5 and LUsol5—see Art. 2.4. Recall that these functions work with the vectors **d**, **e** and **f** that form the diagonals of upper the half of the coefficient matrix. The program that sets up and solves the equations is

```
function fDiff8
% Finite difference method for the 4th-order,
% linear boundary value problem in Example 8.8.

xStart = 0; xStop = 0.5;   % Range of integration.
n = 21;                    % Number of mesh points.
freq = 1;                  % Printout frequency.
h = (xStop - xStart)/(n-1);
x = linspace(xStart,xStop,n)';
[d,e,f,b] = fDiffEqs(x,h,n);
[d,e,f] = LUdec5(d,e,f);
printSol(x,LUsol5(d,e,f,b),freq)

function [d,e,f,b] = fDiffEqs(x,h,n)
% Sets up the pentadiagonal coefficient matrix and the
% constant vector of the finite difference equations.
d = ones(n,1)*6;
```

```
e = ones(n-1,1)*(-4);
f = ones(n-2,1);
b = zeros(n,1);
d(1) = 1; d(2) = 7; d(n-1) = 7; d(n) = 3;
e(1) = 0; f(1) = 0; b(n) = 0.5*h^3;
```

The last two lines of the output are

```
>>      x               y1
     4.7500e-001    5.1953e-003
     5.0000e-001    5.2344e-003
```

Thus at the mid-span we have

$$v|_{x=0.5L} = \frac{PL^3}{EI}\, y|_{\xi=0.5} = 5.2344 \times 10^{-3}\frac{PL^3}{EI}$$

$$\frac{d^2v}{dx^2}\bigg|_{x=0.5L} = \frac{PL^3}{EI}\left(\frac{1}{L^2}\frac{d^2y}{d\xi^2}\bigg|_{\xi=0.5}\right) \approx \frac{PL}{EI}\frac{y_{m-1} - 2y_m + y_{m+1}}{h^2}$$

$$= \frac{PL}{EI}\frac{(5.1953 - 2(5.2344) + 5.1953) \times 10^{-3}}{0.025^2}$$

$$= -0.125\,12\frac{PL}{EI}$$

$$M|_{x=0.5L} = -EI\,\frac{d^2v}{dx^2}\bigg|_{\xi=0.5} = 0.125\,12\,PL$$

In comparison, the exact solution yields

$$v|_{x=0.5L} = 5.208\,3 \times 10^{-3}\frac{PL^3}{EI}$$

$$M|_{x=0.5L} = = 0.125\,00\,PL$$

PROBLEM SET 8.2

Problems 1–5 Use first central difference approximations to transform the boundary value problem shown into simultaneous equations $\mathbf{Ay} = \mathbf{b}$.

1. $y'' = (2 + x)y$, $y(0) = 0$, $y'(1) = 5$.

2. $y'' = y + x^2$, $y(0) = 0$, $y(1) = 1$.

3. $y'' = e^{-x}y'$, $y(0) = 1$, $y(1) = 0$.

4. $y^{(4)} = y'' - y$, $y(0) = 0$, $y'(0) = 1$, $y(1) = 0$, $y'(1) = -1$.

5. $y^{(4)} = -9y + x$, $y(0) = y''(0) = 0$, $y'(1) = y'''(1) = 0$.

Problems 6–10 Solve the given boundary value problem with the finite difference method using $n = 21$.

6. ■ $y'' = xy$, $y(1) = 1.5$ $y(2) = 3$.

7. ■ $y'' + 2y' + y = 0$, $y(0) = 0$, $y(1) = 1$. Exact solution is $y = xe^{1-x}$.

8. ■ $x^2 y'' + xy' + y = 0$, $y(1) = 0$, $y(2) = 0.638961$. Exact solution is $y = \sin(\ln x)$.

9. ■ $y'' = y^2 \sin y$, $y'(0) = 0$, $y(\pi) = 1$.

10. ■ $y'' + 2y(2xy' + y) = 0$, $y(0) = 1/2$, $y'(1) = -2/9$. Exact solution is $y = (2 + x^2)^{-1}$.

11. ■

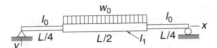

The simply supported beam consists of three segments with the moments of inertia I_0 and I_1 as shown. A uniformly distributed load of intensity w_0 acts over the middle segment. Modeling only the left half of the beam, we can show that the differential equation

$$\frac{d^2 v}{dx^2} = -\frac{M}{EI}$$

for the displacement v is

$$\frac{d^2 v}{dx^2} = -\frac{w_0 L^2}{4 E I_0} \times \begin{cases} \dfrac{x}{L} & \text{in } 0 < x < \dfrac{L}{4} \\[2ex] \dfrac{I_0}{I_1}\left[\dfrac{x}{L} - 2\left(\dfrac{x}{L} - \dfrac{1}{4}\right)^2\right] & \text{in } \dfrac{L}{4} < x < \dfrac{L}{2} \end{cases}$$

Introducing the dimensionless variables

$$\xi = \frac{x}{L} \qquad y = \frac{E I_0}{w_0 L^4} v \qquad \gamma = \frac{I_1}{I_0}$$

changes the differential equation to

$$\frac{d^2 y}{d\xi^2} = \begin{cases} -\dfrac{1}{4}\xi & \text{in } 0 < \xi < \dfrac{1}{4} \\[2ex] -\dfrac{1}{4\gamma}\left[\xi - 2\left(\xi - \dfrac{1}{4}\right)^2\right] & \text{in } \dfrac{1}{4} < \xi < \dfrac{1}{2} \end{cases}$$

with the boundary conditions

$$y\big|_{\xi=0} = \frac{dy}{d\xi}\bigg|_{\xi=1/2} = 0$$

Use the finite difference method to determine the maximum displacement of the beam using $n = 21$ and $\gamma = 1.5$ and compare it with the exact solution

$$v_{\max} = \frac{61}{9216} \frac{w_0 L^4}{E I_0}$$

12. ■

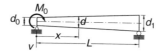

The simply supported, tapered beam has a circular cross section. A couple of magnitude M_0 is applied to the left end of the beam. The differential equation for the displacement v is

$$\frac{d^2 v}{dx^2} = -\frac{M}{EI} = -\frac{M_0(1 - x/L)}{E I_0 (d/d_0)^4}$$

where

$$d = d_0 \left[1 + \left(\frac{d_1}{d_0} - 1 \right) \frac{x}{L} \right] \qquad I_0 = \frac{\pi d_0^4}{64}$$

Substituting

$$\xi = \frac{x}{L} \qquad y = \frac{E I_0}{M_0 L^2} v \qquad \delta = \frac{d_1}{d_0}$$

changes the differential equation to

$$\frac{d^2 y}{d\xi^2} = -\frac{1 - \xi}{[1 + (\delta - 1)\xi]^4}$$

with the boundary conditions

$$y|_{\xi=0} = y|_{\xi=1} = 0$$

Solve the problem with the finite difference method using $\delta = 1.5$ and $n = 21$; plot y vs. ξ. The exact solution is

$$y = -\frac{(3 + 2\delta\xi - 3\xi)\xi^2}{6(1 + \delta\xi - \xi)^2} + \frac{\xi}{3\delta}$$

13. ■ Solve Example 8.4 by the finite difference method with $n = 21$. *Hint*: Compute the end slopes from the second noncentral differences in Tables 5.3.

14. ■ Solve Prob. 20 in Problem Set 8.1 with the finite difference method. Use $n = 21$.

15. ■

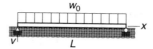

The simply supported beam of length L is resting on an elastic foundation of stiffness $k\,\mathrm{N/m^2}$. The displacement v of the beam due to the uniformly distributed load of intensity w_0 N/m is given by the solution of the boundary value problem

$$EI\frac{d^4v}{dx^4} + kv = w_0, \quad v|_{x=0} = \left.\frac{d^2y}{dx^2}\right|_{x=0} = v|_{x=L} = \left.\frac{d^2v}{dx^2}\right|_{x=L} = 0$$

The nondimensional form of the problem is

$$\frac{d^4y}{d\xi^4} + \gamma y = 1, \quad y|_{\xi=0} = \left.\frac{d^2y}{dx^2}\right|_{\xi=0} = y|_{\xi=1} = \left.\frac{d^2y}{dx^2}\right|_{\xi=1} = 0$$

where

$$\xi = \frac{x}{L} \qquad y = \frac{EI}{w_0 L^4}v \qquad \gamma = \frac{kL^4}{EI}$$

Solve this problem by the finite difference method with $\gamma = 10^5$ and plot y vs. ξ.

16. ■ Solve Prob. 15 if the ends of the beam are free and the load is confined to the middle half of the beam. Consider only the left half of the beam, in which case the nondimensional form of the problem is

$$\frac{d^4y}{d\xi^4} + \gamma y = \begin{cases} 0 \text{ in } 0 < \xi < 1/4 \\ 1 \text{ in } 1/4 < \xi < 1/2 \end{cases}$$

$$\left.\frac{d^2y}{d\xi^2}\right|_{\xi=0} = \left.\frac{d^3y}{d\xi^3}\right|_{\xi=0} = \left.\frac{dy}{d\xi}\right|_{\xi=1/2} = \left.\frac{d^3y}{d\xi^3}\right|_{\xi=1/2} = 0$$

17. ■ The general form of a linear, second-order boundary value problem is

$$y'' = r(x) + s(x)y + t(x)y'$$
$$y(a) = \alpha \text{ or } y'(a) = \alpha$$
$$y(b) = \beta \text{ or } y'(b) = \beta$$

Write a program that solves this problem with the finite difference method for any user-specified $r(x)$, $s(x)$ and $t(x)$. Test the program by solving Prob. 8.

MATLAB Functions

MATLAB has only the following function for solution of boundary value problems:

`sol = bvp4c(@dEqs,@residual,solinit)` uses a high-order finite difference method with an adaptive mesh to solve boundary value problems. The output `sol` is a *structure* (a MATLAB data type) created by `bvp4c`. The first two input arguments are handles to the following user-supplied functions:

F = dEqs(x,y) specifies the first-order differential equations $\mathbf{F}(x, \mathbf{y}) = \mathbf{y}'$. Both F and y are column vectors.

r = residual(ya,yb) specifies all the applicable the boundary residuals $y_i(a) -$ α_i and $y_i(b) - \beta_i$ in a column vector r, where α_i and β_i are the prescribed boundary values.

The third input argument solinit is a structure that contains the x and y-values at the nodes of the initial mesh. This structure can be generated with MATLAB's function bvpinit:

solinit = bvpinit(xinit,@yguess) where xinit is a vector containing the x-coordinates of the nodes; yguess(x) is a user-supplied function that returns a column vector containing the trial solutions for the components of y.

The numerical solution at user-defined mesh points can be extracted from the structure sol with the MATLAB function deval:

y = deval(sol,xmesh) where xmesh is an array containing the x-coordinates of the mesh points. The function returns a matrix with the ith row containing the values of y_i at the mesh points.

The following program illustrates the use of the above functions in solving Example 8.1:

```
function shoot2_matlab
% Solution of Example 8.1 with MATLAB's function bvp4c.

xinit = linspace(0,2,11)';
solinit = bvpinit(xinit,@yguess);
sol = bvp4c(@dEqs,@residual,solinit);
y = deval(sol,xinit)';
printSol(xinit,y,1)            % This is our own func.

function F = dEqs(x,y)         % Differential eqs.
F = [y(2); -3*y(1)*y(2)];

function r = residual(ya,yb) % Boundary residuals.
r = [ya(1); yb(1) - 1];

function yinit = yguess(x)     % Initial guessses for
yinit = [0.5*x; 0.5];          % y1 and y2.
```

9 Symmetric Matrix Eigenvalue Problems

> Find λ for which nontrivial solutions of $\mathbf{Ax} = \lambda\mathbf{x}$ exist

9.1 Introduction

The *standard form* of the matrix eigenvalue problem is

$$\mathbf{Ax} = \lambda\mathbf{x} \tag{9.1}$$

where $\mathbf{A}$ is a given $n \times n$ matrix. The problem is to find the scalar λ and the vector $\mathbf{x}$. Rewriting Eq. (9.1) in the form

$$(\mathbf{A} - \lambda\mathbf{I})\mathbf{x} = \mathbf{0} \tag{9.2}$$

it becomes apparent that we are dealing with a system of n homogeneous equations. An obvious solution is the trivial one $\mathbf{x} = \mathbf{0}$. A nontrivial solution can exist only if the determinant of the coefficient matrix vanishes; that is, if

$$|\mathbf{A} - \lambda\mathbf{I}| = 0 \tag{9.3}$$

Expansion of the determinant leads to the polynomial equation known as the *characteristic equation*

$$a_1\lambda^n + a_2\lambda^{n-1} + \cdots + a_n\lambda + a_{n+1} = 0$$

which has the roots λ_i, $i = 1, 2, \ldots, n$, called the *eigenvalues* of the matrix $\mathbf{A}$. The solutions $\mathbf{x}_i$ of $(\mathbf{A} - \lambda_i\mathbf{I})\mathbf{x} = \mathbf{0}$ are known as the *eigenvectors*.

As an example, consider the matrix

$$\mathbf{A} = \begin{bmatrix} 1 & -1 & 0 \\ -1 & 2 & -1 \\ 0 & -1 & 1 \end{bmatrix} \tag{a}$$

The characteristic equation is

$$|\mathbf{A} - \lambda\mathbf{I}| = \begin{vmatrix} 1-\lambda & -1 & 0 \\ -1 & 2-\lambda & -1 \\ 0 & -1 & 1-\lambda \end{vmatrix} = -3\lambda + 4\lambda^2 - \lambda^3 = 0 \tag{b}$$

The roots of this equation are $\lambda_1 = 0$, $\lambda_2 = 1$, $\lambda_3 = 3$. To compute the eigenvector corresponding the λ_3, we substitute $\lambda = \lambda_3$ into Eq. (9.2), obtaining

$$\begin{bmatrix} -2 & -1 & 0 \\ -1 & -1 & -1 \\ 0 & -1 & -2 \end{bmatrix} \begin{bmatrix} x_1 \\ x_2 \\ x_3 \end{bmatrix} = \begin{bmatrix} 0 \\ 0 \\ 0 \end{bmatrix} \tag{c}$$

We know that the determinant of the coefficient matrix is zero, so that the equations are not linearly independent. Therefore, we can assign an arbitrary value to any one component of $\mathbf{x}$ and use two of the equations to compute the other two components. Choosing $x_1 = 1$, the first equation of Eq. (c) yields $x_2 = -2$ and from the third equation we get $x_3 = 1$. Thus the eigenvector associated with λ_3 is

$$\mathbf{x}_3 = \begin{bmatrix} 1 \\ -2 \\ 1 \end{bmatrix}$$

The other two eigenvectors

$$\mathbf{x}_2 = \begin{bmatrix} 1 \\ 0 \\ -1 \end{bmatrix} \qquad \mathbf{x}_1 = \begin{bmatrix} 1 \\ 1 \\ 1 \end{bmatrix}$$

can be obtained in the same manner.

It is sometimes convenient to display the eigenvectors as columns of a matrix $\mathbf{X}$. For the problem at hand, this matrix is

$$\mathbf{X} = \begin{bmatrix} \mathbf{x}_1 & \mathbf{x}_2 & \mathbf{x}_3 \end{bmatrix} = \begin{bmatrix} 1 & 1 & 1 \\ 1 & 0 & -2 \\ 1 & -1 & 1 \end{bmatrix}$$

It is clear from the above example that the magnitude of an eigenvector is indeterminate; only its direction can be computed from Eq. (9.2). It is customary to *normalize* the eigenvectors by assigning a unit magnitude to each vector. Thus the normalized eigenvectors in our example are

$$\mathbf{X} = \begin{bmatrix} 1/\sqrt{3} & 1/\sqrt{2} & 1/\sqrt{6} \\ 1/\sqrt{3} & 0 & -2/\sqrt{6} \\ 1/\sqrt{3} & -1/\sqrt{2} & 1/\sqrt{6} \end{bmatrix}$$

Throughout this chapter we assume that the eigenvectors are normalized.

Here are some useful properties of eigenvalues and eigenvectors, given without proof:

- All eigenvalues of a symmetric matrix are real.
- All eigenvalues of a symmetric, positive-definite matrix are real and positive.
- The eigenvectors of a symmetric matrix are orthonormal; that is, $\mathbf{X}^T\mathbf{X} = \mathbf{I}$.
- If the eigenvalues of $\mathbf{A}$ are λ_i, then the eigenvalues of $\mathbf{A}^{-1}$ are λ_i^{-1}.

Eigenvalue problems that originate from physical problems often end up with a symmetric $\mathbf{A}$. This is fortunate, because symmetric eigenvalue problems are much easier to solve than their nonsymmetric counterparts. In this chapter we largely restrict our discussion to eigenvalues and eigenvectors of symmetric matrices.

Common sources of eigenvalue problems are the analysis of vibrations and stability. These problems often have the following characteristics:

- The matrices are large and sparse (e.g., have a banded structure).
- We need to know only the eigenvalues; if eigenvectors are required, only a few of them are of interest.

A useful eigenvalue solver must be able to utilize these characteristics to minimize the computations. In particular, it should be flexible enough to compute only what we need and no more.

9.2   Jacobi Method

Similarity Transformation and Diagonalization

Consider the standard matrix eigenvalue problem

$$\mathbf{A}\mathbf{x} = \lambda\mathbf{x} \tag{9.4}$$

where $\mathbf{A}$ is symmetric. Let us now apply the transformation

$$\mathbf{x} = \mathbf{P}\mathbf{x}^* \tag{9.5}$$

where $\mathbf{P}$ is a nonsingular matrix. Substituting Eq. (9.5) into Eq. (9.4) and premultiplying each side by $\mathbf{P}^{-1}$, we get

$$\mathbf{P}^{-1}\mathbf{A}\mathbf{P}\mathbf{x}^* = \lambda\mathbf{P}^{-1}\mathbf{P}\mathbf{x}^*$$

or

$$\mathbf{A}^*\mathbf{x}^* = \lambda\mathbf{x}^* \tag{9.6}$$

where $\mathbf{A}^* = \mathbf{P}^{-1}\mathbf{AP}$. Because λ was untouched by the transformation, the eigenvalues of $\mathbf{A}$ are also the eigenvalues of $\mathbf{A}^*$. Matrices that have the same eigenvalues are deemed to be *similar*, and the transformation between them is called a *similarity transformation*.

Similarity transformations are frequently used to change an eigenvalue problem to a form that is easier to solve. Suppose that we managed by some means to find a $\mathbf{P}$ that diagonalizes $\mathbf{A}^*$, so that Eqs. (9.6) are

$$
\begin{bmatrix}
A_{11}^* - \lambda & 0 & \cdots & 0 \\
0 & A_{22}^* - \lambda & \cdots & 0 \\
\vdots & \vdots & \ddots & \vdots \\
0 & 0 & \cdots & A_{nn}^* - \lambda
\end{bmatrix}
\begin{bmatrix}
x_1^* \\
x_2^* \\
\vdots \\
x_n^*
\end{bmatrix}
=
\begin{bmatrix}
0 \\
0 \\
\vdots \\
0
\end{bmatrix}
$$

The solution of these equations is

$$
\lambda_1 = A_{11}^* \qquad \lambda_2 = A_{22}^* \qquad \cdots \qquad \lambda_n = A_{nn}^* \tag{9.7}
$$

$$
\mathbf{x}_1^* =
\begin{bmatrix}
1 \\
0 \\
\vdots \\
0
\end{bmatrix}
\qquad
\mathbf{x}_2^* =
\begin{bmatrix}
0 \\
1 \\
\vdots \\
0
\end{bmatrix}
\qquad \cdots \qquad
\mathbf{x}_n^* =
\begin{bmatrix}
0 \\
0 \\
\vdots \\
1
\end{bmatrix}
$$

or

$$
\mathbf{X}^* = \begin{bmatrix} \mathbf{x}_1^* & \mathbf{x}_2^* & \cdots & \mathbf{x}_n^* \end{bmatrix} = \mathbf{I}
$$

According to Eq. (9.5) the eigenvector matrix of $\mathbf{A}$ is

$$
\mathbf{X} = \mathbf{PX}^* = \mathbf{PI} = \mathbf{P} \tag{9.8}
$$

Hence the transformation matrix $\mathbf{P}$ is the eigenvector matrix of $\mathbf{A}$ and the eigenvalues of $\mathbf{A}$ are the diagonal terms of $\mathbf{A}^*$.

Jacobi Rotation

A special transformation is the plane rotation

$$
\mathbf{x} = \mathbf{Rx}^* \tag{9.9}
$$

where

$$
\mathbf{R} =
\begin{bmatrix}
1 & 0 & 0 & 0 & 0 & 0 & 0 & 0 \\
0 & 1 & 0 & 0 & 0 & 0 & 0 & 0 \\
0 & 0 & c & 0 & 0 & s & 0 & 0 \\
0 & 0 & 0 & 1 & 0 & 0 & 0 & 0 \\
0 & 0 & 0 & 0 & 1 & 0 & 0 & 0 \\
0 & 0 & -s & 0 & 0 & c & 0 & 0 \\
0 & 0 & 0 & 0 & 0 & 0 & 1 & 0 \\
0 & 0 & 0 & 0 & 0 & 0 & 0 & 1
\end{bmatrix}
\begin{matrix} \\ \\ k \\ \\ \\ \ell \\ \\ \\ \end{matrix}
\tag{9.10}
$$

is called the *Jacobi rotation matrix*. Note that $\mathbf{R}$ is an identity matrix modified by the terms $c = \cos\theta$ and $s = \sin\theta$ appearing at the intersections of columns/rows k and ℓ, where θ is the rotation angle. The rotation matrix has the useful property of being *orthogonal*, or *unitary*, meaning that

$$
\mathbf{R}^{-1} = \mathbf{R}^T \tag{9.11}
$$

One consequence of orthogonality is that the transformation in Eq. (9.9) has the essential characteristic of a rotation: it preserves the magnitude of the vector; that is, $|\mathbf{x}| = |\mathbf{x}^*|$.

The similarity transformation corresponding to the plane rotation in Eq. (9.9) is

$$
\mathbf{A}^* = \mathbf{R}^{-1}\mathbf{A}\mathbf{R} = \mathbf{R}^T\mathbf{A}\mathbf{R} \tag{9.12}
$$

The matrix $\mathbf{A}^*$ not only has the same eigenvalues as the original matrix $\mathbf{A}$, but due to orthogonality of $\mathbf{R}$ it is also symmetric. The transformation in Eq. (9.12) changes only the rows/columns k and ℓ of $\mathbf{A}$. The formulas for these changes are

$$
A_{kk}^* = c^2 A_{kk} + s^2 A_{\ell\ell} - 2cs A_{k\ell}
$$
$$
A_{\ell\ell}^* = c^2 A_{\ell\ell} + s^2 A_{kk} + 2cs A_{k\ell}
$$
$$
A_{k\ell}^* = A_{\ell k}^* = (c^2 - s^2) A_{k\ell} + cs(A_{kk} - A_{\ell\ell}) \tag{9.13}
$$
$$
A_{ki}^* = A_{ik}^* = c A_{ki} - s A_{\ell i}, \quad i \neq k, \quad i \neq \ell
$$
$$
A_{\ell i}^* = A_{i\ell}^* = c A_{\ell i} + s A_{ki}, \quad i \neq k, \quad i \neq \ell
$$

Jacobi Diagonalization

The angle θ in the Jacobi rotation matrix can be chosen so that $A_{k\ell}^* = A_{\ell k}^* = 0$. This suggests the following idea: why not diagonalize $\mathbf{A}$ by looping through all the off-diagonal terms and eliminate them one by one? This is exactly what Jacobi diagonalization does.

However, there is a major snag—the transformation that annihilates an off-diagonal term also undoes some of the previously created zeroes. Fortunately, it turns out that the off-diagonal terms that reappear will be smaller than before. Thus Jacobi method is an iterative procedure that repeatedly applies Jacobi rotations until the off-diagonal terms have virtually vanished. The final transformation matrix $\mathbf{P}$ is the accumulation of individual rotations $\mathbf{R}_i$:

$$\mathbf{P} = \mathbf{R}_1 \cdot \mathbf{R}_2 \cdot \mathbf{R}_3 \cdots \tag{9.14}$$

The columns of $\mathbf{P}$ finish up being the eigenvectors of $\mathbf{A}$ and the diagonal elements of $\mathbf{A}^* = \mathbf{P}^T \mathbf{A} \mathbf{P}$ become the eigenvectors.

Let us now look at the details of a Jacobi rotation. From Eq. (9.13) we see that $A_{k\ell}^* = 0$ if

$$(c^2 - s^2)A_{k\ell} + cs(A_{kk} - A_{\ell\ell}) = 0 \tag{a}$$

Using the trigonometric identities $c^2 - s^2 = \cos 2\theta$ and $cs = (1/2)\sin 2\theta$, we obtain from Eq. (a)

$$\tan 2\theta = -\frac{2A_{k\ell}}{A_{kk} - A_{\ell\ell}} \tag{b}$$

which could be solved for θ, followed by computation of $c = \cos\theta$ and $s = \sin\theta$. However, the procedure described below leads to a better algorithm.[20]

Introducing the notation

$$\phi = \cot 2\theta = -\frac{A_{kk} - A_{\ell\ell}}{2A_{k\ell}} \tag{9.15}$$

and utilizing the trigonometric identity

$$\tan 2\theta = \frac{2t}{(1 - t^2)}$$

where $t = \tan\theta$, we can write Eq. (b) as

$$t^2 + 2\phi t - 1 = 0$$

which has the roots

$$t = -\phi \pm \sqrt{\phi^2 + 1}$$

It has been found that the root $|t| \leq 1$, which corresponds to $|\theta| \leq 45°$, leads to the more stable transformation. Therefore, we choose the plus sign if $\phi > 0$ and the minus

---

[20] The procedure is adapted from W.H. Press *et al.*, *Numerical Recipes in Fortran*, 2nd ed. (1992), Cambridge University Press.

sign if $\phi \leq 0$, which is equivalent to using

$$t = \text{sgn}(\phi) \left(-|\phi| + \sqrt{\phi^2 + 1} \right)$$

To forestall excessive roundoff error if ϕ is large, we multiply both sides of the equation by $|\phi| + \sqrt{\phi^2 + 1}$ and solve for t, which yields

$$t = \frac{\text{sgn}(\phi)}{|\phi| + \sqrt{\phi^2 + 1}} \tag{9.16a}$$

In the case of very large ϕ, we should replace Eq. (9.16a) by the approximation

$$t = \frac{1}{2\phi} \tag{9.16b}$$

to prevent overflow in the computation of ϕ^2. Having computed t, we can use the trigonometric relationship $\tan\theta = \sin\theta / \cos\theta = \sqrt{1 - \cos^2\theta} / \cos\theta$ to obtain

$$c = \frac{1}{\sqrt{1 + t^2}} \qquad s = tc \tag{9.17}$$

We now improve the computational properties of the transformation formulas in Eqs. (9.13). Solving Eq. (a) for $A_{\ell\ell}$, we obtain

$$A_{\ell\ell} = A_{kk} + A_{k\ell} \frac{c^2 - s^2}{cs} \tag{c}$$

Replacing all occurrences of $A_{\ell\ell}$ by Eq. (c) and simplifying, we can write the transformation formulas in Eqs. (9.13) as

$$A_{kk}^* = A_{kk} - t A_{k\ell}$$

$$A_{\ell\ell}^* = A_{\ell\ell} + t A_{k\ell}$$

$$A_{k\ell}^* = A_{\ell k}^* = 0 \tag{9.18}$$

$$A_{ki}^* = A_{ik}^* = A_{ki} - s(A_{\ell i} + \tau A_{ki}), \quad i \neq k, \quad i \neq \ell$$

$$A_{\ell i}^* = A_{i\ell}^* = A_{\ell i} + s(A_{ki} - \tau A_{\ell i}), \quad i \neq k, \quad i \neq \ell$$

where

$$\tau = \frac{s}{1 + c} \tag{9.19}$$

The introduction of τ allowed us to express each formula in the form (original value) + (change), which is helpful in reducing the roundoff error.

At the start of Jacobi's diagonalization process the transformation matrix $\mathbf{P}$ is initialized to the identity matrix. Each Jacobi rotation changes this matrix from $\mathbf{P}$ to $\mathbf{P}^* = \mathbf{PR}$. The corresponding changes in the elements of $\mathbf{P}$ can be shown to be

(only the columns k and ℓ are affected)

$$P_{ik}^* = P_{ik} - s(P_{i\ell} + \tau P_{ik}) \tag{9.20}$$

$$P_{i\ell}^* = P_{i\ell} + s(P_{ik} - \tau P_{i\ell})$$

We still have to decide the order in which the off-diagonal elements of $\mathbf{A}$ are to be eliminated. Jacobi's original idea was to attack the largest element since this results in fewest number of rotations. The problem here is that $\mathbf{A}$ has to be searched for the largest element after every rotation, which is a time-consuming process. If the matrix is large, it is faster to sweep through it by rows or columns and annihilate every element above some threshold value. In the next sweep the threshold is lowered and the process repeated. We adopt Jacobi's original scheme because of its simpler implementation.

In summary, the Jacobi diagonalization procedure, which uses only the upper half of the matrix, is

1. Find the largest (absolute value) off-diagonal element $A_{k\ell}$ in the upper half of $\mathbf{A}$.
2. Compute ϕ, t, c and s from Eqs. (9.15)–(9.17).
3. Compute τ from Eq. (9.19).
4. Modify the elements in the upper half of $\mathbf{A}$ according to Eqs. (9.18).
5. Update the transformation matrix $\mathbf{P}$ using Eqs. (9.20).

Repeat the procedure until the $A_{k\ell} < \varepsilon$, where ε is the error tolerance.

■ jacobi

The function jacobi computes all eigenvalues λ_i and eigenvectors $\mathbf{x}_i$ of a symmetric, $n \times n$ matrix $\mathbf{A}$ by the Jacobi method. The algorithm works exclusively with the upper triangular part of $\mathbf{A}$, which is destroyed in the process. The principal diagonal of $\mathbf{A}$ is replaced by the eigenvalues, and the columns of the transformation matrix $\mathbf{P}$ become the normalized eigenvectors.

```
function [eVals,eVecs] = jacobi(A,tol)
% Jacobi method for computing eigenvalues and
% eigenvectors of a symmetric matrix A.
% USAGE: [eVals,eVecs] = jacobi(A,tol)
% tol = error tolerance (default is 1.0e-9).

if nargin < 2; tol = 1.0e-9; end
n = size(A,1);
maxRot = 5*(n^2);          % Limit number of rotations
```

```
P = eye(n);                  % Initialize rotation matrix
for i = 1:maxRot             % Begin Jacobi rotations
    [Amax,k,L] = maxElem(A);
    if Amax < tol;
        eVals = diag(A); eVecs = P;
        return
    end
    [A,P] = rotate(A,P,k,L);
end
error('Too many Jacobi rotations')

function [Amax,k,L] = maxElem(A)
% Finds Amax = A(k,L) (largest off-diag. elem. of A).
n = size(A,1);
Amax = 0;
for i = 1:n-1
    for j = i+1:n
        if abs(A(i,j)) >= Amax
            Amax = abs(A(i,j));
            k = i; L = j;
        end
    end
end

function [A,P] = rotate(A,P,k,L)
% zeros A(k,L) by a Jacobi rotation and updates
% transformation matrix P.
n = size(A,1);
diff = A(L,L) - A(k,k);
if abs(A(k,L)) < abs(diff)*1.0e-36
    t = A(k,L);
else
    phi = diff/(2*A(k,L));
    t = 1/(abs(phi) + sqrt(phi^2 + 1));
    if phi < 0; t = -t; end;
end
c = 1/sqrt(t^2 + 1); s = t*c;
tau = s/(1 + c);
temp = A(k,L); A(k,L) = 0;
A(k,k) = A(k,k) - t*temp;
```

```
A(L,L) = A(L,L) + t*temp;
for i = 1:k-1                       % For i < k
    temp = A(i,k);
    A(i,k) = temp -s*(A(i,L) + tau*temp);
    A(i,L) = A(i,L) + s*(temp - tau*A(i,L));
end
for i = k+1:L-1                     % For k < i < L
    temp = A(k,i);
    A(k,i) = temp - s*(A(i,L) + tau*A(k,i));
    A(i,L) = A(i,L) + s*(temp - tau*A(i,L));
end
for i = L+1:n                       % For i > L
    temp = A(k,i);
    A(k,i) = temp - s*(A(L,i) + tau*temp);
    A(L,i) = A(L,i) + s*(temp - tau*A(L,i));
end
for i = 1:n        % Update transformation matrix
    temp = P(i,k);
    P(i,k) = temp - s*(P(i,L) + tau*P(i,k));
    P(i,L) = P(i,L) + s*(temp - tau*P(i,L));
end
```

■ sortEigen

The eigenvalues/eigenvectors returned by jacobi are not ordered. The function listed below can be used to sort the results into ascending order of eigenvalues.

```
function [eVals,eVecs] = sortEigen(eVals,eVecs)
% Sorts eigenvalues & eigenvectors into ascending
% order of eigenvalues.
% USAGE: [eVals,eVecs] = sortEigen(eVals,eVecs)

n = length(eVals);
for i = 1:n-1
    index = i; val = eVals(i);
    for j = i+1:n
        if eVals(j) < val
            index = j; val = eVals(j);
        end
    end
```

```
    if index ~= i
        eVals = swapRows(eVals,i,index);
        eVecs = swapCols(eVecs,i,index);
    end
end
```

Transformation to Standard Form

Physical problems often give rise to eigenvalue problems of the form

$$\mathbf{Ax} = \lambda\mathbf{Bx} \qquad (9.21)$$

where $\mathbf{A}$ and $\mathbf{B}$ are symmetric $n \times n$ matrices. We assume that $\mathbf{B}$ is also positive definite. Such problems must be transformed into the standard form before they can be solved by Jacobi diagonalization.

As $\mathbf{B}$ is symmetric and positive definite, we can apply Choleski's decomposition $\mathbf{B} = \mathbf{LL}^T$, where $\mathbf{L}$ is a lower-triangular matrix (see Art. 2.3). Then we introduce the transformation

$$\mathbf{x} = (\mathbf{L}^{-1})^T\mathbf{z} \qquad (9.22)$$

Substituting into Eq. (9.21), we get

$$\mathbf{A}(\mathbf{L}^{-1})^T\mathbf{z} = \lambda\mathbf{LL}^T(\mathbf{L}^{-1})^T\mathbf{z}$$

Premultiplying both sides by $\mathbf{L}^{-1}$ results in

$$\mathbf{L}^{-1}\mathbf{A}(\mathbf{L}^{-1})^T\mathbf{z} = \lambda\mathbf{L}^{-1}\mathbf{LL}^T(\mathbf{L}^{-1})^T\mathbf{z}$$

Because $\mathbf{L}^{-1}\mathbf{L} = \mathbf{L}^T(\mathbf{L}^{-1})^T = \mathbf{I}$, the last equation reduces to the standard form

$$\mathbf{Hz} = \lambda\mathbf{z} \qquad (9.23)$$

where

$$\mathbf{H} = \mathbf{L}^{-1}\mathbf{A}(\mathbf{L}^{-1})^T \qquad (9.24)$$

An important property of this transformation is that it does not destroy the symmetry of the matrix; i.e., a symmetric $\mathbf{A}$ results in a symmetric $\mathbf{H}$.

Here is the general procedure for solving eigenvalue problems of the form $\mathbf{Ax} = \lambda\mathbf{Bx}$:

1. Use Choleski's decomposition $\mathbf{B} = \mathbf{LL}^T$ to compute $\mathbf{L}$.
2. Compute $\mathbf{L}^{-1}$ (a triangular matrix can be inverted with relatively small computational effort).
3. Compute $\mathbf{H}$ from Eq. (9.24).

4. Solve the standard eigenvalue problem $\mathbf{Hz} = \lambda\mathbf{z}$ (e.g., using the Jacobi method).
5. Recover the eigenvectors of the original problem from Eq. (9.22): $\mathbf{x} = (\mathbf{L}^{-1})^T\mathbf{z}$. Note that the eigenvalues were untouched by the transformation.

An important special case is where $\mathbf{B}$ is a diagonal matrix:

$$\mathbf{B} = \begin{bmatrix} \beta_1 & 0 & \cdots & 0 \\ 0 & \beta_2 & \cdots & 0 \\ \vdots & \vdots & \ddots & \vdots \\ 0 & 0 & \cdots & \beta_n \end{bmatrix} \tag{9.25}$$

Here

$$\mathbf{L} = \begin{bmatrix} \beta_1^{1/2} & 0 & \cdots & 0 \\ 0 & \beta_2^{1/2} & \cdots & 0 \\ \vdots & \vdots & \ddots & \vdots \\ 0 & 0 & \cdots & \beta_n^{1/2} \end{bmatrix} \qquad \mathbf{L}^{-1} = \begin{bmatrix} \beta_1^{-1/2} & 0 & \cdots & 0 \\ 0 & \beta_2^{-1/2} & \cdots & 0 \\ \vdots & \vdots & \ddots & \vdots \\ 0 & 0 & \cdots & \beta_n^{-1/2} \end{bmatrix} \tag{9.26a}$$

and

$$H_{ij} = A_{ij}\left(\beta_i\beta_j\right)^{-1/2} \tag{9.26b}$$

■ stdForm

Given the matrices $\mathbf{A}$ and $\mathbf{B}$, the function `stdForm` returns $\mathbf{H}$ and the transformation matrix $\mathbf{T} = (\mathbf{L}^{-1})^T$. The inversion of $\mathbf{L}$ is carried out by the subfunction `invert` (the triangular shape of $\mathbf{L}$ allows this to be done by back substitution).

```
function [H,T] = stdForm(A,B)
% Transforms A*x = lambda*B*x to H*z = lambda*z
% and computes transformation matrix T in x = T*z.
% USAGE: [H,T] = stdForm(A,B)

n = size(A,1);
L = choleski(B); Linv = invert(L);
H = Linv*(A*Linv'); T = Linv';

function Linv = invert(L)
% Inverts lower triangular matrix L.
n = size(L,1);
for j = 1:n-1
    L(j,j) = 1/L(j,j);
```

```
        for i = j+1:n
            L(i,j) = -dot(L(i,j:i-1), L(j:i-1,j)/L(i,i));
        end
    end
    L(n,n) = 1/L(n,n); Linv = L;
```

EXAMPLE 9.1

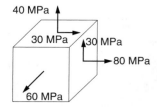

The stress matrix (tensor) corresponding to the state of stress shown is

$$\mathbf{S} = \begin{bmatrix} 80 & 30 & 0 \\ 30 & 40 & 0 \\ 0 & 0 & 60 \end{bmatrix} \text{MPa}$$

(each row of the matrix consists of the three stress components acting on a coordinate plane). It can be shown that the eigenvalues of $\mathbf{S}$ are the *principal stresses* and the eigenvectors are normal to the *principal planes*. (1) Determine the principal stresses by diagonalizing $\mathbf{S}$ with a Jacobi rotation and (2) compute the eigenvectors.

Solution of Part (1) To eliminate S_{12} we must apply a rotation in the 1–2 plane. With $k = 1$ and $\ell = 2$ Eq. (9.15) is

$$\phi = -\frac{S_{11} - S_{22}}{2 S_{12}} = -\frac{80 - 40}{2(30)} = -\frac{2}{3}$$

Equation (9.16a) then yields

$$t = \frac{\text{sgn}(\phi)}{|\phi| + \sqrt{\phi^2 + 1}} = \frac{-1}{2/3 + \sqrt{(2/3)^2 + 1}} = -0.53518$$

According to Eqs. (9.18), the changes in $\mathbf{S}$ due to the rotation are

$$S_{11}^* = S_{11} - t S_{12} = 80 - (-0.53518)(30) = 96.055 \text{ MPa}$$

$$S_{22}^* = S_{22} + t S_{12} = 40 + (-0.53518)(30) = 23.945 \text{ MPa}$$

$$S_{12}^* = S_{21}^* = 0$$

Hence the diagonalized stress matrix is

$$\mathbf{S}^* = \begin{bmatrix} 96.055 & 0 & 0 \\ 0 & 23.945 & 0 \\ 0 & 0 & 60 \end{bmatrix}$$

where the diagonal terms are the principal stresses.

Solution of Part (2) To compute the eigenvectors, we start with Eqs. (9.17) and (9.19), which yield

$$c = \frac{1}{\sqrt{1 + t^2}} = \frac{1}{\sqrt{1 + (-0.53518)^2}} = 0.88168$$

$$s = tc = (-0.53518)(0.88168) = -0.47186$$

$$\tau = \frac{s}{1 + c} = \frac{-0.47186}{1 + 0.88168} = -0.25077$$

We obtain the changes in the transformation matrix $\mathbf{P}$ from Eqs. (9.20). Because $\mathbf{P}$ is initialized to the identity matrix ($P_{ii} = 1$ and $P_{ij} = 0$, $i \neq j$) the first equation gives us

$$P_{11}^* = P_{11} - s(P_{12} + \tau P_{11})$$
$$= 1 - (-0.47186)[0 + (-0.25077)(1)] = 0.88167$$
$$P_{21}^* = P_{21} - s(P_{22} + \tau P_{21})$$
$$= 0 - (-0.47186)[1 + (-0.25077)(0)] = 0.47186$$

Similarly, the second equation of Eqs. (9.20) yields

$$P_{12}^* = -0.47186 \qquad P_{22}^* = 0.88167$$

The third row and column of $\mathbf{P}$ are not affected by the transformation. Thus

$$\mathbf{P}^* = \begin{bmatrix} 0.88167 & -0.47186 & 0 \\ 0.47186 & 0.88167 & 0 \\ 0 & 0 & 1 \end{bmatrix}$$

The columns of $\mathbf{P}^*$ are the eigenvectors of $\mathbf{S}$.

EXAMPLE 9.2

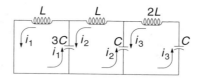

(1) Show that the analysis of the electric circuit shown leads to a matrix eigenvalue problem. (2) Determine the circular frequencies and the relative amplitudes of the currents.

Solution of Part (1) Kirchoff's equations for the three loops are

$$L\frac{di_1}{dt} + \frac{q_1 - q_2}{3C} = 0$$

$$L\frac{di_2}{dt} + \frac{q_2 - q_1}{3C} + \frac{q_2 - q_3}{C} = 0$$

$$2L\frac{di_3}{dt} + \frac{q_3 - q_2}{C} + \frac{q_3}{C} = 0$$

Differentiating and substituting $dq_k/dt = i_k$, we get

$$\frac{1}{3}i_1 - \frac{1}{3}i_2 = -LC\frac{d^2i_1}{dt^2}$$

$$-\frac{1}{3}i_1 + \frac{4}{3}i_2 - i_3 = -LC\frac{d^2i_2}{dt^2}$$

$$-i_2 + 2i_3 = -2LC\frac{d^2i_3}{dt^2}$$

These equations admit the solution

$$i_k(t) = u_k \sin \omega t$$

where ω is the circular frequency of oscillation (measured in rad/s) and u_k are the relative amplitudes of the currents. Substitution into Kirchoff's equations yields $\mathbf{Au} = \lambda\mathbf{Bu}$ ($\sin \omega t$ cancels out), where

$$\mathbf{A} = \begin{bmatrix} 1/3 & -1/3 & 0 \\ -1/3 & 4/3 & -1 \\ 0 & -1 & 2 \end{bmatrix} \qquad \mathbf{B} = \begin{bmatrix} 1 & 0 & 0 \\ 0 & 1 & 0 \\ 0 & 0 & 2 \end{bmatrix} \qquad \lambda = LC\omega^2$$

which represents an eigenvalue problem of the nonstandard form.

Solution of Part (2) Since $\mathbf{B}$ is a diagonal matrix, we can readily transform the problem into the standard form $\mathbf{Hz} = \lambda\mathbf{z}$. From Eq. (9.26a) we get

$$\mathbf{L}^{-1} = \begin{bmatrix} 1 & 0 & 0 \\ 0 & 1 & 0 \\ 0 & 0 & 1/\sqrt{2} \end{bmatrix}$$

and Eq. (9.26b) yields

$$\mathbf{H} = \begin{bmatrix} 1/3 & -1/3 & 0 \\ -1/3 & 4/3 & -1/\sqrt{2} \\ 0 & -1/\sqrt{2} & 1 \end{bmatrix}$$

The eigenvalues and eigenvectors of H can now be obtained with the Jacobi method. Skipping the details, we obtain the following results:

$$\lambda_1 = 0.14779 \qquad \lambda_2 = 0.58235 \qquad \lambda_3 = 1.93653$$

$$\mathbf{z}_1 = \begin{bmatrix} 0.81027 \\ 0.45102 \\ 0.37423 \end{bmatrix} \qquad \mathbf{z}_2 = \begin{bmatrix} 0.56274 \\ -0.42040 \\ -0.71176 \end{bmatrix} \qquad \mathbf{z}_3 = \begin{bmatrix} 0.16370 \\ -0.78730 \\ 0.59444 \end{bmatrix}$$

The eigenvectors of the original problem are recovered from Eq. (9.22): $\mathbf{y}_i = (\mathbf{L}^{-1})^T \mathbf{z}_i$, which yields

$$\mathbf{u}_1 = \begin{bmatrix} 0.81027 \\ 0.45102 \\ 0.26462 \end{bmatrix} \qquad \mathbf{u}_2 = \begin{bmatrix} 0.56274 \\ -0.42040 \\ -0.50329 \end{bmatrix} \qquad \mathbf{u}_3 = \begin{bmatrix} 0.16370 \\ -0.78730 \\ 0.42033 \end{bmatrix}$$

These vectors should now be normalized (each $\mathbf{z}_i$ was normalized, but the transformation to $\mathbf{u}_i$ does not preserve the magnitudes of vectors). The circular frequencies are $\omega_i = \sqrt{\lambda_i/(LC)}$, so that

$$\omega_1 = \frac{0.3844}{\sqrt{LC}} \qquad \omega_2 = \frac{0.7631}{\sqrt{LC}} \qquad \omega_3 = \frac{1.3916}{\sqrt{LC}}$$

EXAMPLE 9.3

The propped cantilever beam carries a compressive axial load P. The lateral displacement $u(x)$ of the beam can be shown to satisfy the differential equation

$$u^{(4)} + \frac{P}{EI} u'' = 0 \tag{a}$$

where EI is the bending rigidity. The boundary conditions are

$$u(0) = u''(0) = 0 \qquad u(L) = u'(L) = 0 \tag{b}$$

(1) Show that displacement analysis of the beam results in a matrix eigenvalue problem if the derivatives are approximated by finite differences. (2) Use the Jacobi method to compute the lowest three buckling loads and the corresponding eigenvectors.

Solution of Part (1) We divide the beam into $n+1$ segments of length $L/(n+1)$ each as shown and enforce the differential equation at nodes 1 to n. Replacing the derivatives of u in Eq. (a) by central finite differences of $\mathcal{O}(h^2)$ at the interior nodes

(nodes 1 to n), we obtain

$$\frac{u_{i-2} - 4u_{i-1} + 6u_i - 4u_{i+1} + u_{i+2}}{h^4}$$

$$= \frac{P}{EI} \frac{-u_{i-1} + 2u_i - u_{i+1}}{h^2}, \quad i = 1, 2, \ldots, n$$

After multiplication by h^4, the equations become

$$u_{-1} - 4u_0 + 6u_1 - 4u_2 + u_3 = \lambda(-u_0 + 2u_1 - u_2)$$

$$u_0 - 4u_1 + 6u_2 - 4u_3 + u_4 = \lambda(-u_1 + 2u_2 - u_3)$$

$$\vdots \qquad\qquad\qquad (c)$$

$$u_{n-3} - 4u_{n-2} + 6u_{n-1} - 4u_n + u_{n+1} = \lambda(-u_{n-2} + 2u_{n-1} - u_n)$$

$$u_{n-2} - 4u_{n-1} + 6u_n - 4u_{n+1} + u_{n+2} = \lambda(-u_{n-1} + 2u_n - u_{n+1})$$

where

$$\lambda = \frac{Ph^2}{EI} = \frac{PL^2}{(n+1)^2 EI}$$

The displacements u_{-1}, u_0, u_{n+1} and u_{n+2} can be eliminated by using the prescribed boundary conditions. Referring to Table 8.1, we obtain the finite difference approximations to the boundary conditions in Eqs. (b):

$$u_0 = 0 \qquad u_{-1} = -u_1 \qquad u_{n+1} = 0 \qquad u_{n+2} = u_n$$

Substitution into Eqs. (c) yields the matrix eigenvalue problem $\mathbf{Ax} = \lambda \mathbf{Bx}$, where

$$\mathbf{A} = \begin{bmatrix} 5 & -4 & 1 & 0 & 0 & \cdots & 0 \\ -4 & 6 & -4 & 1 & 0 & \cdots & 0 \\ 1 & -4 & 6 & -4 & 1 & \cdots & 0 \\ \vdots & \ddots & \ddots & \ddots & \ddots & \ddots & \vdots \\ 0 & \cdots & 1 & -4 & 6 & -4 & 1 \\ 0 & \cdots & 0 & 1 & -4 & 6 & -4 \\ 0 & \cdots & 0 & 0 & 1 & -4 & 7 \end{bmatrix}$$

$$\mathbf{B} = \begin{bmatrix} 2 & -1 & 0 & 0 & 0 & \cdots & 0 \\ -1 & 2 & -1 & 0 & 0 & \cdots & 0 \\ 0 & -1 & 2 & -1 & 0 & \cdots & 0 \\ \vdots & \ddots & \vdots & \ddots & \ddots & \ddots & \vdots \\ 0 & \cdots & 0 & -1 & 2 & -1 & 0 \\ 0 & \cdots & 0 & 0 & -1 & 2 & -1 \\ 0 & \cdots & 0 & 0 & 0 & -1 & 2 \end{bmatrix}$$

Solution of Part (2) The problem with the Jacobi method is that it insists on finding *all* the eigenvalues and eigenvectors. It is also incapable of exploiting banded structures of matrices. Thus the program listed below does much more work than necessary for the problem at hand. More efficient methods of solution will be introduced later in this chapter.

```
% Example 9.3 (Jacobi method)
n = 10;                               % Number of interior nodes.
A = zeros(n); B = zeros(n);           % Start constructing A and B.
for i = 1:n
    A(i,i) = 6; B(i,i) = 2;
end
A(1,1) = 5; A(n,n) = 7;
for i = 1:n-1
    A(i,i+1) = -4; A(i+1,i) = -4;
    B(i,i+1) = -1; B(i+1,i) = -1;
end
for i = 1:n-2
    A(i,i+2) = 1; A(i+2,i) = 1;
end
[H,T] = stdForm(A,B);                 % Convert to std. form.
[eVals,Z] = jacobi(H);                % Solve by Jacobi method.
X = T*Z;                              % Eigenvectors of orig. prob.
for i = 1:n                           % Normalize eigenvectors.
    xMag = sqrt(dot(X(:,i),X(:,i)));
    X(:,i) = X(:,i)/xMag;
end
[eVals,X] = sortEigen(eVals,X); % Sort in ascending order.
eigenvalues = eVals(1:3)'            % Extract 3 smallest
eigenvectors = X(:,1:3)              % eigenvalues & vectors.
```

Running the program resulted in the following output:

```
>> eigenvalues =
    0.1641    0.4720    0.9022
eigenvectors =
    0.1641   -0.1848    0.3070
    0.3062   -0.2682    0.3640
    0.4079   -0.1968    0.1467
    0.4574    0.0099   -0.1219
```

```
0.4515      0.2685     -0.1725
0.3961      0.4711      0.0677
0.3052      0.5361      0.4089
0.1986      0.4471      0.5704
0.0988      0.2602      0.4334
0.0270      0.0778      0.1486
```

The first three mode shapes, which represent the relative displacements of the buckled beam, are plotted below (we appended the zero end displacements to the eigenvectors before plotting the points).

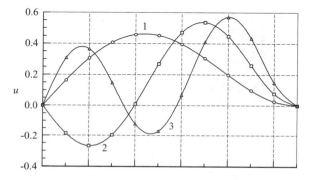

The buckling loads are given by $P_i = (n+1)^2 \lambda_i EI/L^2$. Thus

$$P_1 = \frac{(11)^2 (0.1641) EI}{L^2} = 19.86 \frac{EI}{L^2}$$

$$P_2 = \frac{(11)^2 (0.4720) EI}{L^2} = 57.11 \frac{EI}{L^2}$$

$$P_3 = \frac{(11)^2 (0.9022) EI}{L^2} = 109.2 \frac{EI}{L^2}$$

The analytical values are $P_1 = 20.19 EI/L^2$, $P_2 = 59.68 EI/L^2$ and $P_3 = 118.9 EI/L^2$. It can be seen that the error introduced by the finite element approximation increases with the mode number (the error in P_{i+1} is larger than in P_i). Of course, the accuracy of the finite difference model can be improved by using larger n, but beyond $n = 20$ the cost of computation with the Jacobi method becomes rather high.

9.3   Inverse Power and Power Methods

Inverse Power Method

The inverse power method is a simple iterative procedure for finding the smallest eigenvalue λ_1 and the corresponding eigenvector $\mathbf{x}_1$ of

$$\mathbf{A}\mathbf{x} = \lambda\mathbf{x} \qquad\qquad (9.27)$$

The method works like this:

1. Let $\mathbf{v}$ be an approximation to $\mathbf{x}_1$ (a random vector of unit magnitude will do).
2. Solve

$$\mathbf{A}\mathbf{z} = \mathbf{v} \tag{9.28}$$

for the vector $\mathbf{z}$.
3. Compute $|\mathbf{z}|$.
4. Let $\mathbf{v} = \mathbf{z}/|\mathbf{z}|$ and repeat steps 2–4 until the change in $\mathbf{v}$ is negligible.

At the conclusion of the procedure, $|\mathbf{z}| = \pm 1/\lambda_1$ and $\mathbf{v} = \mathbf{x}_1$. The sign of λ_1 is determined as follows: if $\mathbf{z}$ changes sign between successive iterations, λ_1 is negative; otherwise, λ_1 is positive.

Let us now investigate why the method works. Since the eigenvectors $\mathbf{x}_i$ of Eq. (9.27) are orthonormal, they can be used as the basis for any n-dimensional vector. Thus $\mathbf{v}$ and $\mathbf{z}$ admit the unique representations

$$\mathbf{v} = \sum_{i=1}^{n} v_i \mathbf{x}_i \qquad \mathbf{z} = \sum_{i=1}^{n} z_i \mathbf{x}_i \tag{a}$$

Note that v_i and z_i are not the elements of $\mathbf{v}$ and $\mathbf{z}$, but the components with respect to the eigenvectors $\mathbf{x}_i$. Substitution into Eq. (9.28) yields

$$\mathbf{A}\sum_{i=1}^{n} z_i \mathbf{x}_i - \sum_{i=1}^{n} v_i \mathbf{x}_i = \mathbf{0}$$

But $\mathbf{A}\mathbf{x}_i = \lambda_i \mathbf{x}_i$, so that

$$\sum_{i=1}^{n} (z_i \lambda_i - v_i)\, \mathbf{x}_i = \mathbf{0}$$

Hence

$$z_i = \frac{v_i}{\lambda_i}$$

It follows from Eq. (a) that

$$\mathbf{z} = \sum_{i=1}^{n} \frac{v_i}{\lambda_i} \mathbf{x}_i = \frac{1}{\lambda_1} \sum_{i=1}^{n} v_i \frac{\lambda_1}{\lambda_i} \mathbf{x}_i$$

$$= \frac{1}{\lambda_1} \left(v_1 \mathbf{x}_1 + v_2 \frac{\lambda_1}{\lambda_2} \mathbf{x}_2 + v_3 \frac{\lambda_1}{\lambda_3} \mathbf{x}_3 + \cdots \right) \tag{9.29}$$

Since $|\lambda_1/\lambda_i| < 1$ ($i \neq 1$), we observe that the coefficient of $\mathbf{x}_1$ has become more prominent in $\mathbf{z}$ than it was in $\mathbf{v}$; hence $\mathbf{z}$ is a better approximation to $\mathbf{x}_1$. This completes the first iterative cycle.

In subsequent cycles we set $\mathbf{v} = \mathbf{z}/|\mathbf{z}|$ and repeat the process. Each iteration will increase the dominance of the first term in Eq. (9.29) so that the process converges to

$$\mathbf{z} = \frac{1}{\lambda_1}v_1\mathbf{x}_1 = \frac{1}{\lambda_1}\mathbf{x}_1$$

(at this stage $\mathbf{v} = \mathbf{x}_1$, so that $v_1 = 1, v_2 = v_3 = \cdots = 0$).

The inverse power method also works with the nonstandard eigenvalue problem

$$\mathbf{Ax} = \lambda\mathbf{Bx} \tag{9.30}$$

provided that Eq. (9.28) is replaced by

$$\mathbf{Az} = \mathbf{Bv} \tag{9.31}$$

The alternative is, of course, to transform the problem to standard form before applying the power method.

Eigenvalue Shifting

By inspection of Eq. (9.29) we see that the rate of convergence is determined by the strength of the inequality $|\lambda_1/\lambda_2| < 1$ (the second term in the equation). If $|\lambda_2|$ is well separated from $|\lambda_1|$, the inequality is strong and the convergence is rapid. On the other hand, close proximity of these two eigenvalues results in very slow convergence.

The rate of convergence can be improved by a technique called *eigenvalue shifting*. If we let

$$\lambda = \lambda^* + s \tag{9.32}$$

where s is a predetermined "shift," the eigenvalue problem in Eq. (9.27) is transformed to

$$\mathbf{Ax} = (\lambda^* + s)\mathbf{x}$$

or

$$\mathbf{A}^*\mathbf{x} = \lambda^*\mathbf{x} \tag{9.33}$$

where

$$\mathbf{A}^* = \mathbf{A} - s\mathbf{I} \tag{9.34}$$

Solving the transformed problem in Eq. (9.33) by the inverse power method yields λ_1^* and $\mathbf{x}_1$, where λ_1^* is the smallest eigenvalue of $\mathbf{A}^*$. The corresponding eigenvalue of the original problem, $\lambda = \lambda_1^* + s$, is thus the *eigenvalue closest to s*.

Eigenvalue shifting has two applications. An obvious one is the determination of the eigenvalue closest to a certain value s. For example, if the working speed of a shaft is s rpm, it is imperative to ensure that there are no natural frequencies (which are related to the eigenvalues) close to that speed.

Eigenvalue shifting is also be used to speed up convergence. Suppose that we are computing the smallest eigenvalue λ_1 of the matrix $\mathbf{A}$. The idea is to introduce a shift s that makes λ_1^*/λ_2^* as small as possible. Since $\lambda_1^* = \lambda_1 - s$, we should choose $s \approx \lambda_1$ ($s = \lambda_1$ should be avoided to prevent division by zero). Of course, this method works only if we have a prior estimate of λ_1.

The inverse power method with eigenvalue shifting is a particularly powerful tool for finding eigenvectors if the eigenvalues are known. By shifting very close to an eigenvalue, the corresponding eigenvector can be computed in one or two iterations.

Power Method

The power method converges to the eigenvalue *farthest from zero* and the associated eigenvector. It is very similar to the inverse power method; the only difference between the two methods is the interchange of $\mathbf{v}$ and $\mathbf{z}$ in Eq. (9.28). The outline of the procedure is:

1. Let $\mathbf{v}$ be an approximation to $\mathbf{x}_n$ (a random vector of unit magnitude will do).
2. Compute the vector

$$\mathbf{z} = \mathbf{A}\mathbf{v} \qquad (9.35)$$

3. Compute $|\mathbf{z}|$.
4. Let $\mathbf{v} = \mathbf{z}/|\mathbf{z}|$ and repeat steps 2–4 until the change in $\mathbf{v}$ is negligible.

At the conclusion of the procedure, $|\mathbf{z}| = \pm\lambda_n$ and $\mathbf{v} = \mathbf{x}_n$ (the sign of λ_n is determined in the same way as in the inverse power method).

■ invPower

Given the matrix $\mathbf{A}$ and the scalar s, the function `invPower` returns the eigenvalue of $\mathbf{A}$ closest to s and the corresponding eigenvector. The matrix $\mathbf{A}^* = \mathbf{A} - s\mathbf{I}$ is decomposed as soon as it is formed, so that only the solution phase (forward and back substitution) is needed in the iterative loop. If $\mathbf{A}$ is banded, the efficiency of the program can be improved by replacing `LUdec` and `LUsol` by functions that specialize in banded matrices—see Example 9.6. The program line that forms $\mathbf{A}^*$ must also be modified to be compatible with the storage scheme used for $\mathbf{A}$.

```
function [eVal,eVec] = invPower(A,s,maxIter,tol)
% Inverse power mehod for finding the eigenvalue of A
% closest to s & the correstponding eigenvector.
```

```
% USAGE: [eVal,eVec] = invPower(A,s,maxIter,tol)
% maxIter = limit on number of iterations (default is 50).
% tol = error tolerance (default is 1.0e-6).

if nargin < 4; tol = 1.0e-6; end
if nargin < 3; maxIter = 50; end
n = size(A,1);
A = A - eye(n)*s;     % Form A* = A - sI
A = LUdec(A);         % Decompose A*
x = rand(n,1);        % Seed eigenvecs. with random numbers
xMag = sqrt(dot(x,x)); x = x/xMag;       % Normalize x
for i = 1:maxIter
    xOld = x;         % Save current eigenvecs.
    x = LUsol(A,x); % Solve A*x = xOld
    xMag = sqrt(dot(x,x)); x = x/xMag;  % Normalize x
    xSign = sign(dot(xOld,x)); % Detect sign change of x
    x = x*xSign;
    % Check for convergence
    if sqrt(dot(xOld - x,xOld - x)) < tol
        eVal = s + xSign/xMag; eVec = x;
        return
    end
end
error('Too many iterations')
```

EXAMPLE 9.4
The stress matrix describing the state of stress at a point is

$$S = \begin{bmatrix} -30 & 10 & 20 \\ 10 & 40 & -50 \\ 20 & -50 & -10 \end{bmatrix} \text{MPa}$$

Determine the largest principal stress (the eigenvalue of S farthest from zero) by the power method.

Solution First iteration:

Let $v = \begin{bmatrix} 1 & 0 & 0 \end{bmatrix}^T$ be the initial guess for the eigenvector. Then

$$\mathbf{z} = \mathbf{Sv} = \begin{bmatrix} -30 & 10 & 20 \\ 10 & 40 & -50 \\ 20 & -50 & -10 \end{bmatrix} \begin{bmatrix} 1 \\ 0 \\ 0 \end{bmatrix} = \begin{bmatrix} -30.0 \\ 10.0 \\ 20.0 \end{bmatrix}$$

$$|\mathbf{z}| = \sqrt{30^2 + 10^2 + 20^2} = 37.417$$

$$\mathbf{v} = \frac{\mathbf{z}}{|\mathbf{z}|} = \begin{bmatrix} -30.0 \\ 10.0 \\ 20.0 \end{bmatrix} \frac{1}{37.417} = \begin{bmatrix} -0.801\,77 \\ 0.267\,26 \\ 0.534\,52 \end{bmatrix}$$

Second iteration:

$$\mathbf{z} = \mathbf{Sv} = \begin{bmatrix} -30 & 10 & 20 \\ 10 & 40 & -50 \\ 20 & -50 & -10 \end{bmatrix} \begin{bmatrix} -0.801\,77 \\ 0.267\,26 \\ 0.534\,52 \end{bmatrix} = \begin{bmatrix} 37.416 \\ -24.053 \\ -34.744 \end{bmatrix}$$

$$|\mathbf{z}| = \sqrt{37.416^2 + 24.053^2 + 34.744^2} = 56.442$$

$$\mathbf{v} = \frac{\mathbf{z}}{|\mathbf{z}|} = \begin{bmatrix} 37.416 \\ -24.053 \\ -34.744 \end{bmatrix} \frac{1}{56.442} = \begin{bmatrix} 0.662\,91 \\ -0.426\,15 \\ -0.615\,57 \end{bmatrix}$$

Third iteration:

$$\mathbf{z} = \mathbf{Sv} = \begin{bmatrix} -30 & 10 & 20 \\ 10 & 40 & -50 \\ 20 & -50 & -10 \end{bmatrix} \begin{bmatrix} 0.66291 \\ -0.42615 \\ -0.61557 \end{bmatrix} = \begin{bmatrix} -36.460 \\ 20.362 \\ 40.721 \end{bmatrix}$$

$$|\mathbf{z}| = \sqrt{36.460^2 + 20.362^2 + 40.721^2} = 58.328$$

$$\mathbf{v} = \frac{\mathbf{z}}{|\mathbf{z}|} = \begin{bmatrix} -36.460 \\ 20.362 \\ 40.721 \end{bmatrix} \frac{1}{58.328} = \begin{bmatrix} -0.62509 \\ 0.34909 \\ 0.69814 \end{bmatrix}$$

At this point the approximation of the eigenvalue we seek is $\lambda = -58.328$ MPa (the negative sign is determined by the sign reversal of $\mathbf{z}$ between iterations). This is actually close to the second-largest eigenvalue $\lambda_2 = -58.39$ MPa! By continuing the iterative process we would eventually end up with the largest eigenvalue $\lambda_3 = 70.94$ MPa. But since $|\lambda_2|$ and $|\lambda_3|$ are rather close, the convergence is too slow from this point on for manual labor. Here is a program that does the calculations for us:

```
% Example 9.4 (Power method)
S = [-30 10 20; 10 40 -50; 20 -50 -10];
v = [1; 0; 0];
for i = 1:100
    vOld = v; z = S*v; zMag = sqrt(dot(z,z));
    v = z/zMag; vSign = sign(dot(vOld,v));
```

```
        v = v*vSign;
        if sqrt(dot(vOld - v,vOld - v)) < 1.0e-6
            eVal = vSign*zMag
            numIter = i
            return
        end
    end
end
error('Too many iterations')
```

The results are:

```
>> eVal =
    70.9435
numIter =
    93
```

Note that it took 93 iterations to reach convergence.

EXAMPLE 9.5
Determine the smallest eigenvalue λ_1 and the corresponding eigenvector of

$$
\mathbf{A} = \begin{bmatrix}
11 & 2 & 3 & 1 & 4 \\
2 & 9 & 3 & 5 & 2 \\
3 & 3 & 15 & 4 & 3 \\
1 & 5 & 4 & 12 & 4 \\
4 & 2 & 3 & 4 & 17
\end{bmatrix}
$$

Use the inverse power method with eigenvalue shifting knowing that $\lambda_1 \approx 5$.

Solution

```
% Example 9.5 (Inverse power method)
s = 5;
A = [11   2   3   1   4;
      2   9   3   5   2;
      3   3  15   4   3;
      1   5   4  12   4;
      4   2   3   4  17];
[eVal,eVec] = invPower(A,s)
```

Here is the output:

```
>> eVal =
    4.8739
eVec =
    0.2673
```

```
     -0.7414
     -0.0502
      0.5949
     -0.1497
```

Convergence was achieved with 4 iterations. Without the eigenvalue shift 26 iterations would be required.

EXAMPLE 9.6

Unlike Jacobi diagonalization, the inverse power method lends itself to eigenvalue problems of banded matrices. Write a program that computes the smallest buckling load of the beam described in Example 9.3, making full use of the banded forms. Run the program with 100 interior nodes ($n = 100$).

Solution The function `invPower5` listed below returns the smallest eigenvalue and the corresponding eigenvector of $\mathbf{Ax} = \lambda\mathbf{Bx}$, where $\mathbf{A}$ is a pentadiagonal matrix and $\mathbf{B}$ is a sparse matrix (in this problem it is tridiagonal). The matrix $\mathbf{A}$ is input by its diagonals $\mathbf{d}$, $\mathbf{e}$ and $\mathbf{f}$ as was done in Art. 2.4 in conjunction with the LU decomposition. The algorithm for `invPower5` does not use $\mathbf{B}$ directly, but calls the function `func(v)` that supplies the product $\mathbf{Bv}$. Eigenvalue shifting is not used.

```
function [eVal,eVec] = invPower5(func,d,e,f)
% Finds smallest eigenvalue of A*x = lambda*B*x by
% the inverse power method.
% USAGE: [eVal,eVec] = invPower5(func,d,e,f)
% Matrix A must be pentadiagonal and stored in form
%     A = [f\e\d\e\f].
% func = handle of function that returns B*v.

n = length(d);
[d,e,f] = LUdec5(d,e,f);                % Decompose A
x = rand(n,1);                          % Seed x with random numbers
xMag = sqrt(dot(x,x)); x = x/xMag;      % Normalize x
for i = 1:50
    xOld = x;                               % Save current x
    x = LUsol5(d,e,f,feval(func,x));        % Solve [A]{x} = [B]{xOld}
    xMag = sqrt(dot(x,x)); x = x/xMag;      % Normalize x
    xSign = sign(dot(xOld,x));              % Detect sign change of x
    x = x*xSign;
    % Check for convergence
    if sqrt(dot(xOld - x,xOld - x)) < 1.0e-6
```

```
        eVal = xSign/xMag; eVec = x;
        return
    end
end
error('Too many iterations')
```

The function that computes **Bv** is

```
function Bv = fex9_6(v)
% Computes the product B*v in Example 9.6.

n = length(v);
Bv = zeros(n,1);
for i = 2:n-1
    Bv(i) = -v(i-1) + 2*v(i) - v(i+1);
end
Bv(1) = 2*v(1) - v(2);
Bv(n) = -v(n-1) + 2*v(n);
```

Here is the program that calls `invPower5`:

```
% Example 9.6 (Inverse power method for pentadiagonal A)
n = 100;
d = ones(n,1)*6;
d(1) = 5; d(n) = 7;
e = ones(n-1,1)*(-4);
f = ones(n-2,1);
[eVal,eVec] = invPower5(@fex9_6,d,e,f);
fprintf('PL^2/EI =')
fprintf('%9.4f',eVal*(n+1)^2)
```

The output, shown below, is in excellent agreement with the analytical value.

```
>> PL^2/EI =   20.1867
```

PROBLEM SET 9.1

1. Given

$$A = \begin{bmatrix} 7 & 3 & 1 \\ 3 & 9 & 6 \\ 1 & 6 & 8 \end{bmatrix} \qquad B = \begin{bmatrix} 4 & 0 & 0 \\ 0 & 9 & 0 \\ 0 & 0 & 4 \end{bmatrix}$$

convert the eigenvalue problem $\mathbf{Ax} = \lambda\mathbf{Bx}$ to the standard form $\mathbf{Hz} = \lambda\mathbf{z}$. What is the relationship between $\mathbf{x}$ and $\mathbf{z}$?

2. Convert the eigenvalue problem $\mathbf{Ax} = \lambda\mathbf{Bx}$, where

$$
\mathbf{A} = \begin{bmatrix} 4 & -1 & 0 \\ -1 & 4 & -1 \\ 0 & -1 & 4 \end{bmatrix} \qquad \mathbf{B} = \begin{bmatrix} 2 & -1 & 0 \\ -1 & 2 & -1 \\ 0 & -1 & 1 \end{bmatrix}
$$

to the standard form.

3. An eigenvalue of the problem in Prob. 2 is roughly 2.5. Use the inverse power method with eigenvalue shifting to compute this eigenvalue to four decimal places. Start with $\mathbf{x} = \begin{bmatrix} 1 & 0 & 0 \end{bmatrix}^T$. *Hint*: two iterations should be sufficient.

4. The stress matrix at a point is

$$
\mathbf{S} = \begin{bmatrix} 150 & -60 & 0 \\ -60 & 120 & 0 \\ 0 & 0 & 80 \end{bmatrix} \text{MPa}
$$

Compute the principal stresses (eigenvalues of $\mathbf{S}$).

5.

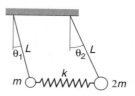

The two pendulums are connected by a spring which is undeformed when the pendulums are vertical. The equations of motion of the system can be shown to be

$$
kL(\theta_2 - \theta_1) - mg\theta_1 = mL\ddot{\theta}_1
$$

$$
-kL(\theta_2 - \theta_1) - 2mg\theta_2 = 2mL\ddot{\theta}_2
$$

where θ_1 and θ_2 are the angular displacements and k is the spring stiffness. Determine the circular frequencies of vibration and the relative amplitudes of the angular displacements. Use $m = 0.25$ kg, $k = 20$ N/m, $L = 0.75$ m and $g = 9.80665$ m/s^2.

6.

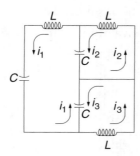

Kirchoff's laws for the electric circuit are

$$3i_1 - i_2 - i_3 = -LC\frac{d^2 i_1}{dt^2}$$

$$-i_1 + i_2 = -LC\frac{d^2 i_2}{dt^2}$$

$$-i_1 + i_3 = -LC\frac{d^2 i_3}{dt^2}$$

Compute the circular frequencies of the circuit and the relative amplitudes of the loop currents.

7. Compute the matrix A^* that results from annihilation of A_{14} and A_{41} in the matrix

$$A = \begin{bmatrix} 4 & -1 & 0 & 1 \\ -1 & 6 & -2 & 0 \\ 0 & -2 & 3 & 2 \\ 1 & 0 & 2 & 4 \end{bmatrix}$$

by a Jacobi rotation.

8. ■ Use the Jacobi method to determine the eigenvalues and eigenvectors of

$$A = \begin{bmatrix} 4 & -1 & -2 \\ -1 & 3 & 3 \\ -2 & 3 & 1 \end{bmatrix}$$

9. ■ Find the eigenvalues and eigenvectors of

$$A = \begin{bmatrix} 4 & -2 & 1 & -1 \\ -2 & 4 & -2 & 1 \\ 1 & -2 & 4 & -2 \\ -1 & 1 & -2 & 4 \end{bmatrix}$$

with the Jacobi method.

10. ■ Use the power method to compute the largest eigenvalue and the corresponding eigenvector of the matrix **A** given in Prob. 9.

11. ■ Find the smallest eigenvalue and the corresponding eigenvector of the matrix **A** in Prob. 9. Use the inverse power method.

12. ■ Let

$$
\mathbf{A} = \begin{bmatrix} 1.4 & 0.8 & 0.4 \\ 0.8 & 6.6 & 0.8 \\ 0.4 & 0.8 & 5.0 \end{bmatrix} \qquad \mathbf{B} = \begin{bmatrix} 0.4 & -0.1 & 0.0 \\ -0.1 & 0.4 & -0.1 \\ 0.0 & -0.1 & 0.4 \end{bmatrix}
$$

Find the eigenvalues and eigenvectors of $\mathbf{Ax} = \lambda\mathbf{Bx}$ by the Jacobi method.

13. ■ Use the inverse power method to compute the smallest eigenvalue in Prob. 12.

14. ■ Use the Jacobi method to compute the eigenvalues and eigenvectors of the matrix

$$
\mathbf{A} = \begin{bmatrix} 11 & 2 & 3 & 1 & 4 & 2 \\ 2 & 9 & 3 & 5 & 2 & 1 \\ 3 & 3 & 15 & 4 & 3 & 2 \\ 1 & 5 & 4 & 12 & 4 & 3 \\ 4 & 2 & 3 & 4 & 17 & 5 \\ 2 & 1 & 2 & 3 & 5 & 8 \end{bmatrix}
$$

15. ■ Find the eigenvalues of $\mathbf{Ax} = \lambda\mathbf{Bx}$ by the Jacobi method, where

$$
\mathbf{A} = \begin{bmatrix} 6 & -4 & 1 & 0 \\ -4 & 6 & -4 & 1 \\ 1 & -4 & 6 & -4 \\ 0 & 1 & -4 & 7 \end{bmatrix} \qquad \mathbf{B} = \begin{bmatrix} 1 & -2 & 3 & -1 \\ -2 & 6 & -2 & 3 \\ 3 & -2 & 6 & -2 \\ -1 & 3 & -2 & 9 \end{bmatrix}
$$

Warning: **B** is not positive definite.

16. ■

The figure shows a cantilever beam with a superimposed finite difference mesh. If $u(x, t)$ is the lateral displacement of the beam, the differential equation of motion governing bending vibrations is

$$
u^{(4)} = -\frac{\gamma}{EI}\ddot{u}
$$

where γ is the mass per unit length and EI is the bending rigidity. The boundary conditions are $u(0, t) = u'(0, t) = u''(L, t) = u'''(L, t) = 0$. With $u(x, t) =$

$y(x) \sin \omega t$ the problem becomes

$$y^{(4)} = \frac{\omega^2 \gamma}{EI} y \qquad y(0) = y'(0) = y''(L) = y'''(L) = 0$$

The corresponding finite difference equations are

$$
\begin{bmatrix}
7 & -4 & 1 & 0 & 0 & \cdots & 0 \\
-4 & 6 & -4 & 1 & 0 & \cdots & 0 \\
1 & -4 & 6 & -4 & 1 & \cdots & 0 \\
\vdots & \ddots & \ddots & \ddots & \ddots & \ddots & \vdots \\
0 & \cdots & 1 & -4 & 6 & -4 & 1 \\
0 & \cdots & 0 & 1 & -4 & 5 & -2 \\
0 & \cdots & 0 & 0 & 1 & -2 & 1
\end{bmatrix}
\begin{bmatrix}
y_1 \\ y_2 \\ y_3 \\ \vdots \\ y_{n-2} \\ y_{n-1} \\ y_n
\end{bmatrix}
= \lambda
\begin{bmatrix}
y_1 \\ y_2 \\ y_3 \\ \vdots \\ y_{n-2} \\ y_{n-1} \\ y_n/2
\end{bmatrix}
$$

where

$$\lambda = \frac{\omega^2 \gamma}{EI} \left(\frac{L}{n} \right)^4$$

(a) Write down the matrix $\mathbf{H}$ of the standard form $\mathbf{Hz} = \lambda \mathbf{z}$ and the transformation matrix $\mathbf{P}$ as in $\mathbf{y} = \mathbf{Pz}$. (b) Write a program that computes the lowest two circular frequencies of the beam and the corresponding mode shapes (eigenvectors) using the Jacobi method. Run the program with $n = 10$. *Note*: the analytical solution for the lowest circular frequency is $\omega_1 = (3.515/L^2) \sqrt{EI/\gamma}$.

17. ■

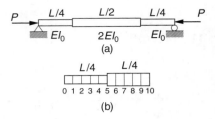

(a)

(b)

The simply supported column in Fig. (a) consists of three segments with the bending rigidities shown. If only the first buckling mode is of interest, it is sufficient to model half of the beam as shown in Fig. (b). The differential equation for the lateral displacement $u(x)$ is

$$u'' = -\frac{P}{EI} u$$

with the boundary conditions $u(0) = u'(L/2) = 0$. The corresponding finite difference equations are

$$
\begin{bmatrix}
2 & -1 & 0 & 0 & 0 & 0 & 0 & \cdots & 0 \\
-1 & 2 & -1 & 0 & 0 & 0 & 0 & \cdots & 0 \\
0 & -1 & 2 & -1 & 0 & 0 & 0 & \cdots & 0 \\
0 & 0 & -1 & 2 & -1 & 0 & 0 & \cdots & 0 \\
0 & 0 & 0 & -1 & 2 & -1 & 0 & \cdots & 0 \\
0 & 0 & 0 & 0 & -1 & 2 & -1 & \cdots & 0 \\
\vdots & \vdots & \vdots & \vdots & \vdots & \ddots & \ddots & \ddots & \vdots \\
0 & \cdots & 0 & 0 & 0 & 0 & -1 & 2 & -1 \\
0 & \cdots & 0 & 0 & 0 & 0 & 0 & -1 & 1
\end{bmatrix}
\begin{bmatrix}
u_1 \\ u_2 \\ u_3 \\ u_4 \\ u_5 \\ u_6 \\ \vdots \\ u_9 \\ u_{10}
\end{bmatrix}
= \lambda
\begin{bmatrix}
u_1 \\ u_2 \\ u_3 \\ u_4 \\ u_5/1.5 \\ u_6/2 \\ \vdots \\ u_9/2 \\ u_{10}/4
\end{bmatrix}
$$

where

$$
\lambda = \frac{P}{E I_0} \left(\frac{L}{20} \right)^2
$$

Write a program that computes the lowest buckling load P of the column with the inverse power method. Utilize the banded forms of the matrices.

18. ■

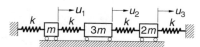

The springs supporting the three-bar linkage are undeformed when the linkage is horizontal. The equilibrium equations of the linkage in the presence of the horizontal force P can be shown to be

$$
\begin{bmatrix}
6 & 5 & 3 \\
3 & 3 & 2 \\
1 & 1 & 1
\end{bmatrix}
\begin{bmatrix}
\theta_1 \\ \theta_2 \\ \theta_3
\end{bmatrix}
= \frac{P}{kL}
\begin{bmatrix}
1 & 1 & 1 \\
0 & 1 & 1 \\
0 & 0 & 1
\end{bmatrix}
\begin{bmatrix}
\theta_1 \\ \theta_2 \\ \theta_3
\end{bmatrix}
$$

where k is the spring stiffness. Determine the smallest buckling load P and the corresponding mode shape. *Hint*: the equations can easily rewritten in the standard form $\mathbf{A}\theta = \lambda\theta$, where $\mathbf{A}$ is symmetric.

19. ■

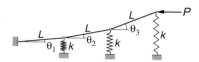

The differential equations of motion for the mass–spring system are

$$k(-2u_1 + u_2) = m\ddot{u}_1$$

$$k(u_1 - 2u_2 + u_3) = 3m\ddot{u}_2$$

$$k(u_2 - 2u_3) = 2m\ddot{u}_3$$

where $u_i(t)$ is the displacement of mass i from its equilibrium position and k is the spring stiffness. Determine the circular frequencies of vibration and the corresponding mode shapes.

20. ■

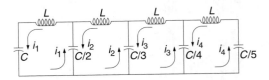

Kirchoff's equations for the circuit are

$$L\frac{d^2 i_1}{dt^2} + \frac{1}{C}i_1 + \frac{2}{C}(i_1 - i_2) = 0$$

$$L\frac{d^2 i_2}{dt^2} + \frac{2}{C}(i_2 - i_1) + \frac{3}{C}(i_2 - i_3) = 0$$

$$L\frac{d^2 i_3}{dt^2} + \frac{3}{C}(i_3 - i_2) + \frac{4}{C}(i_3 - i_4) = 0$$

$$L\frac{d^2 i_4}{dt^2} + \frac{4}{C}(i_4 - i_3) + \frac{5}{C}i_4 = 0$$

Find the circular frequencies of the currents.

21. ■

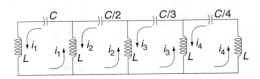

Determine the circular frequencies of oscillation for the circuit shown, given the Kirchoff equations

$$L\frac{d^2 i_1}{dt^2} + L\left(\frac{d^2 i_1}{dt^2} - \frac{d^2 i_2}{dt^2}\right) + \frac{1}{C}i_1 = 0$$

$$L\left(\frac{d^2 i_2}{dt^2} - \frac{d^2 i_1}{dt^2}\right) + L\left(\frac{d^2 i_2}{dt^2} - \frac{d^2 i_3}{dt^2}\right) + \frac{2}{C}i_2 = 0$$

$$L\left(\frac{d^2 i_3}{dt^2} - \frac{d^2 i_2}{dt^2}\right) + L\left(\frac{d^2 i_3}{dt^2} - \frac{d^2 i_4}{dt^2}\right) + \frac{3}{C} i_3 = 0$$

$$L\left(\frac{d^2 i_4}{dt^2} - \frac{d^2 i_3}{dt^2}\right) + L\frac{d^2 i_4}{dt^2} + \frac{4}{C} i_4 = 0$$

22. ■ Several iterative methods exist for finding the eigenvalues of a matrix $\mathbf{A}$. One of these is the *LR method*, which requires the matrix to be symmetric and positive definite. Its algorithm very simple:

> Let $\mathbf{A}_0 = \mathbf{A}$
> do with $i = 0, 1, 2, \ldots$
> > Use Choleski's decomposition $\mathbf{A}_i = \mathbf{L}_i \mathbf{L}_i^T$ to compute $\mathbf{L}_i$
> > Form $\mathbf{A}_{i+1} = \mathbf{L}_i^T \mathbf{L}_i$
> end do

It can be shown that the diagonal elements of $\mathbf{A}_{i+1}$ converge to the eigenvalues of $\mathbf{A}$. Write a program that implements the LR method and test it with

$$\mathbf{A} = \begin{bmatrix} 4 & 3 & 1 \\ 3 & 4 & 2 \\ 1 & 2 & 3 \end{bmatrix}$$

9.4 Householder Reduction to Tridiagonal Form

It was mentioned before that similarity transformations can be used to transform an eigenvalue problem to a form that is easier to solve. The most desirable of the "easy" forms is, of course, the diagonal form that results from the Jacobi method. However, the Jacobi method requires about $10n^3$ to $20n^3$ multiplications, so that the amount of computation increases very rapidly with n. We are generally better off by reducing the matrix to the tridiagonal form, which can be done in precisely $n - 2$ transformations by the Householder method. Once the tridiagonal form is achieved, we still have to extract the eigenvalues and the eigenvectors, but there are effective means of dealing with that, as we see in the next article.

Householder Matrix

Householder's transformation utilizes the *Householder matrix*

$$\mathbf{Q} = \mathbf{I} - \frac{\mathbf{u}\mathbf{u}^T}{H} \tag{9.36}$$

where **u** is a vector and

$$H = \frac{1}{2}\mathbf{u}^T\mathbf{u} = \frac{1}{2}|\mathbf{u}|^2 \qquad (9.37)$$

Note that $\mathbf{u}\mathbf{u}^T$ in Eq. (9.36) is the outer product; that is, a matrix with the elements $(\mathbf{u}\mathbf{u}^T)_{ij} = u_i u_j$. Since **Q** is obviously symmetric ($\mathbf{Q}^T = \mathbf{Q}$), we can write

$$\mathbf{Q}^T\mathbf{Q} = \mathbf{Q}\mathbf{Q} = \left(\mathbf{I} - \frac{\mathbf{u}\mathbf{u}^T}{H}\right)\left(\mathbf{I} - \frac{\mathbf{u}\mathbf{u}^T}{H}\right) = \mathbf{I} - 2\frac{\mathbf{u}\mathbf{u}^T}{H} + \frac{\mathbf{u}\left(\mathbf{u}^T\mathbf{u}\right)\mathbf{u}^T}{H^2}$$

$$= \mathbf{I} - 2\frac{\mathbf{u}\mathbf{u}^T}{H} + \frac{\mathbf{u}\left(2H\right)\mathbf{u}^T}{H^2} = \mathbf{I}$$

which shows that **Q** is also orthogonal.

Now let **x** be an arbitrary vector and consider the transformation **Qx**. Choosing

$$\mathbf{u} = \mathbf{x} + k\mathbf{e}_1 \qquad (9.38)$$

where

$$k = \pm|\mathbf{x}| \qquad \mathbf{e}_1 = \begin{bmatrix} 1 & 0 & 0 & \cdots & 0 \end{bmatrix}^T$$

we get

$$\mathbf{Q}\mathbf{x} = \left(\mathbf{I} - \frac{\mathbf{u}\mathbf{u}^T}{H}\right)\mathbf{x} = \left[\mathbf{I} - \frac{\mathbf{u}\left(\mathbf{x} + k\mathbf{e}_1\right)^T}{H}\right]\mathbf{x}$$

$$= \mathbf{x} - \frac{\mathbf{u}\left(\mathbf{x}^T\mathbf{x} + k\mathbf{e}_1^T\mathbf{x}\right)}{H} = \mathbf{x} - \frac{\mathbf{u}\left(k^2 + kx_1\right)}{H}$$

But

$$2H = \left(\mathbf{x} + k\mathbf{e}_1\right)^T\left(\mathbf{x} + k\mathbf{e}_1\right) = |\mathbf{x}|^2 + k\left(\mathbf{x}^T\mathbf{e}_1 + \mathbf{e}_1^T\mathbf{x}\right) + k^{pt2}\mathbf{e}_1^T\mathbf{e}_1$$

$$= k^2 + 2kx_1 + k^2 = 2\left(k^2 + kx_1\right)$$

so that

$$\mathbf{Q}\mathbf{x} = \mathbf{x} - \mathbf{u} = -k\mathbf{e}_1 = \begin{bmatrix} -k & 0 & 0 & \cdots & 0 \end{bmatrix}^T \qquad (9.39)$$

Hence the transformation eliminates all elements of **x** except the first one.

Householder Reduction of a Symmetric Matrix

Let us now apply the following transformation to a symmetric $n \times n$ matrix **A**:

$$\mathbf{P}_1\mathbf{A} = \begin{bmatrix} 1 & \mathbf{0}^T \\ \mathbf{0} & \mathbf{Q} \end{bmatrix}\begin{bmatrix} A_{11} & \mathbf{x}^T \\ \mathbf{x} & \mathbf{A}' \end{bmatrix} = \begin{bmatrix} A_{11} & \mathbf{x}^T \\ \mathbf{Q}\mathbf{x} & \mathbf{Q}\mathbf{A}' \end{bmatrix} \qquad (9.40)$$

Here **x** is represents the first column of **A** with the first element omitted, and **A**' is simply **A** with its first row and column removed. The matrix **Q** of dimensions

$(n-1) \times (n-1)$ is constructed using Eqs. (9.36)–(9.38). Referring to Eq. (9.39), we see that the transformation reduces the first column of $\mathbf{A}$ to

$$
\begin{bmatrix} A_{11} \\ \mathbf{Qx} \end{bmatrix} = \begin{bmatrix} A_{11} \\ -k \\ 0 \\ \vdots \\ 0 \end{bmatrix}
$$

The transformation

$$
\mathbf{A} \leftarrow \mathbf{P}_1 \mathbf{A} \mathbf{P}_1 = \begin{bmatrix} A_{11} & (\mathbf{Qx})^T \\ \mathbf{Qx} & \mathbf{QA'Q} \end{bmatrix} \tag{9.41}
$$

thus tridiagonalizes the first row as well as the first column of $\mathbf{A}$. Here is a diagram of the transformation for a 4×4 matrix:

1	0	0	0
0			
0		$\mathbf{Q}$	
0			

$\cdot$

A_{11}	A_{12}	A_{13}	A_{14}
A_{21}			
A_{31}		$\mathbf{A'}$	
A_{41}			

$\cdot$

1	0	0	0
0			
0		$\mathbf{Q}$	
0			

$=$

A_{11}	$-k$	0	0
$-k$			
0		$\mathbf{QA'Q}$	
0			

The second row and column of $\mathbf{A}$ are reduced next by applying the transformation to the 3×3 lower right portion of the matrix. This transformation can be expressed as $\mathbf{A} \leftarrow \mathbf{P}_2 \mathbf{A} \mathbf{P}_2$, where now

$$
\mathbf{P}_2 = \begin{bmatrix} \mathbf{I}_2 & \mathbf{0}^T \\ \mathbf{0} & \mathbf{Q} \end{bmatrix} \tag{9.42}
$$

In Eq. (9.42) $\mathbf{I}_2$ is a 2×2 identity matrix and $\mathbf{Q}$ is a $(n-2) \times (n-2)$ matrix constructed by choosing for $\mathbf{x}$ the bottom $n-2$ elements of the second column of $\mathbf{A}$. It takes a total of $n-2$ transformations with

$$
\mathbf{P}_i = \begin{bmatrix} \mathbf{I}_i & \mathbf{0}^T \\ \mathbf{0} & \mathbf{Q} \end{bmatrix}, \quad i = 1, 2, \ldots, n-2
$$

to attain the tridiagonal form.

It is wasteful to form $\mathbf{P}_i$ and then carry out the matrix multiplication $\mathbf{P}_i \mathbf{A} \mathbf{P}_i$. We note that

$$\mathbf{A}'\mathbf{Q} = \mathbf{A}'\left(\mathbf{I} - \frac{\mathbf{u}\mathbf{u}^T}{H}\right) = \mathbf{A}' - \frac{\mathbf{A}'\mathbf{u}}{H}\mathbf{u}^T = \mathbf{A}' - \mathbf{v}\mathbf{u}^T$$

where

$$\mathbf{v} = \frac{\mathbf{A}'\mathbf{u}}{H} \tag{9.43}$$

Therefore,

$$\mathbf{Q}\mathbf{A}'\mathbf{Q} = \left(\mathbf{I} - \frac{\mathbf{u}\mathbf{u}^T}{H}\right)(\mathbf{A}' - \mathbf{v}\mathbf{u}^T) = \mathbf{A}' - \mathbf{v}\mathbf{u}^T - \frac{\mathbf{u}\mathbf{u}^T}{H}(\mathbf{A}' - \mathbf{v}\mathbf{u}^T)$$

$$= \mathbf{A}' - \mathbf{v}\mathbf{u}^T - \frac{\mathbf{u}\left(\mathbf{u}^T\mathbf{A}'\right)}{H} + \frac{\mathbf{u}\left(\mathbf{u}^T\mathbf{v}\right)\mathbf{u}^T}{H}$$

$$= \mathbf{A}' - \mathbf{v}\mathbf{u}^T - \mathbf{u}\mathbf{v}^T + 2g\mathbf{u}\mathbf{u}^T$$

where

$$g = \frac{\mathbf{u}^T\mathbf{v}}{2H} \tag{9.44}$$

Letting

$$\mathbf{w} = \mathbf{v} - g\mathbf{u} \tag{9.45}$$

it can be easily verified that the transformation can be written as

$$\mathbf{Q}\mathbf{A}'\mathbf{Q} = \mathbf{A}' - \mathbf{w}\mathbf{u}^T - \mathbf{u}\mathbf{w}^T \tag{9.46}$$

which gives us the following computational procedure which is to be carried out with $i = 1, 2, \ldots, n - 2$:

1. Let $\mathbf{A}'$ be the $(n - i) \times (n - i)$ lower right-hand portion of $\mathbf{A}$.
2. Let $\mathbf{x} = \begin{bmatrix} A_{i+1,i} & A_{i+2,i} & \cdots & A_{n,i} \end{bmatrix}^T$ (the column of length $n - i$ just to the left of $\mathbf{A}'$).
3. Compute $|\mathbf{x}|$. Let $k = |\mathbf{x}|$ if $x_1 > 0$ and $k = -|\mathbf{x}|$ if $x_1 < 0$ (this choice of sign minimizes the roundoff error).
4. Let $\mathbf{u} = \begin{bmatrix} k+x_1 & x_2 & x_3 & \cdots & x_{n-i} \end{bmatrix}^T$.
5. Compute $H = |\mathbf{u}|^2/2$.
6. Compute $\mathbf{v} = \mathbf{A}'\mathbf{u}/H$.
7. Compute $g = \mathbf{u}^T\mathbf{v}/(2H)$.
8. Compute $\mathbf{w} = \mathbf{v} - g\mathbf{u}$.
9. Compute the transformation $\mathbf{A} \leftarrow \mathbf{A}' - \mathbf{w}^T\mathbf{u} - \mathbf{u}^T\mathbf{w}$.
10. Set $A_{i,i+1} = A_{i+1,i} = -k$.

Accumulated Transformation Matrix

Since we used similarity transformations, the eigenvalues of the tridiagonal matrix are the same as those of the original matrix. However, to determine the eigenvectors $\mathbf{X}$ of original $\mathbf{A}$ we must use the transformation

$$\mathbf{X} = \mathbf{P}\mathbf{X}_{\text{tridiag}}$$

where $\mathbf{P}$ is the accumulation of the individual transformations:

$$\mathbf{P} = \mathbf{P}_1\mathbf{P}_2\cdots\mathbf{P}_{n-2}$$

We build up the accumulated transformation matrix by initializing $\mathbf{P}$ to a $n \times n$ identity matrix and then applying the transformation

$$\mathbf{P} \leftarrow \mathbf{P}\mathbf{P}_i = \begin{bmatrix} \mathbf{P}_{11} & \mathbf{P}_{12} \\ \mathbf{P}_{21} & \mathbf{P}_{22} \end{bmatrix}\begin{bmatrix} \mathbf{I}_i & \mathbf{0}^T \\ \mathbf{0} & \mathbf{Q} \end{bmatrix} = \begin{bmatrix} \mathbf{P}_{11} & \mathbf{P}_{21}\mathbf{Q} \\ \mathbf{P}_{12} & \mathbf{P}_{22}\mathbf{Q} \end{bmatrix} \tag{b}$$

with $i = 1, 2, \ldots, n-2$. It can be seen that each multiplication affects only the rightmost $n-i$ columns of $\mathbf{P}$ (since the first row of $\mathbf{P}_{12}$ contains only zeroes, it can also be omitted in the multiplication). Using the notation

$$\mathbf{P}' = \begin{bmatrix} \mathbf{P}_{12} \\ \mathbf{P}_{22} \end{bmatrix}$$

we have

$$\begin{bmatrix} \mathbf{P}_{12}\mathbf{Q} \\ \mathbf{P}_{22}\mathbf{Q} \end{bmatrix} = \mathbf{P}'\mathbf{Q} = \mathbf{P}'\left(\mathbf{I} - \frac{\mathbf{u}\mathbf{u}^T}{H}\right) = \mathbf{P}' - \frac{\mathbf{P}'\mathbf{u}}{H}\mathbf{u}^T = \mathbf{P}' - \mathbf{y}\mathbf{u}^T \tag{9.47}$$

where

$$\mathbf{y} = \frac{\mathbf{P}'\mathbf{u}}{H} \tag{9.48}$$

The procedure for carrying out the matrix multiplication in Eq. (b) is

- Retrieve $\mathbf{u}$ (in our triangularization procedure the $\mathbf{u}$'s are stored in the columns of the lower triangular portion of $\mathbf{A}$).
- Compute $H = |\mathbf{u}|^2/2$.
- Compute $\mathbf{y} = \mathbf{P}'\mathbf{u}/H$.
- Compute the transformation $\mathbf{P}' \leftarrow \mathbf{P}' - \mathbf{y}\mathbf{u}^T$.

■ householder

This function performs the Householder reduction on the matrix $\mathbf{A}$. Upon return, $\mathbf{d}$ occupies the principal diagonal of $\mathbf{A}$ and $\mathbf{c}$ forms the upper subdiagonal; that is, $\mathbf{d} = \text{diag}(A)$ and $\mathbf{c} = \text{diag}(A,1)$. The portion of A below the principal diagonal is utilized to store the vectors **u** that are needed in the computation of the transformation matrix **P**.

```
function A = householder(A)
% Housholder reduction of A to tridiagonal form A = [c\d\c].
% Extract c and d by d = diag(A), c = diag(A,1).
% USAGE: A = householder(A)

n = size(A,1);
for k = 1:n-2
    u = A(k+1:n,k);
    uMag = sqrt(dot(u,u));
    if u(1) < 0; uMag = -uMag; end
    u(1) = u(1) + uMag;
    A(k+1:n,k) = u;                  % Save u in lower part of A.
    H = dot(u,u)/2;
    v = A(k+1:n,k+1:n)*u/H;
    g = dot(u,v)/(2*H);
    v = v - g*u;
    A(k+1:n,k+1:n) = A(k+1:n,k+1:n) - v*u' - u*v';
    A(k,k+1) = -uMag;
end
```

■ householderP

The function `householderP` returns the accumulated transformation matrix **P**. There is no need to call it if only the eigenvalues are to be computed. Note that the input parameter A is not the original matrix, but the matrix returned by `householder`.

```
function P = householderP(A)
% Computes transformation matrix P after
% householder reduction of A is carried out.
% USAGE: P = householderP(A).

n = size(A,1);
P = eye(n);
for k = 1:n-2
    u = A(k+1:n,k);
    H = dot(u,u)/2;
    v = P(1:n,k+1:n)*u/H;
```

```
        P(1:n,k+1:n) = P(1:n,k+1:n) - v*u';
end
```

EXAMPLE 9.7
Transform the matrix

$$A = \begin{bmatrix} 7 & 2 & 3 & -1 \\ 2 & 8 & 5 & 1 \\ 3 & 5 & 12 & 9 \\ -1 & 1 & 9 & 7 \end{bmatrix}$$

into tridiagonal form using Householder reduction.

Solution Reduce the first row and column:

$$A' = \begin{bmatrix} 8 & 5 & 1 \\ 5 & 12 & 9 \\ 1 & 9 & 7 \end{bmatrix} \qquad x = \begin{bmatrix} 2 \\ 3 \\ -1 \end{bmatrix} \qquad k = |\mathbf{x}| = 3.7417$$

$$\mathbf{u} = \begin{bmatrix} k + x_1 \\ x_2 \\ x_3 \end{bmatrix} = \begin{bmatrix} 5.7417 \\ 3 \\ -1 \end{bmatrix} \qquad H = \frac{1}{2}|\mathbf{u}|^2 = 21.484$$

$$\mathbf{uu}^T = \begin{bmatrix} 32.967 & 17\,225 & -5.7417 \\ 17.225 & 9 & -3 \\ -5.7417 & -3 & 1 \end{bmatrix}$$

$$Q = I - \frac{\mathbf{uu}^T}{H} = \begin{bmatrix} -0.53450 & -0.80176 & 0.26725 \\ -0.80176 & 0.58108 & 0.13964 \\ 0.26725 & 0.13964 & 0.95345 \end{bmatrix}$$

$$QA'Q = \begin{bmatrix} 10.642 & -0.1388 & -9.1294 \\ -0.1388 & 5.9087 & 4.8429 \\ -9.1294 & 4.8429 & 10.4480 \end{bmatrix}$$

$$A \leftarrow \begin{bmatrix} A_{11} & (\mathbf{Qx})^T \\ \mathbf{Qx} & QA'Q \end{bmatrix} = \begin{bmatrix} 7 & -3.7417 & 0 & 0 \\ -3.7417 & 10.642 & -0.1388 & -9.1294 \\ 0 & -0.1388 & 5.9087 & 4.8429 \\ 0 & -9.1294 & 4.8429 & 10.4480 \end{bmatrix}$$

In the last step we used the formula $\mathbf{Qx} = \begin{bmatrix} -k & 0 & \cdots & 0 \end{bmatrix}^T$.

Reduce the second row and column:

$$\mathbf{A}' = \begin{bmatrix} 5.9087 & 4.8429 \\ 4.8429 & 10.4480 \end{bmatrix} \qquad \mathbf{x} = \begin{bmatrix} -0.1388 \\ -9.1294 \end{bmatrix} \qquad k = -|\mathbf{x}| = -9.1305$$

where the negative sign on k was determined by the sign of x_1.

$$\mathbf{u} = \begin{bmatrix} k + x_1 \\ x_2 \end{bmatrix} = \begin{bmatrix} -9.2693 \\ -9.1294 \end{bmatrix} \qquad H = \frac{1}{2}|\mathbf{u}|^2 = 84.633$$

$$\mathbf{u}\mathbf{u}^T = \begin{bmatrix} 85.920 & 84.623 \\ 84.623 & 83.346 \end{bmatrix}$$

$$\mathbf{Q} = \mathbf{I} - \frac{\mathbf{u}\mathbf{u}^T}{H} = \begin{bmatrix} 0.01521 & -0.99988 \\ -0.99988 & 0.01521 \end{bmatrix}$$

$$\mathbf{Q}\mathbf{A}'\mathbf{Q} = \begin{bmatrix} 10.594 & 4.772 \\ 4.772 & 5.762 \end{bmatrix}$$

$$\mathbf{A} \leftarrow \begin{bmatrix} A_{11} & A_{12} & \mathbf{0}^T \\ A_{21} & A_{22} & (\mathbf{Qx})^T \\ \mathbf{0} & \mathbf{Qx} & \mathbf{QA'Q} \end{bmatrix} = \begin{bmatrix} 7 & -3.742 & 0 & 0 \\ -3.742 & 10.642 & 9.131 & 0 \\ 0 & 9.131 & 10.594 & 4.772 \\ 0 & 0 & 4.772 & 5.762 \end{bmatrix}$$

EXAMPLE 9.8

Use the function `householder` to tridiagonalize the matrix in Example 9.7; also determine the transformation matrix **P**.

Solution

```
% Example 9.8 (Householder reduction)
A = [7   2   3 -1;
     2   8   5   1;
     3   5  12   9;
    -1   1   9   7];
A = householder(A);
d = diag(A)'
c = diag(A,1)'
P = householderP(A)
```

The results of running the above program are:

```
>> d =
    7.0000    10.6429    10.5942    5.7629
```

```
c =
    -3.7417      9.1309       4.7716
P =
     1.0000           0            0            0
          0     -0.5345      -0.2551       0.8057
          0     -0.8018      -0.1484      -0.5789
          0      0.2673      -0.9555      -0.1252
```

9.5   Eigenvalues of Symmetric Tridiagonal Matrices

Sturm Sequence

In principle, the eigenvalues of a matrix $\mathbf{A}$ can be determined by finding the roots of the characteristic equation $|\mathbf{A} - \lambda\mathbf{I}| = 0$. This method is impractical for large matrices since the evaluation of the determinant involves $n^3/3$ multiplications. However, if the matrix is tridiagonal (we also assume it to be symmetric), its characteristic polynomial

$$P_n(\lambda) = |\mathbf{A} - \lambda\mathbf{I}| = \begin{vmatrix} d_1 - \lambda & c_1 & 0 & 0 & \cdots & 0 \\ c_1 & d_2 - \lambda & c_2 & 0 & \cdots & 0 \\ 0 & c_2 & d_3 - \lambda & c_3 & \cdots & 0 \\ 0 & 0 & c_3 & d_4 - \lambda & \cdots & 0 \\ \vdots & \vdots & \vdots & \vdots & \ddots & \vdots \\ 0 & 0 & \cdots & 0 & c_{n-1} & d_n - \lambda \end{vmatrix}$$

can be computed with only $3(n - 1)$ multiplications using the following sequence of operations:

$$P_0(\lambda) = 1$$
$$P_1(\lambda) = d_1 - \lambda \tag{9.49}$$
$$P_i(\lambda) = (d_i - \lambda)P_{i-1}(\lambda) - c_{i-1}^2 P_{i-2}(\lambda), \quad i = 2, 3, \ldots, n$$

The polynomials $P_0(\lambda)$, $P_1(\lambda)$, ..., $P_n(\lambda)$ form a *Sturm sequence* that has the following property:

- The number of sign changes in the sequence $P_0(a)$, $P_1(a)$, ..., $P_n(a)$ is equal to the number of roots of $P_n(\lambda)$ that are smaller than a. If a member $P_i(a)$ of the sequence is zero, its sign is to be taken opposite to that of $P_{i-1}(a)$.

As we see shortly, Sturm sequence property makes it relatively easy to bracket the eigenvalues of a tridiagonal matrix.

■ sturmSeq

Given the diagonals **c** and **d** of $\mathbf{A} = [\mathbf{c} \backslash \mathbf{d} \backslash \mathbf{c}]$, and the value of λ, this function returns the Sturm sequence $P_0(\lambda), P_1(\lambda), \ldots, P_n(\lambda)$. Note that $P_n(\lambda) = |\mathbf{A} - \lambda \mathbf{I}|$.

```
function p = sturmSeq(c,d,lambda)
% Returns Sturm sequence p associated with
% the tridiagonal matrix A = [c\d\c] and lambda.
% USAGE: p = sturmSeq(c,d,lambda).
% Note that |A - lambda*I| = p(n).

n = length(d) + 1;
p = ones(n,1);
p(2) = d(1) - lambda;
for i = 2:n-1
    p(i+1) = (d(i) - lambda)*p(i) - (c(i-1)^2 )*p(i-1);
end
```

■ count_eVals

This function counts the number of sign changes in the Sturm sequence and returns the number of eigenvalues of the matrix $\mathbf{A} = [\mathbf{c} \backslash \mathbf{d} \backslash \mathbf{c}]$ that are smaller than λ.

```
function num_eVals = count_eVals(c,d,lambda)
% Counts eigenvalues smaller than lambda of matrix
% A = [c\d\c]. Uses the Sturm sequence.
% USAGE: num_eVals = count_eVals(c,d,lambda).

p = sturmSeq(c,d,lambda);
n = length(p);
oldSign = 1; num_eVals = 0;
for i = 2:n
    pSign = sign(p(i));
    if pSign == 0; pSign = -oldSign; end
    if pSign*oldSign < 0
        num_eVals = num_eVals + 1;
    end
    oldSign = pSign;
end
```

EXAMPLE 9.9

Use the Sturm sequence property to show that the smallest eigenvalue of $\mathbf{A}$ is in the interval $(0.25, 0.5)$, where

$$\mathbf{A} = \begin{bmatrix} 2 & -1 & 0 & 0 \\ -1 & 2 & -1 & 0 \\ 0 & -1 & 2 & -1 \\ 0 & 0 & -1 & 2 \end{bmatrix}$$

Solution Taking $\lambda = 0.5$, we have $d_i - \lambda = 1.5$ and $c_{i-1}^2 = 1$ and the Sturm sequence in Eqs. (9.49) becomes

$$P_0(0.5) = 1$$

$$P_1(0.5) = 1.5$$

$$P_2(0.5) = 1.5(1.5) - 1 = 1.25$$

$$P_3(0.5) = 1.5(1.25) - 1.5 = 0.375$$

$$P_4(0.5) = 1.5(0.375) - 1.25 = -0.6875$$

Since the sequence contains one sign change, there exists one eigenvalue smaller than 0.5.

Repeating the process with $\lambda = 0.25$ ($d_i - \lambda = 1.75$, $c_{i-1}^2 = 1$), we get

$$P_0(0.25) = 1$$

$$P_1(0.25) = 1.75$$

$$P_2(0.25) = 1.75(1.75) - 1 = 2.0625$$

$$P_3(0.25) = 1.75(2.0625) - 1.75 = 1.8594$$

$$P_4(0.25) = 1.75(1.8594) - 2.0625 = 1.1915$$

There are no sign changes in the sequence, so that all the eigenvalues are greater than 0.25. We thus conclude that $0.25 < \lambda_1 < 0.5$.

Gerschgorin's Theorem

Gerschgorin's theorem is useful in determining the *global bounds* on the eigenvalues of an $n \times n$ matrix $\mathbf{A}$. The term "global" means the bounds that enclose all the eigenvalues. We give here a simplified version of the theorem for a symmetric matrix.

- If λ is an eigenvalue of $\mathbf{A}$, then

$$a_i - r_i \leq \lambda \leq a_i + r_i, \quad i = 1, 2, \ldots, n$$

where

$$a_i = A_{ii} \qquad r_i = \sum_{\substack{j=1 \\ j \neq i}}^{n} |A_{ij}| \tag{9.50}$$

It follows that the global bounds on the eigenvalues are

$$\lambda_{\min} \geq \min_i (a_i - r_i) \qquad \lambda_{\max} \leq \max_i (a_i + r_i) \tag{9.51}$$

■ gerschgorin

The function gerschgorin returns the lower and the upper global bounds on the eigenvalues of a symmetric tridiagonal matrix $\mathbf{A} = [\mathbf{c}\backslash\mathbf{d}\backslash\mathbf{c}]$.

```
function [eValMin,eValMax]= gerschgorin(c,d)
% Evaluates the global bounds on eigenvalues
% of A = [c\d\c].
% USAGE: [eValMin,eValMax]= gerschgorin(c,d).

n = length(d);
eValMin = d(1) - abs(c(1));
eValMax = d(1) + abs(c(1));
for i = 2:n-1
    eVal = d(i) - abs(c(i)) - abs(c(i-1));
    if eVal < eValMin; eValMin = eVal; end
    eVal = d(i) + abs(c(i)) + abs(c(i-1));
    if eVal > eValMax; eValMax = eVal; end
end
eVal = d(n) - abs(c(n-1));
if eVal < eValMin; eValMin = eVal; end
eVal = d(n) + abs(c(n-1));
if eVal > eValMax; eValMax = eVal; end
```

EXAMPLE 9.10

Use Gerschgorin's theorem to determine the global bounds on the eigenvalues of the matrix

$$\mathbf{A} = \begin{bmatrix} 4 & -2 & 0 \\ -2 & 4 & -2 \\ 0 & -2 & 5 \end{bmatrix}$$

Solution Referring to Eqs. (9.50), we get

$$a_1 = 4 \qquad a_2 = 4 \qquad a_3 = 5$$
$$r_1 = 2 \qquad r_2 = 4 \qquad r_3 = 2$$

Hence

$$\lambda_{\min} \geq \min(a_i - r_i) = 4 - 4 = 0$$
$$\lambda_{\max} \leq \max(a_i + r_i) = 4 + 4 = 8$$

Bracketing Eigenvalues

The Sturm sequence property together with Gerschgorin's theorem provides us convenient tools for bracketing each eigenvalue of a symmetric tridiagonal matrix.

■ eValBrackets

The function eValBrackets brackets the m smallest eigenvalues of a symmetric tridiagonal matrix $\mathbf{A} = [\mathbf{c}\backslash\mathbf{d}\backslash\mathbf{c}]$. It returns the sequence $r_1, r_2, \ldots, r_{m+1}$, where each interval (r_i, r_{i+1}) contains exactly one eigenvalue. The algorithm first finds the global bounds on the eigenvalues by Gerschgorin's theorem. The method of bisection in conjunction with the Sturm sequence property is then used to determine the upper bounds on $\lambda_m, \lambda_{m-1}, \ldots, \lambda_1$ in that order.

```
function r = eValBrackets(c,d,m)
% Brackets each of the m lowest eigenvalues of A = [c\d\c]
% so that there is one eivenvalue in [r(i), r(i+1)].
% USAGE: r = eValBrackets(c,d,m).

[eValMin,eValMax]= gerschgorin(c,d);   % Find global limits
r = ones(m+1,1); r(1) = eValMin;
% Search for eigenvalues in descending order
for k = m:-1:1
    % First bisection of interval (eValMin,eValMax)
    eVal = (eValMax + eValMin)/2;
    h = (eValMax - eValMin)/2;
    for i = 1:100
        % Find number of eigenvalues less than eVal
        num_eVals = count_eVals(c,d,eVal);
        % Bisect again & find the half containing eVal
```

```
        h = h/2;
        if num_eVals < k ; eVal = eVal + h;
        elseif num_eVals > k ; eVal = eVal - h;
        else; break
        end
    end
    % If eigenvalue located, change upper limit of
    % search and record result in {r}
    ValMax = eVal; r(k+1) = eVal;
end
```

EXAMPLE 9.11

Bracket each eigenvalue of the matrix in Example 9.10.

Solution In Example 9.10 we found that all the eigenvalues lie in $(0, 8)$. We now bisect this interval and use the Sturm sequence to determine the number of eigenvalues in $(0, 4)$. With $\lambda = 4$, the sequence is—see Eqs. (9.49)

$$P_0(4) = 1$$

$$P_1(4) = 4 - 4 = 0$$

$$P_2(4) = (4 - 4)(0) - 2^2(1) = -4$$

$$P_3(4) = (5 - 4)(-4) - 2^2(0) = -4$$

Since a zero value is assigned the sign opposite to that of the preceding member, the signs in this sequence are $(+, -, -, -)$. The one sign change shows the presence of one eigenvalue in $(0, 4)$.

Next we bisect the interval $(4, 8)$ and compute the Sturm sequence with $\lambda = 6$:

$$P_0(6) = 1$$

$$P_1(6) = 4 - 6 = -2$$

$$P_2(6) = (4 - 6)(-2) - 2^2(1) = 0$$

$$P_3(6) = (5 - 6)(0) - 2^2(-2) = 8$$

In this sequence the signs are $(+, -, +, +)$, indicating two eigenvalues in $(0, 6)$.

Therefore

$$0 \le \lambda_1 \le 4 \qquad 4 \le \lambda_2 \le 6 \qquad 6 \le \lambda_3 \le 8$$

Computation of Eigenvalues

Once the desired eigenvalues are bracketed, they can be found by determining the roots of $P_n(\lambda) = 0$ with bisection or Brent's method.

■ eigenvals3

The function `eigenvals3` computes the m smallest eigenvalues of a symmetric tridiagonal matrix with the method of Brent.

```
function eVals = eigenvals3(C,D,m)
% Computes the smallest m eigenvalues of A = [C\D\C].
% USAGE: eVals = eigenvals3(C,D,m).
% C and D must be delared 'global' in calling program.

eVals = zeros(m,1);
r = eValBrackets(C,D,m); % Bracket eigenvalues
for i=1:m
    % Solve |A - eVal*I| for eVal by Brent's method
    eVals(i) = brent(@func,r(i),r(i+1));
end

function f = func(eVal);
% Returns |A - eVal*I| (last element of Sturm seq.)
global C D
p = sturmSeq(C,D,eVal);
f = p(length(p));
```

EXAMPLE 9.12

Determine the three smallest eigenvalues of the 100×100 matrix

$$
A = \begin{bmatrix}
2 & -1 & 0 & \cdots & 0 \\
-1 & 2 & -1 & \cdots & 0 \\
0 & -1 & 2 & \cdots & 0 \\
\vdots & \vdots & \ddots & \ddots & \vdots \\
0 & 0 & \cdots & -1 & 2
\end{bmatrix}
$$

Solution

```
% Example 9.12 (Eigenvals. of tridiagonal matrix)
format short e
global C D
m = 3; n = 100;
D = ones(n,1)*2;
C = -ones(n-1,1);
eigenvalues = eigenvals3(C,D,m)'
```

The result is

```
>> eigenvalues =
  9.6744e-004   3.8688e-003   8.7013e-003
```

Computation of Eigenvectors

If the eigenvalues are known (approximate values will be good enough), the best means of computing the corresponding eigenvectors is the inverse power method with eigenvalue shifting. This method was discussed before, but the algorithm did not take advantage of banding. Here we present a version of the method written for symmetric tridiagonal matrices.

■ invPower3

This function is very similar to invPower listed in Art. 9.3, but executes much faster since it exploits the tridiagonal structure of the matrix.

```
function [eVal,eVec] = invPower3(c,d,s,maxIter,tol)
% Computes the eigenvalue of A =[c\d\c] closest to s and
% the associated eigenvector by the inverse power method.
% USAGE: [eVal,eVec] = invPower3(c,d,s,maxIter,tol).
% maxIter = limit on number of iterations (default is 50).
% tol = error tolerance (default is 1.0e-6).

if nargin < 5; tol = 1.0e-6; end
if nargin < 4; maxIter = 50; end
n = length(d);
e = c; d = d - s;          % Apply shift to diag. terms of A
[c,d,e] = LUdec3(c,d,e);   % Decompose A* = A - sI
x = rand(n,1);             % Seed x with random numbers
xMag = sqrt(dot(x,x)); x = x/xMag; % Normalize x
```

```
for i = 1:maxIter
    xOld = x;                       % Save current x
    x = LUsol3(c,d,e,x);            % Solve A*x = xOld
    xMag = sqrt(dot(x,x)); x = x/xMag;   % Normalize x
    xSign = sign(dot(xOld,x));      % Detect sign change of x
    x = x*xSign;
    % Check for convergence
    if sqrt(dot(xOld - x,xOld - x)) < tol
        eVal = s + xSign/xMag; eVec = x;
        return
    end
end
error('Too many iterations')
```

EXAMPLE 9.13

Compute the 10th smallest eigenvalue of the matrix A given in Example 9.12.

Solution The following program extracts the m th eigenvalue of A by the inverse power method with eigenvalue shifting:

```
Example 9.13 (Eigenvals. of tridiagonal matrix)
format short e
m = 10
n = 100;
d = ones(n,1)*2; c = -ones(n-1,1);
r = eValBrackets(c,d,m);
s =(r(m) + r(m+1))/2;
[eVal,eVec] = invPower3(c,d,s);
mth_eigenvalue = eVal
```

   The result is

```
>> m =
    10
mth_eigenvalue =
  9.5974e-002
```

EXAMPLE 9.14

Compute the three smallest eigenvalues and the corresponding eigenvectors of the matrix A in Example 9.5.

Solution

```
% Example 9.14 (Eigenvalue problem)
global C D
m = 3;
A = [11   2   3   1   4;
      2   9   3   5   2;
      3   3  15   4   3;
      1   5   4  12   4;
      4   2   3   4  17];
eVecMat = zeros(size(A,1),m);        % Init. eigenvector matrix.
A = householder(A);                  % Tridiagonalize A.
D = diag(A); C = diag(A,1);          % Extract diagonals of A.
P = householderP(A);                 % Compute tranf. matrix P.
eVals = eigenvals3(C,D,m);           % Find lowest m eigenvals.
for i = 1:m                          % Compute corresponding
    s = eVals(i)*1.0000001;          %    eigenvectors by inverse
    [eVal,eVec] = invPower3(C,D,s);  %    power method with
    eVecMat(:,i) = eVec;             %    eigenvalue shifting.
end
eVecMat = P*eVecMat;                 % Eigenvectors of orig. A.
eigenvalues = eVals'
eigenvectors = eVecMat

>> eigenvalues =
    4.8739      8.6636      10.9368
eigenvectors =
   -0.2673      0.7291       0.5058
    0.7414      0.4139      -0.3188
    0.0502     -0.4299       0.5208
   -0.5949      0.0696      -0.6029
    0.1497     -0.3278      -0.0884
```

PROBLEM SET 9.2

1. Use Gerschgorin's theorem to determine global bounds on the eigenvalues of

(a) $\mathbf{A} = \begin{bmatrix} 10 & 4 & -1 \\ 4 & 2 & 3 \\ -1 & 3 & 6 \end{bmatrix}$ (b) $\mathbf{B} = \begin{bmatrix} 4 & 2 & -2 \\ 2 & 5 & 3 \\ -2 & 3 & 4 \end{bmatrix}$

2. Use the Sturm sequence to show that

$$A = \begin{bmatrix} 5 & -2 & 0 & 0 \\ -2 & 4 & -1 & 0 \\ 0 & -1 & 4 & -2 \\ 0 & 0 & -2 & 5 \end{bmatrix}$$

   has one eigenvalue in the interval $(2, 4)$.

3. Bracket each eigenvalue of

$$A = \begin{bmatrix} 4 & -1 & 0 \\ -1 & 4 & -1 \\ 0 & -1 & 4 \end{bmatrix}$$

4. Bracket each eigenvalue of

$$A = \begin{bmatrix} 6 & 1 & 0 \\ 1 & 8 & 2 \\ 0 & 2 & 9 \end{bmatrix}$$

5. Bracket every eigenvalue of

$$A = \begin{bmatrix} 2 & -1 & 0 & 0 \\ -1 & 2 & -1 & 0 \\ 0 & -1 & 2 & -1 \\ 0 & 0 & -1 & 1 \end{bmatrix}$$

6. Tridiagonalize the matrix

$$A = \begin{bmatrix} 12 & 4 & 3 \\ 4 & 9 & 3 \\ 3 & 3 & 15 \end{bmatrix}$$

   with Householder's reduction.

7. Use Householder's reduction to transform the matrix

$$A = \begin{bmatrix} 4 & -2 & 1 & -1 \\ -2 & 4 & -2 & 1 \\ 1 & -2 & 4 & -2 \\ -1 & 1 & -2 & 4 \end{bmatrix}$$

   to tridiagonal form.

8. ■ Compute all the eigenvalues of

$$A = \begin{bmatrix} 6 & 2 & 0 & 0 & 0 \\ 2 & 5 & 2 & 0 & 0 \\ 0 & 2 & 7 & 4 & 0 \\ 0 & 0 & 4 & 6 & 1 \\ 0 & 0 & 0 & 1 & 3 \end{bmatrix}$$

9. ■ Find the smallest two eigenvalues of

$$A = \begin{bmatrix} 4 & -1 & 0 & 1 \\ -1 & 6 & -2 & 0 \\ 0 & -2 & 3 & 2 \\ 1 & 0 & 2 & 4 \end{bmatrix}$$

10. ■ Compute the three smallest eigenvalues of

$$A = \begin{bmatrix} 7 & -4 & 3 & -2 & 1 & 0 \\ -4 & 8 & -4 & 3 & -2 & 1 \\ 3 & -4 & 9 & -4 & 3 & -2 \\ -2 & 3 & -4 & 10 & -4 & 3 \\ 1 & -2 & 3 & -4 & 11 & -4 \\ 0 & 1 & -2 & 3 & -4 & 12 \end{bmatrix}$$

and the corresponding eigenvectors.

11. ■ Find the two smallest eigenvalues of the 6×6 Hilbert matrix

$$A = \begin{bmatrix} 1 & 1/2 & 1/3 & \cdots & 1/6 \\ 1/2 & 1/3 & 1/4 & \cdots & 1/7 \\ 1/3 & 1/4 & 1/5 & \cdots & 1/8 \\ \vdots & \vdots & \vdots & \ddots & \vdots \\ 1/6 & 1/7 & 1/8 & \cdots & 1/11 \end{bmatrix}$$

Recall that this matrix is ill-conditioned.

12. ■ Rewrite the function `eValBrackets` so that it will bracket the m largest eigenvalues of a tridiagonal matrix. Use this function to bracket the two largest eigenvalues of the Hilbert matrix in Prob. 11.

13. ■

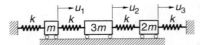

The differential equations of motion of the mass–spring system are

$$k(-2u_1 + u_2) = m\ddot{u}_1$$

$$k(u_1 - 2u_2 + u_3) = 3m\ddot{u}_2$$

$$k(u_2 - 2u_3) = 2m\ddot{u}_3$$

where $u_i(t)$ is the displacement of mass i from its equilibrium position and k is the spring stiffness. Substituting $u_i(t) = y_i \sin \omega t$, we obtain the matrix eigenvalue problem

$$\begin{bmatrix} 2 & -1 & 0 \\ -1 & 2 & -1 \\ 0 & -1 & 2 \end{bmatrix} \begin{bmatrix} y_1 \\ y_2 \\ y_3 \end{bmatrix} = \frac{m\omega^2}{k} \begin{bmatrix} 1 & 0 & 0 \\ 0 & 3 & 0 \\ 0 & 0 & 2 \end{bmatrix} \begin{bmatrix} y_1 \\ y_2 \\ y_3 \end{bmatrix}$$

Determine the circular frequencies ω and the corresponding relative amplitudes y_i of vibration.

14. ■

The figure shows n identical masses connected by springs of different stiffnesses. The equation governing free vibration of the system is $\mathbf{Au} = m\omega^2\mathbf{u}$, where ω is the circular frequency and

$$\mathbf{A} = \begin{bmatrix} k_1 + k_2 & -k_2 & 0 & 0 & \cdots & 0 \\ -k_2 & k_2 + k_3 & -k_3 & 0 & \cdots & 0 \\ 0 & -k_3 & k_3 + k_4 & -k_4 & \cdots & 0 \\ \vdots & \vdots & \ddots & \ddots & \ddots & \vdots \\ 0 & \cdots & 0 & -k_{n-1} & k_{n-1} + k_n & -k_n \\ 0 & \cdots & 0 & 0 & -k_n & k_n \end{bmatrix}$$

Given the spring stiffness array $\mathbf{k} = \begin{bmatrix} k_1 & k_2 & \cdots & k_n \end{bmatrix}^T$, write a program that computes the N lowest eigenvalues $\lambda = m\omega^2$ and the corresponding eigenvectors. Run the program with $N = 4$ and

$$\mathbf{k} = \begin{bmatrix} 400 & 400 & 400 & 0.2 & 400 & 400 & 200 \end{bmatrix}^T \text{ kN/m}$$

Note that the system is weakly coupled, k_4 being small. Do the results make sense?

15. ■

The differential equation of motion of the axially vibrating bar is

$$u'' = \frac{\rho}{E}\ddot{u}$$

where $u(x, t)$ is the axial displacement, ρ represents the mass density and E is the modulus of elasticity. The boundary conditions are $u(0, t) = u'(L, t) = 0$. Letting $u(x, t) = y(x)\sin \omega t$, we obtain

$$y'' = -\omega^2 \frac{\rho}{E}y \qquad y(0) = y'(L) = 0$$

The corresponding finite difference equations are

$$
\begin{bmatrix}
2 & -1 & 0 & 0 & \cdots & 0 \\
-1 & 2 & -1 & 0 & \cdots & 0 \\
0 & -1 & 2 & -1 & \cdots & 0 \\
\vdots & \vdots & \ddots & \ddots & \ddots & \vdots \\
0 & 0 & \cdots & -1 & 2 & -1 \\
0 & 0 & \cdots & 0 & -1 & 1
\end{bmatrix}
\begin{bmatrix}
y_1 \\ y_2 \\ y_3 \\ \vdots \\ y_{n-1} \\ y_n
\end{bmatrix}
= \left(\frac{\omega L}{n}\right)^2 \frac{\rho}{E}
\begin{bmatrix}
y_1 \\ y_2 \\ y_3 \\ \vdots \\ y_{n-1} \\ y_n/2
\end{bmatrix}
$$

(a) If the standard form of these equations is $\mathbf{Hz} = \lambda\mathbf{z}$, write down $\mathbf{H}$ and the transformation matrix $\mathbf{P}$ in $\mathbf{y} = \mathbf{Pz}$. (b) Compute the lowest circular frequency of the bar with $n = 10, 100$ and 1000 utilizing the module `inversePower3`. *Note*: the analytical solution is $\omega_1 = \pi\sqrt{E/\rho}/(2L)$.

16. ∎

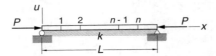

The simply supported column is resting on an elastic foundation of stiffness k (N/m per meter length). An axial force P acts on the column. The differential equation and the boundary conditions for the lateral displacement u are

$$u^{(4)} + \frac{P}{EI}u'' + \frac{k}{EI}u = 0$$

$$u(0) = u''(0) = u(L) = u''(L) = 0$$

Using the mesh shown, the finite difference approximation of these equations is

$$(5 + \alpha)u_1 - 4u_2 + u_3 = \lambda(2u_1 - u_2)$$

$$-4u_1 + (6 + \alpha)u_2 - 4u_3 + u_4 = \lambda(-u_1 + 2u_2 + u_3)$$

$$u_1 - 4u_2 + (6 + \alpha)u_3 - 4u_4 + u_5 = \lambda(-u_2 + 2u_3 - u_4)$$

$$\vdots$$

$$u_{n-3} - 4u_{n-2} + (6+\alpha)u_{n-1} - 4u_n = \lambda(-u_{n-2} + 2u_{n-1} - u_n)$$

$$u_{n-2} - 4u_{n-1} + (5+\alpha)u_n = \lambda(-u_{n-1} + 2u_n)$$

where

$$\alpha = \frac{kh^4}{EI} = \frac{1}{(n+1)^4}\frac{kL^4}{EI} \qquad \lambda = \frac{Ph^2}{EI} = \frac{1}{(n+1)^2}\frac{PL^2}{EI}$$

Write a program that computes the lowest three buckling loads P and the corresponding mode shapes. Run the program with $kL^4/(EI) = 1000$ and $n = 25$.

17. ■ Find smallest five eigenvalues of the 20×20 matrix

$$\mathbf{A} = \begin{bmatrix} 2 & 1 & 0 & 0 & \cdots & 0 & 1 \\ 1 & 2 & 1 & 0 & \cdots & 0 & 0 \\ 0 & 1 & 2 & 1 & \cdots & 0 & 0 \\ \vdots & \vdots & \ddots & \ddots & \ddots & \vdots & \vdots \\ 0 & 0 & \cdots & 1 & 2 & 1 & 0 \\ 0 & 0 & \cdots & 0 & 1 & 2 & 1 \\ 1 & 0 & \cdots & 0 & 0 & 1 & 2 \end{bmatrix}$$

Note: this is a difficult matrix that has many pairs of double eigenvalues.

MATLAB Functions

MATLAB's function for solving eigenvalue problems is `eig`. Its usage for the standard eigenvalue problem $\mathbf{Ax} = \lambda\mathbf{x}$ is

`eVals = eig(A)` returns the eigenvalues of the matrix A (A can be unsymmetric).

`[X,D] = eig(A)` returns the eigenvector matrix X and the diagonal matrix D that contains the eigenvalues on its diagonal; that is, `eVals = diag(D)`.

For the nonstandard form $\mathbf{Ax} = \lambda\mathbf{Bx}$, the calls are

`eVals = eig(A,B)`

`[X,D] = eig(A,B)`

The method of solution is based on *Schur's factorization*: $\mathbf{PAP}^T = \mathbf{T}$, where $\mathbf{P}$ and $\mathbf{T}$ are unitary and triangular matrices, respectively. Schur's factorization is not covered in this text.

10 Introduction to Optimization

> Find **x** that minimizes $F(\mathbf{x})$ subject to $g(\mathbf{x}) = 0$, $h(\mathbf{x}) \geq 0$

10.1 Introduction

Optimization is the term often used for minimizing or maximizing a function. It is sufficient to consider the problem of minimization only; maximization of $F(\mathbf{x})$ is achieved by simply minimizing $-F(\mathbf{x})$. In engineering, optimization is closely related to design. The function $F(\mathbf{x})$, called the *merit function* or *objective function*, is the quantity that we wish to keep as small as possible, such as cost or weight. The components of **x**, known as the *design variables*, are the quantities that we are free to adjust. Physical dimensions (lengths, areas, angles, etc.) are common examples of design variables.

Optimization is a large topic with many books dedicated to it. The best we can do in limited space is to introduce a few basic methods that are good enough for problems that are reasonably well behaved and don't involve too many design variables. By omitting the more sophisticated methods, we may actually not miss all that much. All optimization algorithms are unreliable to a degree—any one of them may work on one problem and fail on another. As a rule of thumb, by going up in sophistication we gain computational efficiency, but not necessarily reliability.

The algorithms for minimization are iterative procedures that require starting values of the design variables **x**. If $F(\mathbf{x})$ has several local minima, the initial choice of **x** determines which of these will be computed. There is no guaranteed way of finding the global optimal point. One suggested procedure is to make several computer runs using different starting points and pick the best result.

More often than not, the design is also subjected to restrictions, or *constraints*, which may have the form of equalities or inequalities. As an example, take the minimum weight design of a roof truss that has to carry a certain loading. Assume that the layout of the members is given, so that the design variables are the cross-sectional areas of the members. Here the design is dominated by inequality constraints that consist of prescribed upper limits on the stresses and possibly the displacements.

The majority of available methods are designed for *unconstrained optimization*, where no restrictions are placed on the design variables. In these problems the minima, if they exit, are stationary points (points where gradient vector of $F(\mathbf{x})$ vanishes). In the more difficult problem of *constrained optimization* the minima are usually located where the $F(\mathbf{x})$ surface meets the constraints. There are special algorithms for constrained optimization, but they are not easily accessible due to their complexity and specialization. One way to tackle a problem with constraints is to use an unconstrained optimization algorithm, but modify the merit function so that any violation of constraints is heavily penalized.

Consider the problem of minimizing $F(\mathbf{x})$ where the design variables are subject to the constraints

$$g_i(\mathbf{x}) = 0, \quad i = 1, 2, \ldots, M \tag{10.1a}$$

$$h_j(\mathbf{x}) \leq 0, \quad j = 1, 2, \ldots, N \tag{10.1b}$$

We choose the new merit function be

$$F^*(\mathbf{x}) = F(\mathbf{x}) + \lambda P(\mathbf{x}) \tag{10.2a}$$

where

$$P(\mathbf{x}) = \sum_{i=1}^{M} [g_i(x)]^2 + \sum_{j=1}^{N} \left\{ \max \left[0, h_j(\mathbf{x}) \right] \right\}^2 \tag{10.2b}$$

is the *penalty function* and λ is a multiplier. The function max(a, b) returns the larger of a and b. It is evident that $P(\mathbf{x}) = 0$ if no constraints are violated. Violation of a constraint imposes a penalty proportional to the square of the violation. Hence the minimization algorithm tends to avoid the violations, the degree of avoidance being dependent on the magnitude of λ. If λ is small, optimization will proceed faster because there is more "space" in which the procedure can operate, but there may be significant violation of constraints. On the other hand, a large λ can result in a poorly conditioned procedure, but the constraints will be tightly enforced. It is advisable to run the optimization program with λ that is on the small side. If the results show unacceptable constraint violation, increase λ and run the program again, starting with the results of the previous run.

An optimization procedure may also become ill-conditioned when the constraints have widely different magnitudes. This problem can be alleviated by *scaling* the offending constraints; that is, multiplying the constraint equations by suitable constants.

10.2 Minimization Along a Line

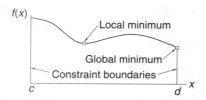

Figure 10.1. Example of local and global minima.

Consider the problem of minimizing a function $f(x)$ of a single variable x with the constraints $c \leq x \leq d$. A hypothetical plot of the function is shown in Fig. 10.1. There are two minimum points: a stationary point characterized by $f'(x) = 0$ that represents a local minimum, and a global minimum at the constraint boundary. It appears that finding the global minimum is simple. All the stationary points could be located by finding the roots of $df/dx = 0$, and each constraint boundary may be checked for a global minimum by evaluating $f(c)$ and $f(d)$. Then why do we need an optimization algorithm? We need it if $f(\mathbf{x})$ is difficult or impossible to differentiate; for example, if f represents a complex computer algorithm.

Bracketing

Before a minimization algorithm can be entered, the minimum point must be bracketed. The procedure of bracketing is simple: start with an initial value of x_0 and move *downhill* computing the function at $x_1, x_2, x_3, \ldots$ until we reach the point x_n where $f(x)$ increases for the first time. The minimum point is now bracketed in the interval (x_{n-2}, x_n). What should the step size $h_i = x_{i+1} - x_i$ be? It is not a good idea have a constant h_i since it often results in too many steps. A more efficient scheme is to increase the size with every step, the goal being to reach the minimum quickly, even if the resulting bracket is wide. We chose to increase the step size by a constant factor; that is, we use $h_{i+1} = ch_i, c > 1$.

Golden Section Search

The golden section search is the counterpart of bisection used in finding roots of equations. Suppose that the minimum of $f(x)$ has been bracketed in the interval (a, b) of length h. To telescope the interval, we evaluate the function at $x_1 = b - Rh$ and $x_2 = a + Rh$, as shown in Fig. 10.2(a). The constant R will be determined shortly. If $f_1 > f_2$ as indicated in the figure, the minimum lies in (x_1, b); otherwise it is located in (a, x_2).

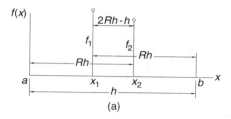

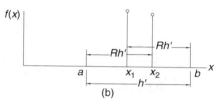

Figure 10.2. Golden section telescoping.

Assuming that $f_1 > f_2$, we set $a \leftarrow x_1$ and $x_1 \leftarrow x_2$, which yields a new interval (a, b) of length $h' = Rh$, as illustrated in Fig. 10.2(b). To carry out the next telescoping operation we evaluate the function at $x_2 = a + Rh'$ and repeat the process.

The procedure works only if Figs. 10.1(a) and (b) are similar; i.e., if the same constant R locates x_1 and x_2 in both figures. Referring to Fig. 10.2(a), we note that $x_2 - x_1 = 2Rh - h$. The same distance in Fig. 10.2(b) is $x_1 - a = h' - Rh'$. Equating the two, we get

$$2Rh - h = h' - Rh'$$

Substituting $h' = Rh$ and cancelling h yields

$$2R - 1 = R(1 - R)$$

the solution of which is the *golden ratio*[21]:

$$R = \frac{-1 + \sqrt{5}}{2} = 0.618\,033\,989\ldots \tag{10.3}$$

Note that each telescoping decreases the interval containing the minimum by the factor R, which is not as good as the factor of 0.5 in bisection. However, the golden search method achieves this reduction with *one function evaluation*, whereas two evaluations would be needed in bisection.

The number of telescopings required to reduce h from $|b - a|$ to an error tolerance ε is given by

$$|b - a| R^n = \varepsilon$$

---

[21] R is the ratio of the sides of a "golden rectangle," considered by ancient Greeks to have the perfect proportions.

which yields

$$n = \frac{\ln(\varepsilon/|b-a|)}{\ln R} = -2.078\,087 \ln \frac{\varepsilon}{|b-a|} \tag{10.4}$$

■ goldBracket

This function contains the bracketing algorithm. For the factor that multiplies successive search intervals we chose $c = 1 + R$.

```
function [a,b] = goldBracket(func,x1,h)
% Brackets the minimum point of f(x).
% USAGE: [a,b] = goldBracket(func,xStart,h)
% INPUT:
% func   = handle of function that returns f(x).
% x1     = starting value of x.
% h      = initial step size used in search.
% OUTPUT:
% a, b = limits on x at the minimum point.

c = 1.618033989;
f1 = feval(func,x1);
x2 = x1 + h; f2 = feval(func,x2);
% Determine downhill direction & change sign of h if needed.
if f2 > f1
    h = -h;
    x2 = x1 + h; f2 = feval(func,x2);
    % Check if minimum is between x1 - h and x1 + h
    if f2 > f1
        a = x2; b = x1 - h; return
    end
end
% Search loop
for i = 1:100
    h = c*h;
    x3 = x2 + h; f3 = feval(func,x3);
    if f3 > f2
        a = x1; b = x3; return
    end
    x1 = x2; f1 = f2; x2 = x3; f2 = f3;
end
error('goldbracket did not find minimum')
```

■ goldSearch

This function implements the golden section search algorithm.

```
function [xMin,fMin] = goldSearch(func,a,b,tol)
% Golden section search for the minimum of f(x).
% The minimum point must be bracketed in a <= x <= b.
% USAGE: [fMin,xMin] = goldSearch(func,xStart,h)
% INPUT:
% func    = handle of function that returns f(x).
% a, b    = limits of the interval containing the minimum.
% tol     = error tolerance (default is 1.0e-6).
% OUTPUT:
% fMin = minimum value of f(x).
% xMin = value of x at the minimum point.

if nargin < 4; tol = 1.0e-6; end
nIter = ceil(-2.078087*log(tol/abs(b-a)));
R = 0.618033989;
C = 1.0 - R;
% First telescoping
x1 = R*a + C*b;
x2 = C*a + R*b;
f1 = feval(func,x1);
f2 = feval(func,x2);
% Main loop
for i =1:nIter
    if f1 > f2
        a = x1; x1 = x2; f1 = f2;
        x2 = C*a + R*b;
        f2 = feval(func,x2);
    else
        b = x2; x2 = x1; f2 = f1;
        x1 = R*a + C*b;
        f1 = feval(func,x1);
    end
end
if f1 < f2; fMin = f1; xMin = x1;
else; fMin = f2; xMin = x2;
end
```

EXAMPLE 10.1

Use goldSearch to find x that minimizes

$$f(x) = 1.6x^3 + 3x^2 - 2x$$

subject to the constraint $x \geq 0$. Compare the result with the analytical solution.

Solution This is a constrained minimization problem. Either the minimum of $f(x)$ is a stationary point in $x \geq 0$, or it is located at the constraint boundary $x = 0$. We handle the constraint with the penalty function method by minimizing $f(x) + \lambda \, [\min(0, x)]^2$.

Starting at $x = 1$ and choosing $h = 0.1$ for the first step size in goldBracket (both choices being rather arbitrary), we arrive at the following program:

```
% Example 10.1 (golden section minimization)
x = 1.0; h = 0.1;
[a,b] = goldBracket(@fex10_1,x,h);
[xMin,fMin] = goldSearch(@fex10_1,a,b)
```

The function to be minimized is

```
function y = fex10_1(x)
% Function used in Example 10.1.
lam = 1.0;          % Penalty function multiplier
c = min(0.0,x);   % Constraint penalty equation
y = 1.6*x^3 + 3.0*x^2 - 2.0*x + lam*c^2;
```

The output from the program is

```
>> xMin =
    0.2735
fMin =
   -0.2899
```

Since the minimum was found to be a stationary point, the constraint was not active. Therefore, the penalty function was superfluous, but we did not know that at the beginning.

The locations of stationary points are obtained analytically by solving

$$f'(x) = 4.8x^2 + 6x - 2 = 0$$

The positive root of this equation is $x = 0.273\,494$. As this is the only positive root, there are no other stationary points in $x \geq 0$ that we must check out. The only other possible location of a minimum is the constraint boundary $x = 0$. But here $f(0) = 0$

is larger than the function at the stationary point, leading to the conclusion that the global minimum occurs at $x = 0.273\,494$.

EXAMPLE 10.2

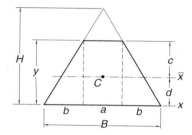

The trapezoid shown is the cross section of a beam. It is formed by removing the top from a triangle of base $B = 48$ mm and height $H = 60$ mm. The problem is to find the height y of the trapezoid that maximizes the section modulus

$$S = I_{\bar{x}}/c$$

where $I_{\bar{x}}$ is the second moment of the cross-sectional area about the axis that passes through the centroid C of the cross section. By optimizing the section modulus, we minimize the maximum bending stress $\sigma_{\max} = M/S$ in the beam, M being the bending moment.

Solution Considering the area of the trapezoid as a composite of a rectangle and two triangles, we find the section modulus through the following sequence of computations:

Base of rectangle	$a = B(H - y)/H$
Base of triangle	$b = (B - a)/2$
Area	$A = (B + a)y/2$
First moment of area about x-axis	$Q_x = (ay)y/2 + 2(by/2)y/3$
Location of centroid	$d = Q_x/A$
Distance involved in S	$c = y - d$
Second moment of area about x-axis	$I_x = ay^3/3 + 2(by^3/12)$
Parallel axis theorem	$I_{\bar{x}} = I_x - Ad^2$
Section modulus	$S = I_{\bar{x}}/c$

We could use the formulas in the table to derive S as an explicit function of y, but that would involve a lot of error-prone algebra and result in an overly complicated expression. It makes more sense to let the computer do the work.

The program we used is listed below. As we wish to maximize S with a minimization algorithm, the merit function is $-S$. There are no constraints in this problem.

```
% Example 10.2 (root finding with golden section)
yStart = 60.0; h = 1.0;
[a,b] = goldBracket(@fex10_2,yStart,h);
[yopt,Sopt] = goldSearch(@fex10_2,a,b);
fprintf('optimal y = %7.4f\n',yopt)
fprintf('optimal S = %7.2f',-Sopt)
```

The function that computes the section modulus is

```
function S = fex10_2(y)
% Function used in Example 10.2
B = 48.0; H = 60.0;
a = B*(H - y)/H; b = (B - a)/2.0;
A = (B + a)*y/2.0;
Q = (a*y^2)/2.0 + (b*y^2)/3.0;
d = Q/A; c = y - d;
I = (a*y^3)/3.0 + (b*y^3)/6.0;
Ibar = I - A*d^2; S = -Ibar/c
```

Here is the output:

```
optimal y = 52.1763
optimal S = 7864.43
```

The section modulus of the original triangle is 7200; thus the optimal section modulus is a 9.2% improvement over the triangle.

10.3   Conjugate Gradient Methods

Introduction

We now look at optimization in n-dimensional design space. The objective is to minimize $F(\mathbf{x})$, where the components of $\mathbf{x}$ are the n independent design variables. One way to tackle the problem is to use a succession of one-dimensional minimizations to close in on the optimal point. The basic strategy is

- Choose a point $\mathbf{x}_0$ in the design space.
- loop with $i = 1, 2, 3, \ldots$

    Choose a vector $\mathbf{v}_i$.
    Minimize $F(\mathbf{x})$ along the line through $\mathbf{x}_{i-1}$ in the direction of $\mathbf{v}_i$. Let the minimum point be $\mathbf{x}_i$.
    if $|\mathbf{x}_i - \mathbf{x}_{i-1}| < \varepsilon$ exit loop
- end loop

The minimization along a line can be accomplished with any one-dimensional optimization algorithm (such as the golden section search). The only question left open is how to choose the vectors $\mathbf{v}_i$.

Conjugate Directions

Consider the quadratic function

$$F(\mathbf{x}) = c - \sum_i b_i x_i + \frac{1}{2} \sum_i \sum_j A_{ij} x_i x_j$$

$$= c - \mathbf{b}^T \mathbf{x} + \frac{1}{2} \mathbf{x}^T \mathbf{A} \mathbf{x} \tag{10.5}$$

Differentiation with respect to x_i yields

$$\frac{\partial F}{\partial x_i} = -b_i + \sum_j A_{ij} x_j$$

which can be written in vector notation as

$$\nabla F = -\mathbf{b} + \mathbf{A}\mathbf{x} \tag{10.6}$$

where ∇F is the *gradient* of F.

Now consider the change in the gradient as we move from point $\mathbf{x}_0$ in the direction of a vector $\mathbf{u}$. The motion takes place along the line

$$\mathbf{x} = \mathbf{x}_0 + s\mathbf{u}$$

where s is the distance moved. Substitution into Eq. (10.6) yields the expression for the gradient along u:

$$\nabla F|_{\mathbf{x}_0 + s\mathbf{u}} = -\mathbf{b} + \mathbf{A}(\mathbf{x}_0 + s\mathbf{u}) = \nabla F|_{\mathbf{x}_0} + s\mathbf{A}\mathbf{u}$$

Note that the change in the gradient is $s\,\mathbf{Au}$. If this change is perpendicular to a vector $\mathbf{v}$; that is, if

$$\mathbf{v}^T\mathbf{Au} = 0 \qquad\qquad (10.7)$$

the directions of $\mathbf{u}$ and $\mathbf{v}$ are said to be mutually *conjugate* (noninterfering). The implication is that once we have minimized $F(\mathbf{x})$ in the direction of $\mathbf{v}$, we can move along $\mathbf{u}$ without ruining the previous minimization.

For a quadratic function of n independent variables it is possible to construct n mutually conjugate directions. Therefore, it would take precisely n line minimizations along these directions to reach the minimum point. If $F(\mathbf{x})$ is not a quadratic function, Eq. (10.5) can be treated as a local approximation of the merit function, obtained by truncating the Taylor series expansion of $F(\mathbf{x})$ about $\mathbf{x}_0$ (see Appendix A1):

$$F(\mathbf{x}) \approx F(\mathbf{x}_0) + \nabla F(\mathbf{x}_0)(\mathbf{x} - \mathbf{x}_0) + \frac{1}{2}(\mathbf{x} - \mathbf{x}_0)^T \mathbf{H}(\mathbf{x}_0)(\mathbf{x} - \mathbf{x}_0)$$

Now the conjugate directions based on the quadratic form are only approximations, valid in the close vicinity of $\mathbf{x}_0$. Consequently, it would take several cycles of n line minimizations to reach the optimal point.

The various conjugate gradient methods use different techniques for constructing conjugate directions. The so-called *zero-order methods* work with $F(\mathbf{x})$ only, whereas the *first-order methods* utilize both $F(\mathbf{x})$ and ∇F. The first-order methods are computationally more efficient, of course, but the input of ∇F (if it is available at all) can be very tedious.

Powell's Method

Powell's method is a zero-order method, requiring the evaluation of $F(\mathbf{x})$ only. If the problem involves n design variables, the basic algorithm is

- Choose a point $\mathbf{x}_0$ in the design space.
- Choose the starting vectors $\mathbf{v}_i,\ i = 1, 2, \ldots, n$ (the usual choice is $\mathbf{v}_i = \mathbf{e}_i$, where $\mathbf{e}_i$ is the unit vector in the x_i-coordinate direction).
- cycle

  do with $i = 1, 2, \ldots, n$

  > Minimize $F(\mathbf{x})$ along the line through $\mathbf{x}_{i-1}$ in the direction of $\mathbf{v}_i$. Let the minimum point be $\mathbf{x}_i$.

  end do

  $\mathbf{v}_{n+1} \leftarrow \mathbf{x}_0 - \mathbf{x}_n$ (this vector is conjugate to $\mathbf{v}_{n+1}$ produced in the previous loop)
  Minimize $F(\mathbf{x})$ along the line through $\mathbf{x}_0$ in the direction of $\mathbf{v}_{n+1}$. Let the minimum point be $\mathbf{x}_{n+1}$.
  if $|\mathbf{x}_{n+1} - \mathbf{x}_0| < \varepsilon$ exit loop

do with $i = 1, 2, \ldots, n$

    $\mathbf{v}_i \leftarrow \mathbf{v}_{i+1}$ (v_1 is discarded, the other vectors are reused)

end do

- end cycle

Powell demonstrated that the vectors $\mathbf{v}_{n+1}$ produced in successive cycles are mutually conjugate, so that the minimum point of a quadratic surface is reached in precisely n cycles. In practice, the merit function is seldom quadratic, but as long as it can be approximated locally by Eq. (10.5), Powell's method will work. Of course, it usually takes more than n cycles to arrive at the minimum of a nonquadratic function. Note that it takes n line minimizations to construct each conjugate direction.

Figure 10.3(a) illustrates one typical cycle of the method in a two dimensional design space ($n = 2$). We start with point $\mathbf{x}_0$ and vectors $\mathbf{v}_1$ and $\mathbf{v}_2$. Then we find the distance s_1 that minimizes $F(\mathbf{x}_0 + s\mathbf{v}_1)$, finishing up at point $\mathbf{x}_1 = \mathbf{x}_0 + s_1\mathbf{v}_1$. Next, we determine s_2 that minimizes $F(\mathbf{x}_1 + s\mathbf{v}_2)$, which takes us to $\mathbf{x}_2 = \mathbf{x}_1 + s_2\mathbf{v}_2$. The last search direction is $\mathbf{v}_3 = \mathbf{x}_2 - \mathbf{x}_0$. After finding s_3 by minimizing $F(\mathbf{x}_0 + s\mathbf{v}_3)$ we get to $\mathbf{x}_3 = \mathbf{x}_0 + s_3\mathbf{v}_3$, completing the cycle.

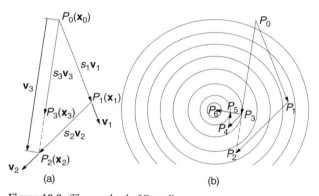

Figure 10.3. The method of Powell.

Figure 10.3(b) shows the moves carried out in two cycles superimposed on the contour map of a quadratic surface. As explained before, the first cycle starts at point P_0 and ends up at P_3. The second cycle takes us to P_6, which is the optimal point. The directions P_0P_3 and P_3P_6 are mutually conjugate.

Powell's method does have a major flaw that has to be remedied—if $F(\mathbf{x})$ is not a quadratic, the algorithm tends to produce search directions that gradually become linearly dependent, thereby ruining the progress towards the minimum. The source of the problem is the automatic discarding of $\mathbf{v}_1$ at the end of each cycle. It has been suggested that it is better to throw out the direction that resulted in the *largest decrease* of $F(\mathbf{x})$, a policy that we adopt. It seems counterintuitive to discard the best direction, but it is likely to be close to the direction added in the next cycle, thereby contributing to linear dependence. As a result of the change, the search directions cease to be mutually conjugate, so that a quadratic form is not minimized in *n* cycles any more. This is not a significant loss since in practice $F(\mathbf{x})$ is seldom a quadratic anyway.

Powell suggested a few other refinements to speed up convergence. Since they complicate the bookkeeping considerably, we did not implement them.

■ powell

The algorithm for Powell's method is listed below. It utilizes two arrays: df contains the decreases of the merit function in the first *n* moves of a cycle, and the matrix u stores the corresponding direction vectors $\mathbf{v}_i$ (one vector per column).

```
function [xMin,fMin,nCyc] = powell(h,tol)
% Powell's method for minimizing f(x1,x2,...,xn).
% USAGE: [xMin,fMin,nCyc] = powell(h,tol)
% INPUT:
% h   = initial search increment (default = 0.1).
% tol = error tolerance (default = 1.0e-6).
% GLOBALS (must be declared GLOBAL in calling program):
% X = starting point
% FUNC = handle of function that returns f.
% OUTPUT:
% xMin = minimum point
% fMin = miminum value of f
% nCyc = number of cycles to convergence

global X FUNC V
if nargin < 2; tol = 1.0e-6; end
if nargin < 1; h = 0.1;  end
if size(X,2) > 1; X = X'; end % X must be column vector
n = length(X);    % Number of design variables
df = zeros(n,1);  % Decreases of f stored here
u = eye(n);       % Columns of u store search directions V
for j = 1:30      % Allow up to 30 cycles
    xOld = X;
    fOld = feval(FUNC,xOld);
    % First n line searches record the decrease of f
    for i = 1:n
        V = u(1:n,i);
        [a,b] = goldBracket(@fLine,0.0,h);
```

```
        [s,fMin] = goldSearch(@fLine,a,b);
        df(i) = fOld - fMin;
        fOld = fMin;
        X = X + s*V;
    end
    % Last line search in the cycle
    V = X - xOld;
    [a,b] = goldBracket(@fLine,0.0,h);
    [s,fMin] = goldSearch(@fLine,a,b);
    X = X + s*V;
    % Check for convergence
    if sqrt(dot(X-xOld,X-xOld)/n) < tol
        xMin = X; nCyc = j; return
    end
    % Identify biggest decrease of f & update search
    % directions
    iMax = 1; dfMax = df(1);
    for i = 2:n
        if df(i) > dfMax
            iMax = i; dfMax = df(i);
        end
    end
    for i = iMax:n-1
        u(1:n,i) = u(1:n,i+1);
    end
    u(1:n,n) = V;
end
error('Powell method did not converge')

function z = fLine(s) % F in the search direction V
global X FUNC V
z = feval(FUNC,X+s*V);
```

EXAMPLE 10.3

Find the minimum of the function[22]

$$F = 100(y - x^2)^2 + (1 - x)^2$$

[22] From Shoup, T. E., and Mistree, F., *Optimization Methods with Applications for Personal Computers,* Prentice-Hall, 1987.

with Powell's method starting at the point $(-1, 1)$. This function has an interesting topology. The minimum value of F occurs at the point $(1, 1)$. As seen in the figure, there is a hump between the starting and minimum points which the algorithm must negotiate.

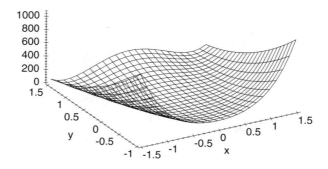

Solution The program that solves this unconstrained optimization problem is

```
% Example 10.3 (Powell's method of minimization)
global X FUNC
FUNC = @fex10_3;
X = [-1.0; 1.0];
[xMin,fMin,numCycles] = powell
```

Note that `powell` receives X and the function handle `FUNC` as `global` variables. The routine for the function to be minimized is

```
function y = fex10_3(X)
y = 100.0*(X(2) - X(1)^2)^2 + (1.0 -X(1))^2;
```

Here are the results:

```
>> xMin =
     1.0000
     1.0000
fMin =
   1.0072e-024
numCycles =
    12
```

EXAMPLE 10.4

Use `powell` to determine the smallest distance from the point $(5, 8)$ to the curve $xy = 5$.

Solution This is a constrained optimization problem: minimize $F(x, y) = (x - 5)^2 + (y - 8)^2$ (the square of the distance) subject to the equality constraint $xy - 5 = 0$. The following program uses Powell's method with penalty function:

```
% Example 10.4 (Powell's method of minimization)
global X FUNC
FUNC = @fex10_4;
X = [1.0; 5.0];
[xMin,fMin,nCyc] = powell;
fprintf('Intersection point = %8.5f %8.5f\n',X(1),X(2))
xy = X(1)*X(2);
fprintf('Constraint x*y = %8.5f\n',xy)
dist = sqrt((X(1) - 5.0)^2 + (X(2) - 8.0)^2);
fprintf('Distance = %8.5f\n',dist)
fprintf('Number of cycles = %2.0f',nCyc)
```

The penalty is incorporated in the M-file of the function to be minimized:

```
function y = fex10_4(X)
% Function used in Example 10.4
lam = 1.0;                        % Penalty multiplier
c = X(1)*X(2) - 5.0;              % Constraint equation
distSq = (X(1) - 5.0)^2 + (X(2) - 8.0)^2;
y = distSq + lam*c^2;
```

As mentioned before, the value of the penalty function multiplier λ (called `lam` in the program) can have profound effects on the result. We chose $\lambda = 1$ (as shown in the listing of `fex10_4`) with the following result:

```
>> Intersection point =  0.73307  7.58776
Constraint x*y =  5.56234
Distance =  4.28680
Number of cycles =  7
```

The small value of λ favored speed of convergence over accuracy. Since the violation of the constraint $xy = 5$ is clearly unacceptable, we ran the program again with

$\lambda = 10\,000$ and changed the starting point to $(0.733\,07, 7.587\,76)$, the end point of the first run. The results shown below are now acceptable.

```
>>Intersection point =   0.65561   7.62654
Constraint x*y =   5.00006
Distance =   4.36041
Number of cycles =   4
```

Could we have used $\lambda = 10\,000$ in the first run? In this case we would be lucky and obtain the minimum in 17 cycles. Hence we save only six cycles by using two runs. However, a large λ often causes the algorithm to hang up, so that it generally wise to start with a small λ.

Fletcher–Reeves Method

Let us assume again that the merit function has the quadratic form in Eq. (10.5). Given a direction $\mathbf{v}$, it took Powell's method n line minimizations to construct a conjugate direction. We can reduce this to a single line minimization with a first-order method. Here is the procedure, known as the Fletcher–Reeves method:

- Choose a starting point $\mathbf{x}_0$.
- $\mathbf{g}_0 \leftarrow -\nabla F(\mathbf{x}_0)$
- $\mathbf{v}_0 \leftarrow \mathbf{g}_0$ (lacking a previous search direction, we choose the steepest descent).
- loop with $i = 0, 1, 2, \ldots$

    Minimize $F(\mathbf{x})$ along $\mathbf{v}_i$; let the minimum point be $\mathbf{x}_{i+1}$.

    $\mathbf{g}_{i+1} \leftarrow -\nabla F(\mathbf{x}_{i+1})$.

    if $\left|\mathbf{g}_{i+1}\right| < \varepsilon$ or $|F(\mathbf{x}_{i+1}) - F(\mathbf{x}_i)| < \varepsilon$ exit loop (convergence criterion).

    $\gamma \leftarrow (\mathbf{g}_{i+1} \cdot \mathbf{g}_{i+1})/(\mathbf{g}_i \cdot \mathbf{g}_i)$.

    $\mathbf{v}_{i+1} \leftarrow \mathbf{g}_{i+1} + \gamma \mathbf{v}_i$.

- end loop

It can be shown that $\mathbf{v}_i$ and $\mathbf{v}_{i+1}$ are mutually conjugate; that is, they satisfy the relationship $\mathbf{v}_i^T \mathbf{A} \mathbf{v}_{i+1} = 0$. Also $\mathbf{g}_i \cdot \mathbf{g}_{i+1} = 0$.

The Fletcher–Reeves method will find the minimum of a quadratic function in n iterations. If $F(\mathbf{x})$ is not quadratic, it is necessary to restart the process after every n iterations. A variant of the Fletcher–Reeves method replaces the expression for γ by

$$\gamma = \frac{(\mathbf{g}_{i+1} - \mathbf{g}_i) \cdot \mathbf{g}_{i+1}}{\mathbf{g}_i \cdot \mathbf{g}_i} \tag{10.6}$$

For a quadratic $F(\mathbf{x})$ this change makes no difference since $\mathbf{g}_i$ and $\mathbf{g}_{i+1}$ are orthogonal. However, for merit functions that are not quadratic, Eq. (10.6) is claimed to eliminate the need for a restart after n iterations.

■ fletcherReeves

```
function [xMin,fMin,nCyc] = fletcherReeves(h,tol)
% Fletcher-Reeves method for minimizing f(x1,x2,...,xn).
% USAGE: [xMin,fMin,nCyc] = fletcherReeves(h,tol)
% INPUT:
% h   = initial search increment (default = 0.1).
% tol = error tolerance (default = 1.0e-6).
% GLOBALS (must be declared GLOBAL in calling program):
% X     = starting point.
% FUNC  = handle of function that returns F.
% DFUNC = handle of function that returns grad(F),
% OUTPUT:
% xMin = minimum point.
% fMin = miminum value of f.
% nCyc = number of cycles to convergence.

global X FUNC DFUNC V
if nargin < 2; tol = 1.0e-6; end
if nargin < 1; h = 0.1; end
if size(X,2) > 1; X = X'; end  % X must be column vector
n = length(X);                 % Number of design variables
g0 = -feval(DFUNC,X);
V = g0;
for i = 1:50
    [a,b] = goldBracket(@fLine,0.0,h);
    [s,fMin] = goldSearch(@fLine,a,b);
    X = X + s*V;
    g1 = -feval(DFUNC,X);
    if sqrt(dot(g1,g1)) <= tol
        xMin = X; nCyc = i; return
    end
    gamma = dot((g1 - g0),g1)/dot(g0,g0);
    V = g1 + gamma*V;
    g0 = g1;
end
error('Fletcher-Reeves method did not converge')

function z = fLine(s) % F in the search direction V
global X FUNC V
z = feval(FUNC,X+s*V);
```

EXAMPLE 10.5

Use the Fletcher–Reeves method to locate the minimum of

$$F(\mathbf{x}) = 10x_1^2 + 3x_2^2 - 10x_1x_2 + 2x_1$$

Start with $\mathbf{x}_0 = \begin{bmatrix} 0 & 0 \end{bmatrix}^T$.

Solution Since $F(\mathbf{x})$ is quadratic, we need only two iterations. The gradient of F is

$$\nabla F(\mathbf{x}) = \begin{bmatrix} 20x_1 - 10x_2 + 2 \\ -10x_1 + 6x_2 \end{bmatrix}$$

First iteration:

$$\mathbf{g}_0 = -\nabla F(\mathbf{x}_0) = \begin{bmatrix} -2 \\ 0 \end{bmatrix} \qquad \mathbf{v}_0 = \mathbf{g}_0 = \begin{bmatrix} -2 \\ 0 \end{bmatrix} \qquad \mathbf{x}_0 + s\mathbf{v}_0 = \begin{bmatrix} -2s \\ 0 \end{bmatrix}$$

$$f(s) = F(\mathbf{x}_0 + s\mathbf{v}_0) = 10(2s)^2 + 3(0)^2 - 10(-2s)(0) + 2(-2s)$$
$$= 40s^2 - 4s$$

$$f'(s) = 80s - 4 = 0 \qquad s = 0.05$$

$$\mathbf{x}_1 = \mathbf{x}_0 + s\mathbf{v}_0 = \begin{bmatrix} 0 \\ 0 \end{bmatrix} + 0.05 \begin{bmatrix} -2 \\ 0 \end{bmatrix} = \begin{bmatrix} -0.1 \\ 0 \end{bmatrix}$$

Second iteration:

$$\mathbf{g}_1 = -\nabla F(\mathbf{x}_1) = \begin{bmatrix} -20(-0.1) + 10(0) - 2 \\ 10(-0.1) - 6(0) \end{bmatrix} = \begin{bmatrix} 0 \\ -1.0 \end{bmatrix}$$

$$\gamma = \frac{\mathbf{g}_1 \cdot \mathbf{g}_1}{\mathbf{g}_0 \cdot \mathbf{g}_0} = \frac{1.0}{4} = 0.25$$

$$\mathbf{v}_1 = \mathbf{g}_1 + \gamma \mathbf{v}_0 = \begin{bmatrix} 0 \\ -1.0 \end{bmatrix} + 0.25 \begin{bmatrix} -2 \\ 0 \end{bmatrix} = \begin{bmatrix} -0.5 \\ -1.0 \end{bmatrix}$$

$$\mathbf{x}_1 + s\mathbf{v}_1 = \begin{bmatrix} -0.1 \\ 0 \end{bmatrix} + s \begin{bmatrix} -0.5 \\ -1.0 \end{bmatrix} = \begin{bmatrix} -0.1 - 0.5s \\ -s \end{bmatrix}$$

$$f(s) = F(\mathbf{x}_1 + s\mathbf{v}_1)$$
$$= 10(-0.1 - 0.5s)^2 + 3(-s)^2 - 10(-0.1 - 0.5s)(-s) + 2(-0.1 - 0.5s)$$
$$= 0.5s^2 - s - 0.1$$

$$f'(s) = s - 1 = 0 \qquad s = 1.0$$

$$\mathbf{x}_2 = \mathbf{x}_1 + s\mathbf{v}_1 = \begin{bmatrix} -0.1 \\ 0 \end{bmatrix} + 1.0 \begin{bmatrix} -0.5 \\ -1.0 \end{bmatrix} = \begin{bmatrix} -0.6 \\ -1.0 \end{bmatrix}$$

We have now reached the minimum point.

EXAMPLE 10.6

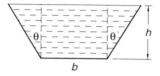

The figure shows the cross section of a channel carrying water. Determine h, b and θ that minimize the length of the wetted perimeter while maintaining a cross-sectional area of 8 m². (Minimizing the wetted perimeter results in least resistance to the flow.) Use the Fletcher–Reeves mathod.

Solution The cross-sectional area of the channel is

$$A = \frac{1}{2}\,[b + (b + 2h\tan\theta)]\,h = (b + h\tan\theta)h$$

and the length of the wetted perimeter is

$$S = b + 2(h\sec\theta)$$

The optimization problem can be cast as

$$\text{minimize } b + 2h\sec\theta$$

$$\text{subject to } (b + h\tan\theta)h = 8$$

Equality constraints can often be used to eliminate some of the design variables. In this case we can solve the area constraint for b, obtaining

$$b = \frac{8}{h} - h\tan\theta$$

Substituting the result into the expression for S, we get

$$S = \frac{8}{h} - h\tan\theta + 2h\sec\theta$$

We have now arrived at an unconstrained optimization problem of finding h and θ that minimize S. The gradient of the merit function is

$$\nabla S = \begin{bmatrix} \partial S/\partial h \\ \partial S/\partial\theta \end{bmatrix} = \begin{bmatrix} -8/h^2 - \tan\theta + 2\sec\theta \\ -h\sec^2\theta + 2h\sec\theta\tan\theta \end{bmatrix}$$

Letting $\mathbf{x} = \begin{bmatrix} h & \theta \end{bmatrix}^T$ and starting with $\mathbf{x}_0 = \begin{bmatrix} 2 & 0 \end{bmatrix}^T$, we arrive at the following program:

```
% Example 10.6 (Minimization with Fletcher-Reeves)
global X FUNC DFUNC
FUNC = @fex10_6; DFUNC = @dfex10_6;
X = [2.0;0.0];
[xMin,fMin,nCyc] = fletcherReeves;
b = 8.0/X(1) - X(1)*tan(X(2));
theta = X(2)*180.0/pi;  % Convert into degrees
fprintf('b = %8.5f\n',b)
fprintf('h = %8.5f\n',X(1))
fprintf('theta = %8.5f\n',theta)
fprintf('perimeter = %8.5f\n',fMin)
fprintf('number of cycles = %2.0f',nCyc)
```

Note that the starting point X and the function handles FUNC (function defining F) and DFUNC (function defining ∇F) are declared global. The M-files for the two functions are

```
function y = fex10_6(X)
% Function defining F in Example 10.6
y = 8.0/X(1) - X(1)*(tan(X(2)) - 2.0/cos(X(2)));


function g = dfex10_6(X)
% Function defining grad(F) in Example 10.6
g = zeros(2,1);
g(1) = -8.0/(X(1)^2) - tan(X(2)) + 2.0/cos(X(2));
g(2) = X(1)*(-1.0/cos(X(2)) + 2.0*tan(X(2)))/cos(X(2));
```

The results are (θ is in degrees):

```
>> b =   2.48161
h =   2.14914
theta = 30.00000
perimeter =   7.44484
number of cycles =   5
```

PROBLEM SET 10.1

1. ■ The Lennard–Jones potential between two molecules is

$$V = 4\varepsilon \left[\left(\frac{\sigma}{r} \right)^{12} - \left(\frac{\sigma}{r} \right)^{6} \right]$$

where ε and σ are constants, and r is the distance between the molecules. Use the functions goldBracket and goldSearch to find σ/r that minimizes the potential and verify the result analytically.

2. ■ One wave function of the hydrogen atom is

$$\psi = C \left(27 - 18\sigma + 2\sigma^2 \right) e^{-\sigma/3}$$

where

$$\sigma = zr/a_0$$

$$C = \frac{1}{81\sqrt{3\pi}} \left(\frac{z}{a_0} \right)^{2/3}$$

$$z = \text{nuclear charge}$$

$$a_0 = \text{Bohr radius}$$

$$r = \text{radial distance}$$

Find σ where ψ is at a minimum. Verify the result analytically.

3. ■ Determine the parameter p that minimizes the integral

$$\int_0^\pi \sin x \cos px \, dx$$

Hint: use numerical quadrature to evaluate the integral.

4. ■

Kirchoff's equations for the two loops of the electrical circuit are

$$R_1 i_1 + R_3 i_1 + R(i_1 - i_2) = E$$

$$R_2 i_2 + R_4 i_2 + R_5 i_2 + R(i_2 - i_1) = 0$$

Find the resistance R that maximizes the power dissipated by R. *Hint*: solve Kirchoff's equations numerically with one of the functions in Chapter 2.

5. ■

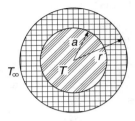

A wire carrying an electric current is surrounded by rubber insulation of outer radius r. The resistance of the wire generates heat, which is conducted through the insulation and convected into the surrounding air. The temperature of the wire can be shown to be

$$T = \frac{q}{2\pi}\left(\frac{\ln(r/a)}{k} + \frac{1}{hr}\right) + T_\infty$$

where

q = rate of heat generation in wire = 50 W/m

a = radius of wire = 5 mm

k = thermal conductivity of rubber = 0.16 W/m · K

h = convective heat-transfer coefficient = 20 W/m^2 · K

T_∞ = ambient temperature = 280 K

Find r that minimizes T.

6. ■ Minimize the function

$$F(x, y) = (x - 1)^2 + (y - 1)^2$$

subject to the constraints $x + y \le 1$ and $x \ge 0.6$.

7. ■ Find the minimum of the function

$$F(x, y) = 6x^2 + y^3 + xy$$

in $y \ge 0$. Verify the result analytically.

8. ■ Solve Prob. 7 if the constraint is changed to $y \ge -2$.

9. ■ Determine the smallest distance from the point $(1, 2)$ to the parabola $y = x^2$.

10. ■

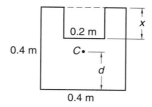

Determine x that minimizes the distance d between the base of the area shown and its centroid C.

11. ■

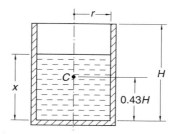

The cylindrical vessel of mass M has its center of gravity at C. The water in the vessel has a depth x. Determine x so that the center of gravity of the vessel–water combination is as low as possible. Use $M = 115$ kg, $H = 0.8$ m and $r = 0.25$ m.

12. ■

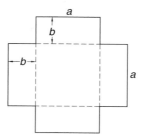

The sheet of cardboard is folded along the dashed lines to form a box with an open top. If the volume of the box is to be 1.0 m³, determine the dimensions a and b that would use the least amount of cardboard. Verify the result analytically.

13. ■

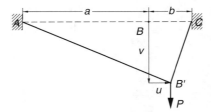

The elastic cord ABC has an extensional stiffness k. When the vertical force P is applied at B, the cord deforms to the shape $AB'C$. The potential energy of the system in the deformed position is

$$V = -Pv + \frac{k(a+b)}{2a}\delta_{AB}^2 + \frac{k(a+b)}{2b}\delta_{BC}^2$$

where

$$\delta_{AB} = \sqrt{(a+u)^2 + v^2} - a$$
$$\delta_{BC} = \sqrt{(b-u)^2 + v^2} - b$$

are the elongations of AB and BC. Determine the displacements u and v by minimizing V (this is an application of the principle of minimum potential energy: a system is in stable equilibrium if its potential energy is at a minimum). Use $a = 150$ mm, $b = 50$ mm, $k = 0.6$ N/mm and $P = 5$ N.

14. ■

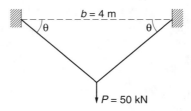

Each member of the truss has a cross-sectional area A. Find A and the angle θ that minimize the volume

$$V = \frac{bA}{\cos\theta}$$

of the material in the truss without violating the constraints

$$\sigma \leq 150 \text{ MPa} \qquad \delta \leq 5 \text{ mm}$$

where

$$\sigma = \frac{P}{2A\sin\theta} = \text{stress in each member}$$

$$\delta = \frac{Pb}{2EA\sin 2\theta\sin\theta} = \text{displacement at the load } P$$

and $E = 200 \times 10^9$ Pa.

15. ■ Solve Prob. 14 if the allowable displacement is changed to 2.5 mm.

16. ■

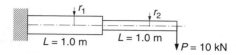

The cantilever beam of circular cross section is to have the smallest volume possible subject to constraints

$$\sigma_1 \le 180 \text{ MPa} \qquad \sigma_2 \le 180 \text{ MPa} \qquad \delta \le 25 \text{ mm}$$

where

$$\sigma_1 = \frac{8PL}{\pi r_1^3} = \text{maximum stress in left half}$$

$$\sigma_2 = \frac{4PL}{\pi r_2^3} = \text{maximum stress in right half}$$

$$\delta = \frac{4PL^3}{3\pi E}\left(\frac{7}{r_1^4} + \frac{1}{r_2^4}\right) = \text{displacement at free end}$$

and $E = 200$ GPa. Determine r_1 and r_2.

17. ■ Find the minimum of the function

$$F(x, y, z) = 2x^2 + 3y^2 + z^2 + xy + xz - 2y$$

and confirm the result analytically.

18. ■

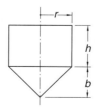

The cylindrical container has a conical bottom and an open top. If the volume V of the container is to be 1.0 m^3, find the dimensions r, h and b that minimize the surface area S. Note that

$$V = \pi r^2 \left(\frac{b}{3} + h \right)$$

$$S = \pi r \left(2h + \sqrt{b^2 + r^2} \right)$$

19. ■

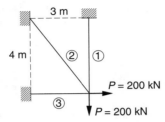

The equilibrium equations of the truss shown are

$$\sigma_1 A_1 + \frac{4}{5}\sigma_2 A_2 = P \qquad \frac{3}{5}\sigma_2 A_2 + \sigma_3 A_3 = P$$

where σ_i is the axial stress in member i and A_i are the cross-sectional areas. The third equation is supplied by compatibility (geometrical constraints on the elongations of the members):

$$\frac{16}{5}\sigma_1 - 5\sigma_2 + \frac{9}{5}\sigma_3 = 0$$

Find the cross-sectional areas of the members that minimize the weight of the truss without the stresses exceeding 150 MPa.

20. ■

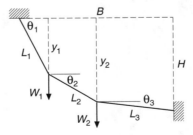

A cable supported at the ends carries the weights W_1 and W_2. The potential energy of the system is

$$V = -W_1 y_1 - W_2 y_2$$

$$= -W_1 L_1 \sin \theta_1 - W_2 (L_1 \sin \theta_1 + L_2 \sin \theta_2)$$

and the geometric constraints are

$$L_1 \cos \theta_1 + L_2 \cos \theta_2 + L_3 \cos \theta_3 = B$$

$$L_1 \sin \theta_1 + L_2 \sin \theta_2 + L_3 \sin \theta_3 = H$$

The principle of minimum potential energy states that the equilibrium configuration of the system is the one that satisfies geometric constraints and minimizes the potential energy. Determine the equilibrium values of θ_1, θ_2 and θ_3 given that $L_1 = 1.2$ m, $L_2 = 1.5$ m, $L_3 = 1.0$ m, $B = 3.5$ m, $H = 0$, $W_1 = 20$ kN and $W_2 = 30$ kN.

MATLAB Functions

x = fmnbnd(@func,a,b) returns x that minimizes the function func of a single variable. The minimum point must be bracketed in (a,b). The algorithm used is *Brent's method* that combines golden section search with quadratic interpolation. It is more efficient than goldSearch that uses just the golden section search.

x = fminsearch(@func,xStart) returns the vector of independent variables that minimizes the multivariate function func. The vector xStart contains the starting values of x. The algorithm is the *Nelder–Mead method*, also known as the *downhill simplex*, which is reliable, but much less efficient than Powell's method.

Both of these functions can be called with various control options that set optimization parameters (e.g., the error tolerance) and control the display of results. There are also additional output parameters that may be used in the function call, as illustrated in the following example (the data is taken from Example 10.4):

```
>> [x,fmin,output] = fminsearch(@fex10_4,[1 5])

x =
    0.7331    7.5878
fmin =
    18.6929
output =
    iterations: 38
    funcCount: 72
    algorithm: 'Nelder-Mead simplex direct search'
```

Appendices

A1   Taylor Series

Function of a Single Variable

The Taylor series expansion of a function $f(x)$ about the point $x = a$ is the infinite series

$$f(x) = f(a) + f'(a)(x - a) + f''(a)\frac{(x - a)^2}{2!} + f'''(a)\frac{(x - a)^3}{3!} + \cdots \tag{A1}$$

In the special case $a = 0$ the series is also known as the *MacLaurin series*. It can be shown that the Taylor series expansion is unique in the sense that no two functions have identical Taylor series.

A Taylor series is meaningful only if all the derivatives of $f(x)$ exist at $x = a$ and the series converges. In general, convergence occurs only if x is sufficiently close to a; i.e., if $|x - a| \leq \varepsilon$, where ε is called the *radius of convergence*. In many cases ε is infinite.

Another useful form of the Taylor series is the expansion about an arbitrary value of x:

$$f(x + h) = f(x) + f'(x)h + f''(x)\frac{h^2}{2!} + f'''(x)\frac{h^3}{3!} + \cdots \tag{A2}$$

Since it is not possible to evaluate all the terms of an infinite series, the effect of truncating the series in Eq. (A2) is of great practical importance. Keeping the first $n + 1$ terms, we have

$$f(x + h) = f(x) + f'(x)h + f''(x)\frac{h^2}{2!} + \cdots + f^{(n)}(x)\frac{h^n}{n!} + E_n \tag{A3}$$

where E_n is the *truncation error* (sum of the truncated terms). The bounds on the truncation error are given by *Taylor's theorem*:

$$E_n = f^{(n+1)}(\xi)\frac{h^{n+1}}{(n + 1)!} \tag{A4}$$

where ξ is some point in the interval $(x, x + h)$. Note that the expression for E_n is identical to the first discarded term of the series, but with x replaced by ξ. Since the value of ξ is undetermined (only its limits are known), the most we can get out of Eq. (A4) are the upper and lower bounds on the truncation error.

If the expression for $f^{(n+1)}(\xi)$ is not available, the information conveyed by Eq. (A4) is reduced to

$$E_n = \mathcal{O}(h^{n+1}) \tag{A5}$$

which is a concise way of saying that the truncation error is *of the order of* h^{n+1}, or behaves as h^{n+1}. If h is within the radius of convergence, then

$$\mathcal{O}(h^n) > \mathcal{O}(h^{n+1})$$

i.e., the error is always reduced if a term is added to the truncated series (this may not be true for the first few terms).

In the special case $n = 1$, Taylor's theorem is known as the *mean value theorem*:

$$f(x + h) = f(x) + f'(\xi)h, \quad x \leq \xi \leq x + h \tag{A6}$$

Function of Several Variables

If f is a function of the m variables $x_1, x_2, \ldots, x_m$, then its Taylor series expansion about the point $\mathbf{x} = [x_1, x_2, \ldots, x_m]^T$ is

$$f(\mathbf{x} + \mathbf{h}) = f(\mathbf{x}) + \sum_{i=1}^{m} \frac{\partial f}{\partial x_i}\bigg|_{\mathbf{x}} h_i + \frac{1}{2!} \sum_{i=1}^{m} \sum_{j=1}^{m} \frac{\partial^2 f}{\partial x_i \partial x_j}\bigg|_{\mathbf{x}} h_i h_j + \cdots \tag{A7}$$

This is sometimes written as

$$f(\mathbf{x} + \mathbf{h}) = f(\mathbf{x}) + \nabla f(\mathbf{x}) \cdot \mathbf{h} + \frac{1}{2}\mathbf{h}^T \mathbf{H}(\mathbf{x})\mathbf{h} + \cdots \tag{A8}$$

The vector ∇f is known as the *gradient* of f and the matrix $\mathbf{H}$ is called the *Hessian matrix* of f.

EXAMPLE A1
Derive the Taylor series expansion of $f(x) = \ln(x)$ about $x = 1$.

Solution The derivatives of f are

$$f'(x) = \frac{1}{x} \quad f''(x) = -\frac{1}{x^2} \quad f'''(x) = \frac{2!}{x^3} \quad f^{(4)} = -\frac{3!}{x^4} \text{ etc.}$$

Evaluating the derivatives at $x = 1$, we get

$$f'(1) = 1 \quad f''(1) = -1 \quad f'''(1) = 2! \quad f^{(4)}(1) = -3! \text{ etc.}$$

which upon substitution into Eq. (A1) together with $a = 1$ yields

$$\ln(x) = 0 + (x - 1) - \frac{(x - 1)^2}{2!} + 2!\frac{(x - 1)^3}{3!} - 3!\frac{(x - 1)^4}{4!} + \cdots$$

$$= (x - 1) - \frac{1}{2}(x - 1)^2 + \frac{1}{3}(x - 1)^3 - \frac{1}{4}(x - 1)^4 + \cdots$$

EXAMPLE A2

Use the first five terms of the Taylor series expansion of e^x about $x = 0$:

$$e^x = 1 + x + \frac{x^2}{2!} + \frac{x^3}{3!} + \frac{x^4}{4!} + \cdots$$

together with the error estimate to find the bounds of e.

Solution

$$e = 1 + 1 + \frac{1}{2} + \frac{1}{6} + \frac{1}{24} + E_4 = \frac{65}{24} + E_4$$

$$E_4 = f^{(4)}(\xi)\frac{h^5}{5!} = \frac{e^\xi}{5!}, \quad 0 \le \xi \le 1$$

The bounds on the truncation error are

$$(E_4)_{min} = \frac{e^0}{5!} = \frac{1}{120} \qquad (E_4)_{max} = \frac{e^1}{5!} = \frac{e}{120}$$

Thus the lower bound on e is

$$e_{min} = \frac{65}{24} + \frac{1}{120} = \frac{163}{60}$$

and the upper bound is given by

$$e_{max} = \frac{65}{24} + \frac{e_{max}}{120}$$

which yields

$$\frac{119}{120}e_{max} = \frac{65}{24} \qquad e_{max} = \frac{325}{119}$$

Therefore,

$$\frac{163}{60} \le e \le \frac{325}{119}$$

EXAMPLE A3

Compute the gradient and the Hessian matrix of

$$f(x, y) = \ln \sqrt{x^2 + y^2}$$

at the point $x = -2$, $y = 1$.

Solution

$$\frac{\partial f}{\partial x} = \frac{1}{\sqrt{x^2 + y^2}} \left(\frac{1}{2} \frac{2x}{\sqrt{x^2 + y^2}} \right) = \frac{x}{x^2 + y^2} \qquad \frac{\partial f}{\partial y} = \frac{y}{x^2 + y^2}$$

$$\nabla f(x, y) = \begin{bmatrix} x/(x^2 + y^2) & y/(x^2 + y^2) \end{bmatrix}^T$$

$$\nabla f(-2, 1) = \begin{bmatrix} -0.4 & 0.2 \end{bmatrix}^T$$

$$\frac{\partial^2 f}{\partial x^2} = \frac{(x^2 + y^2) - x(2x)}{(x^2 + y^2)^2} = \frac{-x^2 + y^2}{(x^2 + y^2)^2}$$

$$\frac{\partial^2 f}{\partial y^2} = \frac{x^2 - y^2}{(x^2 + y^2)^2}$$

$$\frac{\partial^2 f}{\partial x \partial y} = \frac{\partial^2 f}{\partial y \partial x} = \frac{-2xy}{(x^2 + y^2)^2}$$

$$\mathbf{H}(x, y) = \begin{bmatrix} -x^2 + y^2 & -2xy \\ -2xy & x^2 - y^2 \end{bmatrix} \frac{1}{(x^2 + y^2)^2}$$

$$\mathbf{H}(-2, 1) = \begin{bmatrix} -0.12 & 0.16 \\ 0.16 & 0.12 \end{bmatrix}$$

A2   Matrix Algebra

A matrix is a rectangular array of numbers. The *size* of a matrix is determined by the number of rows and columns, also called the *dimensions* of the matrix. Thus a matrix of m rows and n columns is said to have the size $m \times n$ (the number of rows is always listed first). A particularly important matrix is the square matrix, which has the same number of rows and columns.

An array of numbers arranged in a single column is called a *column vector*, or simply a *vector*. If the numbers are set out in a row, the term *row vector* is used. Thus a column vector is a matrix of dimensions $n \times 1$ and a row vector can be viewed as a matrix of dimensions $1 \times n$.

We denote matrices by boldface, upper case letters. For vectors we use boldface, lower case letters. Here are examples of the notation:

$$\mathbf{A} = \begin{bmatrix} A_{11} & A_{12} & A_{13} \\ A_{21} & A_{22} & A_{23} \\ A_{31} & A_{32} & A_{33} \end{bmatrix} \qquad \mathbf{b} = \begin{bmatrix} b_1 \\ b_2 \\ b_3 \end{bmatrix} \qquad \text{(A9)}$$

Indices of the elements of a matrix are displayed in the same order as its dimensions: the row number comes first, followed by the column number. Only one index is needed for the elements of a vector.

Transpose

The transpose of a matrix $\mathbf{A}$ is denoted by $\mathbf{A}^T$ and defined as

$$A_{ij}^T = A_{ji}$$

The transpose operation thus interchanges the rows and columns of the matrix. If applied to vectors, it turns a column vector into a row vector and *vice versa*. For example, transposing $\mathbf{A}$ and $\mathbf{b}$ in Eq. (A9), we get

$$\mathbf{A}^T = \begin{bmatrix} A_{11} & A_{21} & A_{31} \\ A_{12} & A_{22} & A_{32} \\ A_{13} & A_{23} & A_{33} \end{bmatrix} \qquad \mathbf{b}^T = \begin{bmatrix} b_1 & b_2 & b_3 \end{bmatrix}$$

An $n \times n$ matrix is said to be *symmetric* if $\mathbf{A}^T = \mathbf{A}$. This means that the elements in the upper triangular portion (above the diagonal connecting A_{11} and A_{nn}) of a symmetric matrix are mirrored in the lower triangular portion.

Addition

The sum $\mathbf{C} = \mathbf{A} + \mathbf{B}$ of two $m \times n$ matrices $\mathbf{A}$ and $\mathbf{B}$ is defined as

$$C_{ij} = A_{ij} + B_{ij}, \quad i = 1, 2, \ldots, m; \quad j = 1, 2, \ldots, n \tag{A10}$$

Thus the elements of $\mathbf{C}$ are obtained by adding elements of $\mathbf{A}$ to the elements of $\mathbf{B}$. Note that addition is defined only for matrices that have the same dimensions.

Multiplication

The scalar or *dot product* $c = \mathbf{a} \cdot \mathbf{b}$ of the vectors $\mathbf{a}$ and $\mathbf{b}$, each of size m, is defined as

$$c = \sum_{k=1}^{m} a_k b_k \tag{A11}$$

It can also be written in the form $c = \mathbf{a}^T \mathbf{b}$.

The matrix product $\mathbf{C} = \mathbf{AB}$ of an $l \times m$ matrix $\mathbf{A}$ and an $m \times n$ matrix $\mathbf{B}$ is defined by

$$C_{ij} = \sum_{k=1}^{m} A_{ik} B_{kj}, \quad i = 1, 2, \ldots, l; \quad j = 1, 2, \ldots, n \tag{A12}$$

The definition requires the number of columns in A (the dimension m) to be equal to the number of rows in B. The matrix product can also be defined in terms of the dot product. Representing the ith row of A as the vector $\mathbf{a}_i$ and the jth column of B as the vector $\mathbf{b}_j$, we have

$$C_{ij} = \mathbf{a}_i \cdot \mathbf{b}_j \tag{A13}$$

A square matrix of special importance is the identity or *unit matrix*

$$\mathbf{I} = \begin{bmatrix} 1 & 0 & 0 & \cdots & 0 \\ 0 & 1 & 0 & \cdots & 0 \\ 0 & 0 & 1 & \cdots & 0 \\ \vdots & \vdots & \vdots & \ddots & \vdots \\ 0 & 0 & 0 & 0 & 1 \end{bmatrix} \tag{A14}$$

It has the property $\mathbf{AI} = \mathbf{IA} = \mathbf{A}$.

Inverse

The inverse of an $n \times n$ matrix A, denoted by $\mathbf{A}^{-1}$, is defined to be an $n \times n$ matrix that has the property

$$\mathbf{A}^{-1}\mathbf{A} = \mathbf{A}\mathbf{A}^{-1} = \mathbf{I} \tag{A15}$$

Determinant

The determinant of a square matrix A is a scalar denoted by $|A|$ or $\det(A)$. There is no concise definition of the determinant for a matrix of arbitrary size. We start with the determinant of a 2×2 matrix, which is defined as

$$\begin{vmatrix} A_{11} & A_{12} \\ A_{21} & A_{22} \end{vmatrix} = A_{11}A_{22} - A_{12}A_{21} \tag{A16}$$

The determinant of a 3×3 matrix is then defined as

$$\begin{vmatrix} A_{11} & A_{12} & A_{13} \\ A_{21} & A_{22} & A_{23} \\ A_{31} & A_{32} & A_{33} \end{vmatrix} = A_{11}\begin{vmatrix} A_{22} & A_{23} \\ A_{32} & A_{33} \end{vmatrix} - A_{12}\begin{vmatrix} A_{21} & A_{23} \\ A_{31} & A_{33} \end{vmatrix} + A_{13}\begin{vmatrix} A_{21} & A_{22} \\ A_{31} & A_{32} \end{vmatrix}$$

Having established the pattern, we can now define the determinant of an $n \times n$ matrix in terms of the determinant of an $(n-1) \times (n-1)$ matrix:

$$|\mathbf{A}| = \sum_{k=1}^{n} (-1)^{k+1} A_{1k} M_{1k} \tag{A17}$$

where M_{ik} is the determinant of the $(n-1) \times (n-1)$ matrix obtained by deleting the ith row and kth column of **A**. The term $(-1)^{k+i} M_{ik}$ is called a *cofactor* of A_{ik}.

Equation (A17) is known as *Laplace's development* of the determinant on the first row of **A**. Actually Laplace's development can take place on any convenient row. Choosing the ith row, we have

$$|\mathbf{A}| = \sum_{k=1}^{n} (-1)^{k+i} A_{ik} M_{ik} \tag{A18}$$

The matrix **A** is said to be *singular* if $|\mathbf{A}| = 0$.

Positive Definiteness

An $n \times n$ matrix A is said to be positive definite if

$$\mathbf{x}^T \mathbf{A} \mathbf{x} > 0 \tag{A19}$$

for all nonvanishing vectors **x**. It can be shown that a matrix is positive definite if the determinants of all its leading minors are positive. The leading minors of **A** are the n square matrices

$$\begin{bmatrix} A_{11} & A_{12} & \cdots & A_{1k} \\ A_{12} & A_{22} & \cdots & A_{2k} \\ \vdots & \vdots & \ddots & \vdots \\ A_{k1} & A_{k2} & \cdots & A_{kk} \end{bmatrix}, \quad k = 1, 2, \ldots, n$$

Therefore, positive definiteness requires that

$$A_{11} > 0, \quad \begin{vmatrix} A_{11} & A_{12} \\ A_{21} & A_{22} \end{vmatrix} > 0, \quad \begin{vmatrix} A_{11} & A_{12} & A_{13} \\ A_{21} & A_{22} & A_{23} \\ A_{31} & A_{32} & A_{33} \end{vmatrix} > 0, \ldots, |A| > 0 \tag{A20}$$

Useful Theorems

We list without proof a few theorems that are utilized in the main body of the text. Most proofs are easy and could be attempted as exercises in matrix algebra.

$$(\mathbf{AB})^T = \mathbf{B}^T \mathbf{A}^T \tag{A21a}$$

$$(\mathbf{AB})^{-1} = \mathbf{B}^{-1} \mathbf{A}^{-1} \tag{A21b}$$

$$\left| \mathbf{A}^T \right| = |\mathbf{A}| \tag{A21c}$$

$$|\mathbf{AB}| = |\mathbf{A}| \, |\mathbf{B}| \tag{A21d}$$

$$\text{if } \mathbf{C} = \mathbf{A}^T \mathbf{B} \mathbf{A} \text{ where } \mathbf{B} = \mathbf{B}^T, \text{ then } \mathbf{C} = \mathbf{C}^T \tag{A21e}$$

EXAMPLE A4

Letting

$$A = \begin{bmatrix} 1 & 2 & 3 \\ 1 & 2 & 1 \\ 0 & 1 & 2 \end{bmatrix} \qquad u = \begin{bmatrix} 1 \\ 6 \\ -2 \end{bmatrix} \qquad v = \begin{bmatrix} 8 \\ 0 \\ -3 \end{bmatrix}$$

compute $u + v$, $u \cdot v$, Av and $u^T Av$.

Solution

$$u + v = \begin{bmatrix} 1+8 \\ 6+0 \\ -2-3 \end{bmatrix} = \begin{bmatrix} 9 \\ 6 \\ -5 \end{bmatrix}$$

$$u \cdot v = 1(8)) + 6(0) + (-2)(-3) = 14$$

$$Av = \begin{bmatrix} a_1 \cdot v \\ a_2 \cdot v \\ a_3 \cdot v \end{bmatrix} = \begin{bmatrix} 1(8) + 2(0) + 3(-3) \\ 1(8) + 2(0) + 1(-3) \\ 0(8) + 1(0) + 2(-3) \end{bmatrix} = \begin{bmatrix} -1 \\ 5 \\ -6 \end{bmatrix}$$

$$u^T Av = u \cdot (Av) = 1(-1) + 6(5) + (-2)(-6) = 41$$

EXAMPLE A5

Compute $|A|$, where A is given in Example A4. Is A positive definite?

Solution Laplace's development of the determinant on the first row yields

$$|A| = 1 \begin{vmatrix} 2 & 1 \\ 1 & 2 \end{vmatrix} - 2 \begin{vmatrix} 1 & 1 \\ 0 & 2 \end{vmatrix} + 3 \begin{vmatrix} 1 & 2 \\ 0 & 1 \end{vmatrix}$$

$$= 1(3) - 2(2) + 3(1) = 2$$

Development on the third row is somewhat easier due to the presence of the zero element:

$$|A| = 0 \begin{vmatrix} 2 & 3 \\ 2 & 1 \end{vmatrix} - 1 \begin{vmatrix} 1 & 3 \\ 1 & 1 \end{vmatrix} + 2 \begin{vmatrix} 1 & 2 \\ 1 & 2 \end{vmatrix}$$

$$= 0(-4) - 1(-2) + 2(0) = 2$$

To verify positive definiteness, we evaluate the determinants of the leading minors:

$$A_{11} = 1 > 0 \quad \text{O.K.}$$

$$\begin{vmatrix} A_{11} & A_{12} \\ A_{21} & A_{22} \end{vmatrix} = \begin{vmatrix} 1 & 2 \\ 1 & 2 \end{vmatrix} = 0 \quad \text{Not O.K.}$$

A is not positive definite.

EXAMPLE A6

Evaluate the matrix product **AB**, where **A** is given in Example A4 and

$$B = \begin{bmatrix} -4 & 1 \\ 1 & -4 \\ 2 & -2 \end{bmatrix}$$

Solution

$$AB = \begin{bmatrix} \mathbf{a}_1 \cdot \mathbf{b}_1 & \mathbf{a}_1 \cdot \mathbf{b}_2 \\ \mathbf{a}_2 \cdot \mathbf{b}_1 & \mathbf{a}_2 \cdot \mathbf{b}_2 \\ \mathbf{a}_3 \cdot \mathbf{b}_1 & \mathbf{a}_3 \cdot \mathbf{b}_2 \end{bmatrix}$$

$$= \begin{bmatrix} 1(-4) + 2(1) + 3(2) & 1(1) + 2(-4) + 3(-2) \\ 1(-4) + 2(1) + 1(2) & 1(1) + 2(-4) + 1(-2) \\ 0(-4) + 1(1) + 2(2) & 0(1) + 1(-4) + 2(-2) \end{bmatrix} = \begin{bmatrix} 4 & -13 \\ 0 & -9 \\ 5 & -8 \end{bmatrix}$$

Index